T0205917

Safety Differently

Human Factors for a New Era

Second Edition

Sidney Dekker

CRC Press
Taylor & Francis Group
Boca Raton   London   New York

CRC Press is an imprint of the
Taylor & Francis Group, an **informa** business

CRC Press
Taylor & Francis Group
6000 Broken Sound Parkway NW, Suite 300
Boca Raton, FL 33487-2742

© 2015 by Taylor & Francis Group, LLC
CRC Press is an imprint of Taylor & Francis Group, an Informa business

No claim to original U.S. Government works

Printed on acid-free paper
Version Date: 20140313

International Standard Book Number-13: 978-1-4822-4199-0 (Paperback)

Library of Congress Cataloging-in-Publication Data

Dekker, Sidney.
    Safety differently : human factors for a new era / author, Sidney Dekker. -- Second edition.
        pages cm
    Includes bibliographical references and index.
    ISBN 978-1-4822-4199-0 (alk. paper)
    1. Human engineering. I. Title.

T59.7.D44 2014
363.1--dc23

2014000190

Visit the Taylor & Francis Web site at
http://www.taylorandfrancis.com

and the CRC Press Web site at
http://www.crcpress.com

Contents

Contents

Preface

Is the human a problem to control, or a solution to harness? For half a century, safety thinking was dominated by the idea that people are the target for intervention. The "human factor" was about individual people and their "mental, physical or moral deficiencies." People were the problem to control. They had to be carefully selected—not as much for their strengths as for their absence of limitations and defects—and then molded around the fixed features of the technology with which they had to work. Safety problems were addressed by controlling the human factor that went into work. Halfway through the 20th century, a remarkable transition occurred in this thinking. Increasing complexity of technology and a quickening pace of technological change combined with a growing skepticism about the scientific basis and pragmatic value of the behaviorist psychology that had underpinned and driven much of the ideological basis for safety thinking. "Human factors" as a field sprouted from this transition. It was able to demonstrate that the world in which people work is not fixed. Instead, technology could be adapted to fit the strengths and limitations of people—independent of their individual differences. Safety problems were increasingly addressed by controlling the technology, the environment, the system.

Over the last 40 years, recognition has grown that mishaps (a refinery explosion, a commercial aircraft accident) are inextricably linked to the (mal)functioning of that system—the surrounding organizations and institutions. People are not the instigators of failure; they are the recipients of it, the inheritors. The construction and operation of commercial airliners or upstream gas networks or healthcare or space shuttles or passenger ferries spawn vast networks of organizations to support them, to advance and improve them, to control and regulate them. Complex technologies cannot exist without these organizations and institutions—carriers, regulators, government agencies, manufacturers, subcontractors, maintenance facilities, training outfits—that, in principle, are designed to protect and secure their operation. Their very mandate boils down to not having accidents happen. Since the 1978 nuclear accident at Three Mile Island, we have increasingly realized that the very organizations meant to keep a technology safe and stable (human operators, regulators, management, maintenance) are actually among the major contributors to breakdown. Sociotechnical failures are impossible without such contributions.

This recognition has been hugely emancipatory and empowering. Rather than controlling individual human factors deep down in an organization, it has allowed safety resources to be directed at entire systems, at organizations, at the sorts of design issues, and operational and organizational constraints that create problems and safety risk for everybody. But it has the seeds of its own reversion and regression implanted in it. Safety is increasingly seen as a matter of administrative ordering and managerial control. The most popular organizational safety image of the late 20th century was that of the Swiss Cheese Model. This shows an outcome failure as the result of many smaller, prior failures in organizational and administrative layers upstream from the sharp end. Again, it is empowering and emancipatory because it provides a strong reminder of context, of the system surrounding people, not just the person at the sharp end. But it reifies Newtonian ideas about risk (as energy to be contained) and linear cause–effect relationships. The Newtonian worldview actively hampers our ability to create a human factors thinking, a different safety thinking, suited for a new era of complexity and interconnectedness. What is more, it does not escape prehistoric thinking about people as a problem to control. If layers of organizational protection are imperfect and porous, this is the result not only of bad procedures or unsuitable designs but also (in the language of the model) of violations, of unsafe acts, of line management deficiencies, of fallible decisions and deficient supervision.

But most importantly, the image conveys the modernist idea that safety at the sharp end is assured by good administrative and technological ordering of the upstream system. We need to find and fix the holes or failures there to prevent outcome failures. Putting our faith in bureaucracy and administration, as well as science and technology, is a good, modern thing to do. This has, however, helped entrench the idea that safety is generated chiefly through planning, process, paperwork, audit trails, and administrative work. It has resulted in a new focus on safety management systems that count and tabulate negative events, on procedural compliance, and surveillance and monitoring, and it has thus put new limits on the people who do work at the sharp end. The emancipation promised by the transition halfway through the 20th century, and implemented by human factors thinking, has become subjected to a kind of counter-reformation. Safety increasingly means deference to liability concerns, to protocol, to insurance mercantilism, to fear of regulation and litigation. In many industries, safety has morphed from an operational value into an administrative machinery. It has changed from an ethical responsibility *for* people who do the organization's dangerous work, to a bureaucratic accountability *to* those who attempt to control corporate risk. Burgeoning safety bureaucracies, which work at an increasing distance from the operation, are usually organized around lagging, negative indicators. They tend to establish vocabularies of deficit, control, and constraint. People are once again a problem to control. Expanding machineries of safety can make an organization listen less to technical expertise and operational experience, and disempower middle management and supervisory layers. People may no longer feel as able or empowered to think for themselves. This can stifle innovation, choke off initiative, and erode problem ownership.

SAFETY, DIFFERENTLY

A new era for human factors calls for a different kind of safety thinking. A thinking that sees people as the source of diversity, insight, creativity, and wisdom about safety, not as sources of risk that undermine an otherwise safe system. It calls for a kind of thinking that is quicker to trust people and mistrust bureaucracy, and that is more committed to actually preventing harm than to looking good. The key transitions for a new era are

- We need to transition from seeing people as a problem to control, to seeing people as a solution to harness.

- We need to transition from seeing safety as a bureaucratic accountability *up*, to seeing it as an ethical responsibility *down*.
- We need to transition from seeing safety as an absence of negatives to seeing it as the presence of a positive capacity to make things go right. A focus on safety and risk should become a focus on resilience.
- We need to depart from our Cartesian–Newtonian languages of linear cause–effect relationships, of defenses-in-depth and other static metaphors. Instead, we need to embrace a language of complexity, of change and evolution, of holistic relationships rather than individual components.
- We need to trade our vocabularies of control, constraint, and human deficit for new vocabularies of empowerment, diversity, and human opportunity.

In its current form, human factors and safety thinking is not necessarily well equipped to handle such transitions. The field still largely relies on a vocabulary based on a particular conception of the natural sciences. This is derived from its roots in engineering and experimental psychology. This vocabulary, the subtle use of metaphors, images, and ideas, seems more and more at odds with the interpretative demands posed by modern organizational accidents or with complexity and the vast, networked possibilities of the information age. The vocabulary expresses a worldview that is (perhaps) appropriate for simple mechanical or technical failures, or for an individual's interactions with a piece of technology. But it is much less capable of embracing and penetrating the complexity of big sociotechnical issues—those that involve the intertwined effects of technology and the organized, continually evolving social complexity surrounding its use.

Any language, and the worldview it mediates, imposes limitations on our understanding of the world, of the successes and failures in it. Yet these limitations are now becoming increasingly evident and pressing. With growth in system size and complexity, the nature of accidents is changing (system accidents, sociotechnical failures). Resource scarcity and competition mean that systems incrementally push their operations toward the edges of their safety envelopes. They have to do this in order to remain successful in their dynamic environments. Commercial returns at the boundaries are greater, but the difference between having and not having an accident is up to stochastics more than available margins. Open systems are continually adrift within their safety envelopes, and the processes that drive such migration are not easy to recognize or control, nor is the exact location of the boundaries. At the same time, the fast pace of technological change creates new types of hazards, especially those that come with increased reliance on computer technology. Both engineered and social systems (and their interplay) rely to an ever greater extent on information technology. Although computational speed and access to information would seem a safety advantage in principle, our ability to make sense of data is not keeping pace with our ability to collect and generate it. By knowing more, we may actually know less. Managing safety by numbers (incidents, error counts, safety threats), as if safety is just another index of a Harvard business model, can create a false impression of rationality and managerial control. It may ignore higher-order variables that could tell other, more useful stories of the nature and direction of system drift. It may also come at the cost of deeper understandings of sociotechnical functioning.

It becomes really problematic if we do not recognize the other implications of modernism in this. The assumption that easily slips into the investigation of accidents is that the systems that failed are merely complicated, not complex. More about that distinction will be said in the book. For now, complicated systems are believed to have intentional designs, to be relatively closed off to the environment, and to not evolve into new or unknown states. They can be understood, and the parts that failed can be found, by decomposing them. Is that authentic to the systems whose rubble we find at accident sites, though? It seems that the organizations and technologies we help build and understand have more in common with organic systems—open to interaction with their environments and evolving in ways that were not foreseen on the drawing board. What happens in these systems is not neat; it is messy. The environment can reach in, fold in, and touch each participating human or element in its own way. The constantly changing relationships between parts give rise to emergent behaviors and properties that cannot be seen at the level of individual components. In such environments, under pressures of scarcity and competition, with opaque, large, complex sociotechnical systems, where patterns of information reach certain agents but not others, decisions and trade-offs get made that make local sense but can lead to global problems, to a drift into failure. Drift into failure is associated with normal adaptive organizational processes. Organizational failures in safe systems are not preceded by failures, by the breaking or lack of quality of single components. Instead, organizational failure in safe systems is preceded by normal work, by normal people doing normal work in seemingly normal organizations. This appears to challenge the definition of an incident and may undermine the value of incident reporting as a tool for learning beyond a certain safety level. The border between normal work and incident is clearly elastic and subject to incremental revision. With every little step away from previous norms, past success can be taken as a guarantee of future safety. Incrementalism notches the entire system closer to the edge of breakdown, but without compelling empirical indications that it is headed that way.

Current human factors and safety models cannot adequately deal with this. They require failures as a prerequisite for failures. They are still oriented toward finding failures in and of components (e.g., human errors, holes in layers of defense, latent problems, organizational deficiencies, resident pathogens) and rely on externally dictated standards of work and structure, rather than taking insider accounts (of what is a failure vs. normal work) as canonical. Processes of sensemaking, of the creation of local rationality by those who actually make the thousands of little and larger trade-offs that ferry a system along its drifting course, lie outside today's human factors lexicon. Current models typically view organizations as Newtonian–Cartesian machines with components and linkages between them. Mishaps get modeled as a sequence of events (actions and reactions) between a trigger and an outcome. Such models can say nothing about the buildup of latent failures, about the gradual, incremental loosening or loss of control. The processes of erosion of constraints, of attrition of safety, of drift toward margins, cannot be captured because structuralist approaches are static, modernist metaphors for resulting forms, not dynamic models oriented toward processes of formation. In their inability to meaningfully address drift into failure, which intertwines technical, social, institutional, and individual factors, human factors and safety are currently paying for their theoretical exclusion of transactional and social processes between individuals and world. The componentialism and fragmentation of human factors research are still an obstacle to making progress in this respect. An enlargement of the unit of analysis (as done in the ideas of cognitive systems engineering and distributed cognition), and a call to make action central in understanding assessments and thought, have been ways to catch up with new practical developments for which human factors and safety were not prepared.

As a psychology of pragmatics, of constant improvement, human factors and safety have generally retained a Cartesian–Newtonian view of science and scientific method and a belief in the certainty of scientific knowledge. The aim of science was to achieve control by deriving general, and ideally mathematical, laws of nature. A heritage of this can be seen in human factors, particularly in the predominance of experiments, the nomothetic rather than ideographic inclination of its research, and a strong faith in the realism of observed facts. It can also be recognized in the reductive strategies human factors and safety rely on to deal with complexity, just like Taylor once did. Newtonian problem solving is analytic. It consists of breaking up problems into pieces and in arranging these in some logical order. Phenomena are decomposed into more basic parts, and the whole is explained by reference to its constituent components and their interactions. In human factors and safety, *mind* is often understood as a box-like construction with a mechanistic trade in internal representations; *work* is broken into procedural steps through hierarchical task analyses; *organizations* are not organic or dynamic but consist of static layers and compartments and linkages; and *safety* is a structural property that can be understood in terms of its lower-order mechanisms (reporting systems, error rates and audits, safety management function in the organizational chart, quality systems).

Newton and Descartes, with their particular take on natural science, have a firm grip on human factors and safety in other areas, too. The information-processing paradigm, for example, so useful in explaining early information-transfer problems in World War II radar and radio operators, all but colonized human factors research. It is still a dominant force, buttressed by the Spartan laboratory experiments that seem to confirm its utility and validity. The paradigm has mechanized mind, chunked it up into separate components (which Neisser once called an "intra-psychic highway") with linkages in between. Newton would have loved the mechanics of it. Descartes would have liked it, too: A clear separation between mind and world solved (or circumvented, rather) a lot of problems associated with the transactions between the two. A mechanistic model such as information processing of course holds appeal for engineering and other consumers of human factors research results. Pragmatics dictate bridging the gap between practitioner and science, and having a cognitive model that is a simile of a technical device familiar to applied people is one powerful way to do just that. But there is no empirical reason to restrict our understanding of attitudes, memories, or heuristics as mentally encoded deposits, as some contents of consciousness with certain expiry dates. In fact, such a model severely restricts our ability to understand how people use talk and action to construct perceptual and social order; how, through discourse and action, people create the environments that in turn determine further action and possible assessments, and that constrain what will subsequently be seen as acceptable discourse or rational decisions. We cannot begin to understand drift into failure without understanding how groups of people, through assessment and action, assemble versions of the world in which they assess and act.

Information processing fits within a larger, dominant metatheoretical perspective that takes the individual as its central focus. This view, too, is a heritage of the scientific revolution and Enlightenment, which increasingly popularized the humanistic idea of a "self-contained individual." For most of psychology, this has meant that all processes worth studying take place within the boundaries of the body (or mind). This is epitomized by the mentalist focus of information processing, but visible also in how we measure "safety culture" for example. Durkheim argued long ago that any attempt to explain the social and cultural in terms of the individual puts one in harm's way analytically. But that is exactly what we do—ask individuals for their experiences and proclivities in their organization and see the aggregate as "culture." This might be because many human factors researchers come from psychology and not anthropology or ethnography. In those fields, of course, notions of culture as a summation of individual opinions, practices, and choices are ludicrous. And the normative idea that there are "good" cultures and "bad" cultures (as in "a good safety culture") is not only hubristic, but deeply ominous. Perhaps this sort of research is just one more expression of the triumphalism of the individual, the metatheory that is so characteristic of folk or lay social science and so natural for most of us in the West.

We can recognize the prominence of Newtonian–Cartesian deconstruction and componentialism in empiricist notions of a perception of elements as well—elements perceived in the world, which then gradually get converted into meaning through stages of mental processing. These ancient ideas are legitimate theoretical notions in much traditional human factors research to this day. Empiricism was once a force in the history of psychology. Yet buoyed by the information-processing paradigm, its central tenets have made a comeback in, for example, theories of situation awareness. In adopting such a folk model from an applied community and subjecting it to putative scientific scrutiny, human factors of course meets its pragmatist ideal. Folk models fold neatly into the concerns of human factors as an applied discipline. Few theories can close the gap between researcher and practitioner better than those that apply and dissect practitioner vernacular for scientific study. But folk models come with an epistemological price tag. Research that claims to investigate a phenomenon (say, shared situation awareness, or complacency), but that does not define that phenomenon (because, as a folk model, everybody is assumed to know what it means), cannot make falsifiable contact with empirical reality. This leaves such human factors research without the major mechanism for scientific quality control since Karl Popper.

Can the once pragmatic ideas of human factors and system safety still adequately meet the problems that have started to emerge from today's world? We may be in for a repetition of the shifts that came with the technological developments of World War II, where behaviorism was shown to fall short. This time, it may be the turn of human factors and safety. Contemporary developments, however, are not just technical. They are sociotechnical: understanding what makes systems safe or brittle requires more than knowledge of the human–machine interface. As David Meister pointed out (and he has been around for a while), human factors has not made much progress since 1950. "We have had 50 years of research," he wonders rhetorically, "but how much more do we know than we did at the beginning?" (Meister 2003, p. 5). It is not that approaches taken by human factors and system safety are no longer useful, but their usefulness can only really be appreciated when we see their limits. Building on its predecessor (*Ten Questions about Human Error*), this book is an installment in a larger transformation that has begun to identify both deep-rooted constraints and new leverage points in our views of human factors and safety. It raises questions about human factors and system safety as disciplines, about where they stand today and what they stand on top of. This book attempts to show where our current thinking is limited; where our vocabulary, our models, and our ideas are constraining progress. In every chapter, the book tries to provide directions for new ideas and models that could perhaps better cope with the complexity of problems facing us now. If it talks about what we have done so far, then that should not be taken as criticism. It should be taken as critique. There is a difference between criticism and critique. Criticism is an expression of disapproval because you see faults or mistakes. It is usually not very constructive; it can mean the *end* of a conversation. Critique is not as normative. It is more inquisitive. Critique is about revealing and questioning the assumptions underlying an idea. Critique can be empowering. When you start discovering the kinds of analytical trade-offs and premises that hold up some of our most popular, most cherished ideas, then that can be the *beginning* of a great conversation. A great conversation in class, in the literature, with clients, students, colleagues, lay people. I hope it is a book that makes you stop and think. And that it makes you think critically.

Technological changes have traditionally given rise to most of our human factors and safety thinking. The practical demands posed by technological changes endowed human factors and safety with the pragmatic spirit they have to this day. But pragmatic is no longer pragmatic if it does not match the demands created by what is happening around us now. The pace of sociotechnological change is not likely to slow down any time soon, nor is the growth of complexity in our world. If we think that World War II generated a lot of interesting changes, giving birth to human factors as a discipline, then we may be living in even more exciting times today. If we in human factors and safety keep doing what we have been doing, simply because it worked for us in the past, we may become one of those systems that drift into failure. Pragmatics requires that we too adapt to better cope with the complexity of the world facing us now. Our past successes are no guarantee of continued future achievement. This is why we need safety, differently. This is why we need to develop human factors for a new era.

Acknowledgments

For the ideas developed in this book, I am particularly indebted to the work of Ken Gergen, John Flach, Jens Rasmussen, David Woods, Erik Hollnagel, Nancy Leveson, Judith Orasanu, Gene Rochlin, James Reason, Gary Klein, Karl Weick, John Senders, and Diane Vaughan. Among an even larger number of students and colleagues—past and present—I gratefully acknowledge my collaborations with James Nyce, Johan Bergstrom, Kip Smith, Margareta Lutzhoft, Isis Amer Wahlin, Nicklas Dahlstrom, David Capers, Roel van Winsen, Eder Henriqson, Penny Sanderson, Maurice Peters, Caroline Bjorklund, Jens Alfredsson, Gideon Singer, Heather Parker, Rob Robson, Nancy Berlinger, Orjan Goteman, Ivan Pupulidy, Shawn Pruchnicki, Hans Houtman, Erik van der Lely, Monique Mann, Shane Durdin, Daniel Hummerdal, Mike Goddu, and Rick Strycker. Then there are the people who were able, throughout the years, to contribute with the kind of research funding that allows one's imagination to roam; the support that creates the space for new ideas to germinate and grow. I mention especially Arne Axelsson, and more recently Martin O'Neill, Kelvin Genn, Corrie Pitzer, and John Green. Finally, a thank you to my students who persuaded me to produce a second edition of *Ten Questions about Human Error* and to Ebony King, my research assistant, without whose unfailing help there still would be no second edition, no *Safety, Differently*.

Author

Sidney Dekker is currently a professor at Griffith University in Brisbane, Australia, where he runs the Safety Science Innovation Lab. Best-selling author of many books in human factors and safety, he has recently been active flying the Boeing 737NG as a part-time airline pilot.

1 Making the World a Better Place

CONTENTS

KEY POINTS

- During the 20th century, a remarkable transition occurred in how we look at the relationship between safety and the human factor. This transition was behind the emergence of the field we call "human factors."
- In the first half of the 20th century, the human factor was mostly seen as the cause of safety problems. Safety interventions targeted the human. The human had to be picked for and adapted to fixed systems and technologies. Safety problems were addressed by controlling the human.
- In the second half of the 20th century, the human was seen as the recipient of safety problems. Safety interventions targeted the system, where technology was not taken as fixed but adaptable to human strengths and limitations. Safety problems were addressed by controlling the technology.
- The reformation that has thus occurred has recently been met with a counter-reformation, in which workers are increasingly given responsibility for

safety practices and behaviors, and sanctioned if they do not comply. This trend is consistent with the criminalization of human error as well as with the bureaucratization of safety in many industries.

- Human factors has been empowering, as it looks for safety improvements in the system surrounding people rather than just the people. This has, however, reaffirmed the modernist faith that a better hierarchical ordering of organization and work translate into greater safety.

- Modernist assumptions, particularly our faith in the rationality of our technologies, systems, and bureaucracies, are evident in both approaches to safety and the human factor. Many organizations rely on safety management systems, loss prevention systems, or similar bureaucratic machineries to count and tabulate negative events as evidence for the absence or presence of a "safety culture."

- Burgeoning safety bureaucracies preoccupied with counting and tabulating lagging negatives (e.g., "violations," or deviations, incidents) tend to devalue expertise and experience. They turn safety from an ethical responsibility *for* operational people into a bureaucratic accountability *to* non-operational people.

- Moving to a new era in safety will involve seeing people not as a problem to control but as a solution to harness. It will require seeing safety as a presence of capabilities to make things go right, rather than the absence of negatives. And it means we recommit to safety as a responsibility *for* people who do our safety-critical work, rather than as a bureaucratic accountability *to* people above us.

SAFETY AND THE HUMAN FACTOR IN THE 20TH CENTURY

HUMANS OR TECHNOLOGY AS THE TARGET FOR SAFETY WORK

Since the early 20th century, the pursuit of safety has been an integral part of making the world a better place. To pursue progress on safety, most industries, if not all, have had to take the human factor into account. People, after all, make a difference in how safety is created, assured, engineered, and designed for—as well as forgotten and broken. The understanding of the human factor has undergone a remarkable transformation during the 20th century. The shift, roughly, is this (see also Table 1.1):

- In the first half of the 20th century, the human factor was mostly seen as the cause of safety trouble. Safety interventions targeted the human—through aptitude testing, selection, training, reminders, sanctions, and incentives. Technologies and tasks were seen as fixed: the human had to be picked for and molded to them. Individual differences between people were to be exploited to fit the human to the system. Safety problems were addressed by controlling the human.

- In the second half of the 20th century, the human was seen as the recipient of safety trouble—trouble that was created upstream and then handed down by imperfect tools, technologies, or tasks. Safety interventions targeted

TABLE 1.1
How Our Understanding of the Relationship between Safety and the Human Factor Has Undergone a Remarkable Transformation during the 20th Century

First Half of the 20th Century	Second Half of the 20th Century
Human is cause of trouble	Human is recipient of trouble
Safety interventions target the human through selection, training, sanctions, and rewards	Safety interventions target the organizational and technological environment
Technology and tasks are fixed; the human has to be picked for them and adapted to them	Technology and tasks are malleable and should be adapted to human strengths and limitations
Individual differences are key to fitting the right human to the task	Technologies and tasks should be devised to be error-resistant and error-tolerant, independent of individual differences
Safety problems addressed by controlling the human	Safety problems addressed by controlling the technology
Psychology is useful for influencing people's behavior; it allows us to engineer people to fit our systems	Psychology is useful for understanding perception, attention, memory, and decision making, so that we can engineer systems fit for people

the system—with better design and better organization. Technology was not taken as fixed but as malleable, to be adapted to human strengths and limitations. Individual differences were much less important than devising technologies and systems that would resist or tolerate the actions of individuals, independent of their differences. Safety problems were addressed by controlling the technology.

The shift has been accompanied by a change in psychological theorizing as well. Behaviorism was the school of psychology that dominated the first half of the 20th century. It was a psychology that was not interested in mental phenomena but in molding and shaping people's behavior to existing environments using clever systems of selection, rewards, and sanctions. Interrogating the mind to understand why people did what they did was not as important as working with their behavior to get them to do the right thing. The second half of the 20th century, in contrast, saw the growth of cognitive psychology, social psychology, and engineering psychology. Mental and social phenomena once again became important for understanding how best to design and engineer technologies that fit the strengths and limitations of human perception, memory, attention, collaboration, communication, and decision making.

BEHAVIORAL SAFETY AND RESPONSIBILIZATION

Of course, history never splits into two as neatly as this. Taylorism, as we will see, is a movement from the early 20th century. And indeed, it was obsessed with fitting people to the task. Measurement of people and their tasks, timing the execution of work, studying the motions of people, recording, administering, and a host of other bureaucratic machinery made Taylorism into a modernist intervention if ever there was one. But Taylorism also looked critically at the system—at plant and production line design, planning, and supervision—as it realized that just asking people to work harder with imperfect technologies was not very useful.

On the other side of the transformation, the late 20th century shows a reversion to many of the beliefs from the first half of the century. Behavioral safety has made a comeback in many industries, as industry leaders and safety managers are looking for ideas to further reduce their flat-lining incident and injury numbers even *after* they have put all kinds of systems, procedures, treatises, and protections in place. This makes the human, and his or her behavior, once again the target for intervention. Such safety work is driven by vocabularies of human deficit and the imposition of constraint and control by systems. The vocabulary can be focused on what humans should not do, on the things they are not good at—on the need to control their shortcomings in an otherwise great system:

> It is now generally acknowledged that individual human frailties … lie behind the majority of the remaining accidents. Although many of these have been anticipated in safety rules, prescriptive procedures and management treatises, people don't always do what they are supposed to do. Some employees have negative attitudes to safety which adversely affect their behaviours. This undermines the system of multiple defences that an organisation constructs and maintains to guard against injury to its workers and damage to its property (Lee and Harrison 2000, pp. 61–62).

Even legally, there is a neo-liberal trend toward "responsibilization" in many Western and other countries. A recent study shows how workers are assigned ever more responsibility for their own safety at work and are held accountable, judged, and sanctioned as individual instigators of trouble—not collective recipients of it. In some places and industries, workers themselves are increasingly blamed (sanctioned, ticketed) for safety violations, with over two-thirds of all citations handed out by workplace safety inspectors directed at workers or immediate supervisors (Gray 2009). This is a departure from the gradual moral and juridical enshrining of the notion of *employers* as safety offenders (i.e., blaming systems, organizations, and sometimes the people behind them) during the 20th century. Now, individual responsibility once again falls heavier on workers who are "instructed to become prudent subjects who must 'practice legal responsibility'" (Gray 2009, p. 327). Workers, in other words, are enticed (by winning their "hearts and minds") to do the right thing: pay attention, wear protective equipment, ensure machine guarding, use a lifting device, ask questions, speak up. And if they don't, "the failure to practise individual responsibility in the face of workplace dangers is often used to explain why workers who perform unsafe jobs become injured" (Gray 2009, p. 330). The putative scientific legitimation of this new responsibilization was not lost on those on the receiving end:

> There's a lot of dangerous crap at work. You breathe it, you lift it, you touch it, you despair of it. Now, thanks to the Health and Safety Executive, you may have to swallow a lot more BS too. HSE is dusting off the discredited science of "behavioural safety" so

however many hazards you face at work, when things go wrong you can safely assume "it's all your fault." ... Gone are demands for engineering control, toxic use reduction, and ergonomic job design, as attention shifts to workers wearing personal protective equipment and using proper body position. Gone is any focus on how work is organized or being restructured—issues like adequate staffing levels, limits on extended work hours, humane work load and work pace are not even considered. In fact, BS schemes can increase the dangers of work. These programmes and policies have a chilling effect on workers' reporting of symptoms, injuries and illnesses (USW 2010, p. 33).

Such a view of failures, particularly in the face of campaigns to win workers' "hearts" for safety, has deep ethical implications. A failure to comply constitutes not only a citable legal offense, but a moral failing, a lack of heart. As Leape put it, we have "come to view an error as a failure of character—you weren't careful enough, you didn't try hard enough. This kind of thinking lies behind a common reaction…: how can there be an error without negligence?" (Leape 1994, p. 1851). The consequences for progress on safety can be quite regressive. Note the idea of "unsafe acts" (see Reason 1990, 1997), which has been quite a prominent component of safety thinking over the last 30 years, but is seen as problematic and contributing to blaming the worker here:

A behavior-based approach blames workers themselves for job injuries and illnesses, and drives both injury reporting and hazard reporting underground. If injuries aren't reported, the hazards contributing to those injuries go unidentified and unaddressed. Injured workers may not get the care they need, and medical costs get shifted from workers compensation (paid for by employers) to workers' health insurance (where workers can get saddled with increased costs). In addition, if a worker is trained to observe and identify fellow workers' "unsafe acts," he or she will report "you're not lifting properly" rather than "the job needs to be redesigned" (Frederick and Lessin 2000).

At the same time, something as seemingly fundamental as safety by design is making only slow inroads in some industries—even though technologies have putatively been the central target of safety interventions for half a century. Safety by design is an eminent way to address safety problems by controlling technology rather than the human. Standardized design of equipment has long been expected practice in aviation, and there are strong links between the safety record of an industry and the extent to which it has adopted standardized designs that are error resistant and error tolerant (i.e., helping people avoid error and harmlessly absorbing the consequences in case they happen anyway) (Amalberti 2001; Billings 1997). This is not true for all industries. Many tools and technologies available today still pose unnecessary risks by being neither error resistant nor error tolerant—from meat slicers in supermarkets to scissor lifts in construction. Also, some of their guards or protective features can get in the way of doing the job those technologies are meant to support or enable. Some of the very typical interventions today, then, are once again driven by logics from the first half of the 20th century. People are the target. Stick warnings and procedures on the unchanging technology; tell people to watch out and be careful.

SAFETY BY DESIGN IN SINGAPORE

A clear example of safety by design comes from green buildings in Singapore. A city state with a very limited area, its residents have built ever-higher buildings. But they prize green, and no wonder: Singapore is in the tropics. Green buildings, with walls clad in tropical plants, beautify the cityscape and are quite popular. The problem, of course, is maintenance: gardening on a vertical surface that rises up 10 or 20 floors poses unique challenges. There are the normal working-at-height interventions that we see in window cleaning (harnesses, cables, hanging scaffolding). But that still places workers on the outside of the building and creates its own unique risks. In one building, the green walls have been made rotatable inwards. This allows workers to simply take the elevator, go to the appropriate floor, unseat the pin that holds a wall section fixed, swivel it around to face them, do their gardening, then rotate and fix it back in place. That is safety by design. The risk and the opportunities for error have been designed out of the system even before it has become operational.

Our current situation, with the push and pull from reformation and counter-reformation—from humans to systems and back—can make more sense if we take a careful look at where we have come from. This is the main purpose of this chapter: to lay out how we have evolved in our thinking about safety and the role of the human factor in it. It will set the transformation around the middle of the 20th century in the context of intellectual and technological developments, both before and after. It will explore the shifting role of psychology and our view of humanity—ourselves—as we transformed our understanding of accident risk. Dismayingly, this history may end up where it started: with humans as the central problem to control. Today, however, those efforts to control the errant human factor have the full force of modernism behind them: the scientific–technical rationality of surveillance, of monitoring and counting and tabulating and storing workplace behaviors—from loss prevention systems in the offshore industry, to error counting in airline cockpits, to data recorders in hospital operating rooms. And when that is still not effective, we turn increasingly to the judicial system to deal with the human factor. Industries ranging from shipping to aviation, healthcare, and construction are reporting an increase in the criminalization of human error. Italy has a specific criminal category of "causing air disaster" on its legal books, and Sweden recently debated the introduction of a similar "patient safety crime" (Dekker 2011a). Alarmingly, the supposedly rational, fair, and truth-seeking judiciary has begun to co-opt the very concepts we perhaps once innocently coined. As I write this, an operator is in jail for four years after the prosecution successfully argued he had "lost situation awareness" in a bad-weather, automation-related accident that killed two people (see Chapter 4). In the worst cases, this represents a rather dystopian view of humanity; a frightening and unpleasant view, where a supposedly degraded workplace ethic can be handled only by stronger totalitarian and juridical control of the human factor. If ever there was a time to open our minds to a new era in safety, it is now.

IMPROVING WORK, IMPROVING THE WORLD

Let's start with a more utopian, humanist assumption. People involved in safety work want to make the world, or their small part of it, a better place. They want to avert tragedy, enhance security and resilience, and avoid injuries and incidents. These are foremost social and ethical aspirations. They represent a kind of ambition that has, in our not too distant history, played out on a much grander scale—during the Enlightenment and Scientific Revolution. It is from there that our sheer ability to even think in terms of risk, safety, and the human factor springs. Managing safety effectively, after all, takes a few things that were not widespread in the Western world before:

- The sense that bad things are not the result of fate or some god's will, but that humans were in control of their own destiny, by using their own intellect. In a fatalistic or deeply religious society, there is little point in human efforts to control risk: divine predestination cannot be overruled, after all.
- The wealth to develop and build technologies and systems that can improve or ensure safety. Without the enormous growth in capital and its spread from a few feudal, royal, or ecclesiastic hands to entrepreneurs and (eventually) their workers, there would have been no resources to invest in safety protection.
- The large-scale scientific–rational measurement, recording and storing of data, including accident and incident data. This made safety problems more visible and public, something that was accompanied by a growth in risk securitization and the insurance industry.
- The democratization and gradual moral maturing of Western societies, in which accidental death, injury, violence, and risk have all become increasingly illegitimate, and where worker and middle classes have become increasingly vocal with demands for safety protection.

The Enlightenment and the Scientific Revolution (and the resulting Industrial Revolution) were largely responsible for introducing this, which is why we turn to the first of them now and trace the influence of the others later in the chapter.

THE ENLIGHTENMENT

The Enlightenment is vital to our understanding of the deeper assumptions and hopes on which our field is founded. The Enlightenment is the collective label for a series of intellectual, political, cultural, and social movements across Europe and later in the American colonies during the 17th and 18th centuries. The overriding purposes were to try to reform society; to dislodge it from the power of the Church and monarchies; to challenge ideas grounded in tradition, hierarchy, and faith; to give common people a voice. Important Enlightenment thinkers included Spinoza in the Netherlands; Hobbes and Locke in England; Kant in Konigsberg; Smith and Hume in Scotland; Voltaire, Rousseau, and Montesquieu in France; and Franklin in North America. National expressions of Enlightenment varied widely. In France, it morphed into a battle against government domination and censorship; in England, it had significant scientific implications (Isaac Newton was given a knighthood and lucrative government office in charge of the mint); in the German states, it reached deep into the middle classes and became nationalist without threatening governments or churches. In the Netherlands, economic progress during the Golden Age (roughly most of the 17th century) aided liberalization and democratization characteristic of the Enlightenment, and it was further fueled by a Protestant reaction against the Catholic counterreformation of the century before.

The various Enlightenment movements across Europe began in earnest with the Scientific Revolution in the 16th and 17th centuries. New scientific insights began to unseat long-held Christianity-driven ideas about a deistic, geocentric cosmos, and made room for discoveries about a whole range of natural phenomena. These discoveries offered new ways to influence and control nature's vicissitudes, in fields ranging from medicine to chemistry, physics, and engineering. Science could indeed make the world a better place—visibly, systematically, demonstrably so. Newton's ideas, for example, promised that nature really was a simple, ordered domain. We could discover the rules by which it worked if we just studied it long and carefully enough. Laws governed the workings of nature, and human reason could capture those laws. Enlightenment thinkers relied on humans, not a god; and on human reasoning, not the dictates of a king or tradition. They were distrustful of other sources of knowledge and authority—not surprising after centuries of domination by church and crown. Scientific insight and the capacity for human reason and for rationality were powerful enough to literally shape the world, to make the world a better place. The Enlightenment laid the philosophical basis and offered practical schemes for improving the human condition. It had a major, lasting impact on the culture, politics, and governments of the Western world, eventually expressing itself in things such as universal suffrage, abolition of indentured labor and slavery, a greater focus on public health, and access to healthcare, public housing, and education. An essentially optimistic vision for humanity, the Enlightenment promised that, by using reason, progress and improvement were always possible.

FROM CRAFTSMAN TO WORKER

In pre-modern systems, designers and users had considerable autonomy. The builders of printing presses in 11th century China, or 15th century Europe, for example, had ample discretion in the use of materials available to them; in the number, dimension, and shapes of the parts they used; in the size of the paper that would fit into it; and more. There was a craftsmanship in design and building and using the press. Designer, builder, and user were often the same person. Craftsmen were not equivalent: some were better than others and had strong reputations because of it. Variations in quality and product safety were very common. In some ship-building today, this craftsmanship is still visible. It leads to ship bridge designs that are highly variable and often unique, where system operating skills learned on one bridge are not easily transposed onto another bridge, and substitution errors or other confusions easily arise (Lützhoft and Dekker 2002). Fitts and Jones and their colleagues (soon to make their appearance in this chapter) also gathered a number of interesting accounts where (what seemed like) craftsman-like production of aircraft led to predictable problems:

> We had an alert one morning about eleven o'clock, because about 35 Japanese planes had been picked up on the radar screen. In the mad scramble for planes, the one I happened to pick out was a brand new ship which had arrived about two days previously. I climbed in and it seemed the whole cockpit was re-arranged. Finally, I got it started but the Japs hit just about that time. The rest of the gang had gotten off and were climbing up to altitude. I took a look at that instrument panel and viewed the gages around me, sweat falling off my brow. The first bomb dropped just about a hundred yards from operations. I figured then and there I wasn't going to take it off but I sure could run it on the ground. That's exactly what I did—ran it all around the filed [sic], up and down the runway, during the attack (USAF 1947, p. 37).

On an interesting side note, surgery today is sometimes characterized as having a craftsman outlook (Amalberti et al. 2005). Surgeons are often selected and known for their individual reputations, in contrast to, for example, anesthetists. Anesthesia is very safe, and as a field of practice and technology, it is relatively stable and standardized. To be sure, there is still a lot in anesthesia equipment design that is far from stable or standardized (Cook et al. 1991). But the risk of death in otherwise healthy patients undergoing anesthesia is very low: close to 10^{-6} deaths per anesthetic episode. This is not far from the safety level of commercial aviation in most Western countries (Amalberti 2001). Surgery is not as safe as anesthesia: it lies closer to 10^{-4} and unintended nonfatal harm is even more common (Baker et al. 2004). Of course, ultra-standardization and actor equivalency are often eschewed for good reasons: neither the anatomy to be operated on nor the knowledge base that drives such interventions (particularly in cancer surgery, for example) is very standard or stable.

The Industrial Revolution changed this for many fields. Opportunities and calls for mass production introduced a whole different set of requirements. Functionality and uniformity became important means to achieve efficiency and predictability. Take the steam-operated rotary presses that replaced hand-operated Gutenberg-style printing presses in the 19th century, for instance. Their increasing standardization was meant to achieve higher output and linearity in production processes. It called for a particular kind and size of paper and replaceable standard parts. It transformed the printer from an actively involved craftsman who was intimately connected to the design and building of his or her press to a more detached and dispensable industrial worker. The worker had little or no say in the design, building, or placement of the press and had little influence in planning or pacing the work that was done with it. Workplaces became increasingly seen as "machines to work in," just as cars became "machines to travel in" and buildings became "machines to live in" (Harvey 1990). Anybody involved in designing, building, or using printing presses (or pretty much any other means of production) was forced to accept ever-stricter limitations on their individual discretion and autonomy. Where craftsmanship had determined design specifications from the bottom-up, these limitations were imposed from the top-down.

CONTROL FROM ABOVE

Ideas of the Enlightenment thus became an achieved reality in the modernism of the 19th and early 20th centuries. The rational planning, science, and technology of the Industrial Revolution putatively made the world a better place for growing masses of consumers. Mechanization and mass production meant that they could now routinely get access to things that had been out of reach—cloth, sugar, cars, ideas printed in their own language, spare time. It was hugely empowering. But was this only a good thing? Control over the environment, over people's work, was achieved by imposing control hierarchically, from above. This is modernism. It wants to improve the world by making things more rational and more technical, by making them less haphazard and less improvised. It centralizes the planning, modeling, design, and building of things. And it separates that from their use, from the execution of the work that is done with them.

In this, irrational rather than rational consequences became increasingly visible. The Industrial Revolution brought new workplace accidents. It produced diseases from working with toxic materials, led to environmental degradation, and created a new, entrenched urban underclass living largely in poverty. Child labor and new forms of servitude to capitalist bosses were extremely common. In many cities, life expectancy in the new Industrial Age in many cities was shorter than it had been during the Middle Ages (Mokyr 1992). And new forms of authority and knowledge (scientific, managerial, corporate) excluded other voices, just like church and crown had done before. Take Frederick Taylor's "scientific management" as an example. It promised smarter, better, cheaper manufacturing by doing everything that modernism prescribed: rational control, hierarchically imposed from above, standardization of tasks and equipment, the eradication of initiative, improvisation, bottom-up innovation or worker autonomy, the parceling up of work and skills into minute, controllable bits. Scientific management pretty much asserted that managers were smart and workers were dumb. Workers could execute, but not model the work, nor manage it. Planning the work should not be left to them, and so their voices and opinions were largely irrelevant. They were not worth listening to, only worth looking at for the purposes of study (the classic time-and-motion studies), intervention, and better control. With appropriate application of rationality, any skilled craft could be decomposed into the tiniest fragments of work so as to increase its efficiency.

FREDERICK TAYLOR AND SCIENTIFIC MANAGEMENT

Frederick Winslow Taylor (1856–1915) is best known for "Scientific Management," which is also known as Taylorism. His and Frank Gilbreth's innovation was to study work processes "scientifically" (e.g., by time-and-motion studies): how was work performed and how could its efficiency be improved? Although there were substantial philosophical differences between the two men, both had backgrounds in engineering (mechanical and industrial). Both

Taylor and Gilbreth realized that just making people work as hard as they could was not as efficient as optimizing the way that work was done. Careful, scientific study could not only identify the minimum chunks of labor that any task could be reduced to, but also help specify the required skill levels and expected or possible workloads at each station or on each line. Early experiments with bricklaying and loading pig iron (oblong blocks of crude iron obtained from a smelting furnace) showed success: by analyzing, reducing, and then redistributing micro-tasks, efficiencies in laying the bricks and loading the iron went up dramatically. By calculating the time needed for the component elements of a task, Taylor could develop the "best" way to complete that task. This "one best" standard allowed him, in proto-behaviorist fashion, to promote the idea of "a fair day's pay for a fair day's work" (which in itself might have aided the efficiency increases): if a worker didn't achieve enough in a day, he didn't deserve to be paid as much as another worker who was more productive. Scientific management made those productivity differences between workers more visible, and thus a better target for monetary incentives.

Production lines were soon to follow. The problem that motivated them came from the production line—which was not pioneered in automobile manufacturing, even though Henry Ford readily copied the idea. The production line that occupied Taylor and colleagues was that of the Chicago slaughterhouses, and their major issue was "balancing the line." Workers, despite their input, were unable to clear bottlenecks or make up shortfalls between the different meat processing stations along the line. It was a problem not of effort, but of planning and coordination. "Almost every act of the workman should be preceded by one or more preparatory acts of the management," Taylor said, "which enable him to do his work better and quicker than he otherwise could. A worker should not be left to his own unaided devices" (Geller 2001, p. 26). Supervisory control emerged as a new class of work. Supervisors set the initial conditions, received information about the details of activities on their shop floor, and accordingly fine-tuned and adjusted the work assigned (Sheridan 1987).

Taylor developed four principles of scientific management on the basis of his experiments and insights (Geller 2001):

1. Do not allow people to work by heuristic ("rule of thumb"), habit, or common sense. Such human judgment cannot be trusted, because it is plagued by laxity, ambiguity, and unnecessary complexity. Instead, use the scientific method to study work and determine the most efficient way to perform a specific task.
2. Do not just assign workers to any job. Instead, match workers to their jobs based on capability and motivation, and train them to work at maximum efficiency.
3. Monitor worker performance and provide instructions and supervision to ensure that they are complying with the most efficient way of working. Affairs of workers are best guided by experts.

4. Allocate the work between managers and workers so that the managers spend their time planning and training. This allows the workers to perform their assigned tasks most efficiently.

These are principles from more than 100 years ago, but their reverberations are felt every day. Whenever problems in safety or efficiency arise, it is still quite natural to reach for Tayloristic solutions. We might tell people to follow a procedure more carefully, to comply better to the "one best method" that has been determined for their task. We might choose to increase the number of occupational health and safety experts who plan and monitor work, for example. Or we might consider ways of automating parts of the work: if machines do it, then human errors will no longer be possible. Indeed, Taylorism gave rise to the idea of the redundancy of parts. Because scientific management chunked tasks up into their most elemental parts, those could simply become seen as interchangeable parts of an industrial process—whether done by human or machine. Whichever did the task more reliably and efficiently should get to do it. The idea is still with us in what is known as function allocation or MABA/MABA (Men-Are-Better-At/Machines-Are-Better-At) lists (Dekker and Woods 2002). The basis for this is Taylor's redundancy-of-parts idea: by conceiving of work in terms of the most reducible component tasks, it became possible to think about automating parts of those component tasks. If a machine could do that task more reliably and more efficiently than a human, the task should be automated. Today, this is known as the substitution myth—the false idea that humans and machines are interchangeable. This myth, however, has a powerful hold on many ideas (popular as well as scientific) in human factors and safety. More will be said about it in the chapter on automation.

Another very important innovation was Taylor's separation of planning from execution. It represented a major departure from the craftsman phase that had preceded mass industrialization. With it went the image of humans as innovative, creative, and intrinsically motivated. Taylorism became the embodiment of Cartesian ideas from much earlier in the Scientific Revolution and Enlightenment: that of humans as machines, as mere "locomotives capable of working" (more about Descartes later in the book). Taylor was genuinely surprised at the bottom-up resistance that his ideas created. He had not foreseen (or had perhaps no way to see or interpret) how his work dehumanized labor, how the eradication of worker initiative, of local expertise and craftsmanship not only brought efficiency but also hollowed out the meaning of work. The immense personnel turnover in major factories like Ford's bewildered him. Did he not bring those people a crisper, cleaner, more predictable world? Henry Ford had to go to great lengths to "bribe" his workers to stay, making the radical move of offering a $5-a-day wage. This had the added benefit that workers like his could now afford to buy the cars they produced, thus increasing the size of Ford's market dramatically. Ford's initiative helped convert what had been a society of producers (scratching out a living on farms and in workshops) into one of consumers.

The Taylorist principles that separated planning from execution, that standardized tools and equipment, and that rationally analyzed and decomposed work, are visible in the work of human factors and safety in many forms today. The requirement to carefully execute routines that were planned by someone else (in the form of checklists and procedures and using standardized equipment) is common in many fields of practice and industries today. Deviating from such procedures is a "violation." This is a term often used by human factors and safety professionals. As explained above, it implies both normativism and moralism. There is a norm that should be upheld (by following the rules and plans drawn up by someone else) and a professional or even moral deficiency in not doing so. To Taylor, this would all have made good sense. He would have recognized himself and his ideas in both the importance we give to following procedures and the moral undertones of the word *violation*. Implementing his scientific management ideas was more than smart management, he argued. It was a moral imperative—it was the right thing to do, the ethical thing to do. Taylor did not have to make up the philosophical foundations for his ideas, for those were modernism's own. As the next chapter will bear out, the Newtonian notion of "one best method" to achieve a result is everywhere we write procedures or checklists and demand compliance with them. But we are getting ahead of ourselves. Let us first look at how our thinking about accidents and the human factor grew among, and out of, the Industrial Revolution. We will also see what the role of psychology was in this, and how all of this changed with the Second World War.

HUMAN FACTOR AS PROBLEM TO CONTROL

WORKERS AS CAUSE OF ACCIDENT AND INJURY

Pure accidents, said American railway safety expert H. B. Rockwell in 1905, don't happen. There is nothing purely "accidental" about the events we call accidents. Rather, "someone has blundered, someone has disobeyed an order or undertaken to reverse a law of nature" (Burnham 2009, p. 17). Employees' carelessness or negligence, as well as passengers' or pedestrians' thoughtless or risky behavior, were often responsible for safety problems and failures, thought Rockwell. In the early 20th century, mention of the "human factor" became more prevalent. Both Britain and Germany had pioneered many of the safety interventions directed at their workers that would later be characterized by the human factor. In Britain in 1919, the human factor was explained as "the worker [who] through some lack or defect or through the influence of unfavourable surroundings, may contribute to the accident which causes him injury, and [whom] as such may be regarded as a cause" (Burnham 2009, p. 18). There was mention of "unfavourable conditions," but the location of cause was unambiguous: the human factor was ultimately responsible for safety trouble.

Attention to accidents had been growing since the late 19th century. Accidents were slowly moving from the personal domain, into the public. They were no longer just a private concern. Where they had previously been seen as unfortunate but unpredictable (and essentially unpreventable) coincidences of space and time, their number, scope, and cost started mounting as the Industrial Revolution unfolded across Europe and its (former) colonies. Germany had been paying systematic attention to workplace injuries and fatalities since the 1880s. Individual plant owners were starting to tabulate their own statistics across the Western world, which helped develop tallies of the kinds and totals of injuries. This in turn helped lay the basis for interventions and for the examination of individual workers' accident records (Burnham 2009). By 1920, "industrial accidents" were a well-known category of cost, concern, and suffering not only for workers, but for directors, managers, regulators, magistrates, lawyers, courts, and insurers. Oxford psychologist Vernon, having amassed a record of 50,000 personal injury accidents, concluded that "Accidents depend, in the main, on carelessness and lack of attention of the workers" (Burnham 2009, p. 53), despite mitigating factors of fatigue, lighting, and production pressure.

ACCIDENT-PRONE WORKERS

The idea that workers themselves are the cause of accident and injury dominated the first half of the 20th century. As early as 1913, Tolman and Kendall, two pioneers of the American safety movement, strongly recommended managers to be on the lookout for men who always hurt themselves and to take the hard decision to get rid of them. In the long run, that would be cheaper and less arduous than keeping them around the workplace. In 1922, a US management consultant named Boyd Fisher published *Mental Causes of Accidents*, a book that explained how various kinds of individual actions and habits were behind error and accident. This included, he found, ignorance, predispositions, inattention, and preoccupation. Some of these were transient mental states; some were more enduring character traits. Boyd also commented on the connection between age and the tendency to accident, pointing out that the young were more liable to accident (as were very old men). This is echoed today in observations by some managers of Generation Y workers' tendency to be involved in more shortcuts, injuries, incidents, and violations than older colleagues. Apart from the many accidents that had long been occurring in mining, construction, and manufacturing, transportation too, first by rail and then road, became a major and growing producer of accidents. As early as 1906, 374 people died in automobile crashes in the United States alone. By 1919, the number for the year had climbed to 10,000. Concern with industrial accidents, in part because of their sheer numbers, grew again during the Great Depression and the government-sponsored employment projects it spawned (the Hoover dam, for example). The new disparity between the supply and demand for cheap labor in Western economies might have had consequences for the amount of protection afforded to workers.

The counting and tabulating of incident and injury statistics in the first quarter of the 20th century started yielding a pattern. Around 1925, independently from each other, British and German psychologists suggested that there were particularly "accident-prone" workers. The inclination to "accident" (used as a verb back then) was proportional to the number of accidents previously suffered, German psychologist Karl Marbe pointed out. The contrast was even more convincing: those who consistently had no accidents were significantly less likely to have them in the future as well. The English-speaking world, too, realized that personal, individual factors contribute to accidents. In England, Eric Farmer, industrial psychologist at Cambridge, left his work on Taylorist time-and-motion studies and began devising tests to identify people who were likely to have accidents. "The human factor" at the time was thought of as exactly that: not generalizable psychological or environmental conditions (like fatigue or error traps in equipment design), but characteristics specific to the individual, or, in the vocabulary of deficit of the time, "physical, mental, or moral defects" (Burnham 2009, p. 61). These negative characteristics could be identified by carefully testing and screening employees—something for which psychological institutes across the continent, the United Kingdom, and the United States were developing ever-cleverer simulations and contraptions. Some referred back to the putative successes of pilot selection during World War I: the scientific selection of pilot candidates helped eliminate those unsuitable to pilot airplanes and reduced human factor–related accidents from 80% to 20% (Miles 1925). The idea seemed to appeal to, and mirror, common sense as well. The medical director of the Selby shoe company in England said in 1933 that

> Anyone connected with an industrial clinic or hospital soon becomes acquainted with workers who make frequent visits because of accidents or illness. During the past seventeen years I have repeatedly remarked, "Here comes the fellow who is always getting hurt." At short intervals he reports with a cut, bruise, sprain or burn. Interspersed with his numerous minor accidents will be an occasional serious accident. Often you hear him remark, "I don't believe anybody has more bad luck than I have." He is the type of fellow who attracts the attention of safety engineers and safety committees. He and other such workers have been classed as "accident prone." … The accident prone workers are interesting people because they present so many problems and because the best authorities today admit they know so little about them (Burnham 2009, p. 35).

Indeed, if specific workers were the problem, then the target for intervention was pretty clear, Marbe thought. And it should interest not only workers or their employers, but insurance companies and government regulators as well. Marbe proposed that production damage and costs could be minimized by finding the right worker for the job. This should be done by testing workers and carefully vetting and selecting them, excluding the accident-prone ones from the jobs where they could do most harm. Moede, a colleague of Karl Marbe

> …enthusiastically claimed that standard psychotechnology could solve the problem of excesses of accidents by addressing the components of accident proneness. Psychotechnologists, he explained, started by doing a detailed analysis of accidents and errors to find out exactly what was happening. Workers with reaction and attention deficiencies could be detected by psychotechnological testing, either before or during employment. Testing, especially aptitude testing, could also identify people with other personality traits that made them predisposed to have accidents. And at that point, wrote Moede optimistically, psychotechnologists could provide either ways to screen out unsafe job candidates, or they could suggest safety training that would counter undesirable traits. Thus, the problem and the cure were already being handled by psychotechnologists (Burnham 2009, p. 48).

The other idea suggested by Marbe was the workers should have a personal card on which accidents and errors would be noted—a trace for employers and others to base hiring and firing decisions on. This proposal was not followed concretely at the time, but today similar employment records are used across many industries to guide managerial decision making about human resources in the organization. Alternative interventions targeting undesirable workers were available to employers at the time anyway, something that did not seem to ethically bother anyone. The focus was on issues of safety, managerial influence, and cost control, and keeping people out of positions where they could have accidents was the canonical way to do that. The public transport company in Boston, for example, found in the mid-1920s that 27% of its drivers on subways, street cars, or buses were causing 55% of all accidents. Psychologically testing new employees and eliminating those who tested badly on efficiency and power use helped reduce their overall accident rate—something that caught the attention of transportation operators around the world. Accident proneness had become firmly established as an individual and undesirable psychological trait. It kept a growing machinery of *psychotechnik*, of testing and selection, busy well into the Second World War. The tone of the psychologists behind the idea had become bullish, claiming that

> Accident proneness is no longer a theory but an established fact, and must be recognized as an important element in determining accident incidence. This does not mean that knowledge of the subject is complete, or that the liability of any particular individual to accident can with certainty be predicted. What has been shown, so far, is that it is to some extent possible to detect those most liable to sustain accidents (Farmer 1945, p. 224).

PSYCHOLOGICAL BEHAVIORISM

The psychology that carried most of this with concepts and theory was behaviorism. The era of behaviorism (roughly the first half of the 20th century), championed by John Watson (1878–1958) has been called a mental Ice Age (Neisser 1976). Behaviorism saw the study of mind as illegitimate and unscientific. Its rejection of the study of internal mental phenomena was itself a response to what had reigned before. Behaviorism was a psychology of protest, coined in sharp contrast against German Wundtian experimental introspection that had preceded it. Before Wundt, Watson claimed that there was no psychology. And after Wundt, there was only confusion and anarchy. Behaviorism could change that: by refusing to study consciousness and instead dedicating itself to scientifically manipulating and influencing visible behavior, psychology would attain new rigor and a new future.

BEHAVIORISM

"Psychology as the behaviorist views it is a purely objective experimental branch of natural science. Its theoretical goal is the prediction and control of behavior. Introspection forms no essential part of its methods, nor is the scientific value of its data dependent upon the readiness with which they lend themselves to interpretation in terms of consciousness" (Watson 1978, p. 435).

This is how John Watson opened his 1913 broadside against psychology as he saw it: preoccupied with internal, invisible things, using methods that were anything but scientific, and losing credibility and standing as a result. The way to deal with that? Declare that only visible things could be legitimately studied, using objective methods. Mental states could no longer be the object of observation or study: it only produced subjective theorizing. A departure from studying consciousness, or mental states, elevated the status of animal models in psychology. Consciousness would be difficult to study in animals, but behavior could be observed, and even more easily influenced than that of humans. Rats, pigeons, kittens, chickens, dogs—they could all generate valuable insights into how organisms learn and respond to incentives, constraints, and sanctions in their environment. Watson was not alone in this, or the first. Pavlov and Thorndike had gone before him with their studies of animal behavior. The term *trial and error* learning, for example, came from Thorndike, stemming from his animal work in the late 1890s.

The basic psychological unit of analysis in behaviorism was that of stimulus–response, or S–R. The brain was considered a "mystery box" in which the stimulus was connected to the response, but it did not matter how. Two factors responsible for Watson's success, and the impact he had on psychology and beyond, are worth mentioning. One is his own youthful optimism, his tough-mindedness and his uncompromising, self-confident expression in spoken and written word. The other is the era in which his ideas were able to gain traction. Published only two years after Frederick Taylor's Scientific Management and the emergence of the production line (see below), Watson's ideas about influencing human behavior were a welcome source of inspiration for managers seeking to maximize efficiencies of human labor. The world fair in Chicago in 1933 claimed in its motto that *Science finds, Industry applies, Man adapts.* Such was the vision of technological progress at that time and the optimism that had been fed by successful industrialization during the century before. Man adapts to technology—and behaviorism was going to help industry do exactly that.

TECHNOLOGY, ENGINEERING, AND NEW SOCIAL STRATEGIES

As announced at the beginning of the chapter, a remarkable transition occurred in how we look at the relationship between safety and the human factor around the middle of the 20th century. In the years leading up to World War II, the human factor was largely seen as responsible for safety trouble. It was the cause of accidents, injuries, and incidents. The target for intervention, by managers, psychologists, and engineers, was the human. Through aptitude testing, screening, selection, training, incentives, and sanction, humans were increasingly molded to tasks and technologies. Those who were particularly accident-prone were prevented from touching safety-critical technologies altogether. Expert dissent was growing, however, toward the middle of the century. Psychologists started to question the measurements and statistics that had always been the behaviorists' strongest suit. The method of percentages (such as those used by the Boston public transport system: that 27% of its drivers were causing 55% of all accidents) came under fire. It required, after all, that everyone needed to expect the same number of accidents; otherwise, deviations from that norm could never be found. This was an untenable assumption, as the mix of processes and factors (including chance) that went into the causation of accidents was way too complex and random—something perhaps better represented by a Poisson (rather than Gaussian) distribution. Indeed, the "complexities of the uncontrolled world surrounding operational accidents" ruled out that we could at all predict future accidents on the basis of having had past ones (Webb 1956, p. 144). It was concluded that

> the evidence so far available does not enable one to make categorical statements in regard to accident-proneness, either one way or the other, and as long as we choose to deceive ourselves that they do, just so long will we stagnate in our abysmal ignorance of the real factors involved in the personal liability to accidents (Arbous and Kerrich 1951, p. 424).

World War II really spelled the end of accident proneness and some of the psychological behaviorism behind it. Technological developments were so rapid, so pervasive, and so complex, that no amount of intervention toward the human worker alone could solve the emerging safety problems. Personal proneness as an explanatory factor for accident causation paled in comparison to all the other contributors: a worker's proneness to have an accident proved to be much more a function of the tools and tasks he or she was given than the result of any personal characteristics. Roger Green, one of the early aviation psychologists, commented on it like this:

> Closer examination of individual accidents reveals why this is so, for it is very rare for an accident to be caused entirely because of deficiencies or other discrete human factors problems; culpability almost invariably lies with an interaction between these and other factors. The possible number of such interactions are virtually limitless; consequently, the determination of the most likely aetiology of a given accident is likely to remain a non-precise "science" (Green 1977, p. 923).

So instead of trying to change the human so that accidents became less likely, engineers and others realized that they could, and should, change the technologies and tasks so as to make error and accident less likely. Humans were no longer seen as just the cause of trouble—they were the recipients of trouble: trouble that could be engineered and organized away. Organizational resources spent on measuring and tabulating statistics to weed out accident-prone individuals were increasingly seen as waste, hampering safety research and getting in the way of more effective considerations. Safety conferences and professional meetings in the decades after the War started to call attention more to managerial responsibilities and engineering solutions for safe work, leaving the focus on individual workers to human resource professionals or supervisors (Burnham 2009). The moral imperative shone through clearly as well (something that is being reversed in the responsibilization that was talked about in the beginning of this chapter). As one review said, this focus on the individual worker as cause of injury and incident

...may allow managements to escape their responsibilities in machine design, in selection and training of personnel in safe operating procedures, and in meticulous attention to the environment where high energy transfer can impinge on the human (Connolly 1981, p. 474).

Let us look in more detail at the emergence of "human factors" as a discipline around the middle of the 20th century. It represents one of the first and perhaps strongest transformations in our thinking about the sources of safety and risk—and where to direct our efforts to control them.

HUMAN ERROR AS A CONSEQUENCE, NOT A CAUSE

EMERGENCE OF "HUMAN FACTORS"

We made a normal take-off in a heavily loaded F-13. After leaving the ground, I gave the signal to the copilot for gear up. The aircraft began to settle back toward the ground as the flaps were retracted instead of the gear. Elevator control was sufficient to prevent contacting the ground again and flying speed was maintained. The copilot, in a scramble to get the gear up, was unable to operate the two safety latches on the gear switch and I was forced to operate this switch to raise the gear. Everything would have been all right had not the number-one engine quit at this point. Full rudder and aileron would not maintain directional control and the airspeed and altitude would not permit any retarding of power on the opposite side. As the gear retracted, the airspeed built up sufficiently to enable me to maintain directional control and finally to gain a few feet of altitude. The combination of difficulties almost caused a crash. I believe that the error of raising the flaps instead of the gear was caused by inexperience of the copilot and the location of the two switches. Numerous instances of this error were reported by members of my squadron while overseas, although no accidents were known to be caused by it. However, it could result in very dangerous situations especially when combined with some other failure (Fitts and Jones 1947, p. 350).

Thus reports the commander of a WWII era aircraft about his narrow escape from a crash. The narrative comes from one of the first systematic investigations that went beyond the label of "pilot error." In 1947, Paul Fitts and Richard Jones, building on pioneering work by people like Alphonse Chapanis, wanted to get a better understanding of how the design and location of controls influenced the kinds of errors that pilots made. Using recorded interviews and written reports, they built up a large corpus of accounts of errors in using aircraft controls. The question asked of pilots from the Air Materiel Command, the Air Training Command, and the Army Air Force Institute of Technology, as well as former pilots, was this: "Describe in detail an error in the operation of a cockpit control, flight control, engine control, toggle switch, selector switch, trim tab, etc. which was made by yourself or by another person whom you were watching at the time" (Fitts and Jones 1947, p. 332). Practically all Army Air Force pilots, they found, regardless of experience and skill, reported that they sometimes made errors in using cockpit controls.

Others, not involved in these studies, had started to notice the same problem. It was a confluence of evidence and insight that formed the beginning of a remarkable transformation in how we think about the relationship between safety and the human factor:

> It happened this way. In 1943, Lt. Alphonse Chapanis was called on to figure out why pilots and copilots of P-47s, B-17s, and B-25s frequently retracted the wheels instead of the flaps after landing. Chapanis, who was the only psychologist at Wright Field until the end of the war, was not involved in the ongoing studies of human factors in equipment design. Still, he immediately noticed that the side-by-side wheel and flap controls—in most cases identical toggle switches or nearly identical levers—could easily be confused. He also noted that the corresponding controls on the C-47 were not adjacent and their methods of actuation were quite different; hence C-47 copilots never pulled up the wheels after landing. Chapanis realized that the so-called pilot errors were really cockpit design errors and that by coding the shapes and modes of operation of controls, the problem could be solved. As an immediate wartime fix, a small, rubber-tired wheel was attached to the end of the wheel control and a small wedge-shaped end to the flap control on several types of airplanes, and the pilots and copilots of the modified planes stopped retracting their wheels after landing. When the war was over, these mnemonically shape-coded wheel and flap controls were standardized worldwide, as were the tactually discriminable heads of the power control levers found in conventional airplanes today (Roscoe 1997, p. 2).

A study like that of Fitts and Jones confirmed these insights, demonstrating how features of World War II airplane cockpits systematically influenced the way in which pilots made errors. Human factors were born out of an effort to make the work of operators more manageable, to make it more successful and less arduous.

BEYOND BEHAVIORISM

Fitts and Jones opened their 1947 paper with an insight that could hardly be simpler, and hardly more profound. "It should be possible to eliminate a large proportion of so-called 'pilot-error' accidents by designing equipment in accordance with human requirements" (Fitts and Jones 1947, p. 332). "Pilot error" was put in quotation marks—denoting the researchers' suspicion of the term. The point was not the pilot error. That was just the symptom of trouble, not the cause of trouble. The remedy did not lie in telling pilots not to make errors. Rather, Fitts and Jones argued, change the tools, and you can eliminate the errors of people who work with those tools. Skill and experience, after all, had very little influence on error rates: getting people trained better would not have much impact. Rather change the environment, and you change the behavior that goes on inside of it.

> What happened was that in the last decades of the twentieth century, experts dealing with mechanization in many settings all moved away from a focus on the careless or cursed individual who caused accidents. Instead, they now concentrated, to an extent that is remarkable, on devising technologies that would prevent damage no matter how wrongheaded the actions of an individual person, such as a worker or a driver. In the withering away of the idea of accident proneness, it is possible to see the growing dominance of a new social strategy, solving safety problems by controlling the techno-logical environment (Burnham 2009, p. 5).

The insight would be the beginning of a fundamental revision, inversion even, of how we studied and tried to improve human performance. Up to WWII, and for a decade or so beyond, the study of human behavior and cognition was governed by a fundamentally different assumption: the world was a fixed place. The only way to change behavior in it was to change something about the people. Through an intricate system of selection and training, rewards, and sanctions, the right sort of behavior could be picked out and molded so that it became adapted to its environment. As we will see later in the book, this is still a widely held idea in, for example, behavioral safety (Geller 2001) and is the foundation of many safety interventions in fields ranging from healthcare, to aviation, to nuclear power.

If behaviorism had been a psychology of protest, then human factors was a psychology of pragmatics. The Second World War brought such a furious pace of technological development that behaviorism was caught shorthanded. Practical problems in operator vigilance and decision making emerged that were altogether immune against Watson's behaviorist repertoire of motivational exhortations. Up to that point, psychology had largely assumed that the world was fixed and that humans had to adapt to its demands through selection and training. Human factors showed that the world was not fixed. Changes in the environment could easily lead to performance increments not achievable through behaviorist interventions. In a behaviorist intervention, performance had to be shaped after features of the world. In a human factors intervention, features of the world were shaped after the limits and capabilities of performance.

> I was acting as control for basic students shooting a stage. At completion of the day's flying and after all students had been dispatched to home base, I started up my airplane, BT-13, taxied out to the take-off strip, ran through pre-take-off checks and proceeded to advance the throttle. I held the plane down to pick up excess airspeed and as it left the ground, proceeded to pull back the propeller control to low RPM. Immediately, the engine cut out and I could see nothing but fence posts at the end of the field staring me in the face. Luckily, I immediately pushed the prop control forward to high RPM's and the engine caught just in time to keep from plowing into the ground. You guessed it. It wasn't the prop control at all. It was the mixture control* (Fitts and Jones 1947, p. 349).

Indeed, WWII pilots mixed up throttle, mixture, and propeller controls because the locations of levers kept changing across different cockpits. The study of Fitts and Jones showed that such errors were not surprising, random degradations of human performance. Rather, they were actions and assessments that made sense once researchers understood features of the world in which people worked; once they had analyzed the situation surrounding the operator. Their insight was that human errors are systematically connected to features of people's tools and tasks. It may be difficult to predict when or how often errors will occur (though human reliability techniques have certainly tried). With a critical examination of the system in which people work, however, it is not that difficult to anticipate where errors will occur.

---

* The mixture control on this sort of piston engine controls the inflow of the air/fuel mixture into the engine's cylinders. At the backstop of the mixture control, the air/fuel is cut off altogether, sending nothing to the cylinders.

Human factors has worked off this premise ever since: the notion of designing error-tolerant and error-resistant systems is founded on it.

This is empowering. The premise is that by designing things more smartly, human factors can make bad things go away. Greater safety and quality are achieved by reducing the number of negative events in a particular workplace. That, however, has not been enough for researchers recently. There is a growing literature on how particular designs not only make us less dumb or less error-prone but also amplify capabilities and capacities, making us considerably smarter (Norman 1993). It has been augmented by the pursuit of pleasurable products. This deploys knowledge of human factors and ergonomics to actively *add* to the quality and enjoyment of life, not just to reduce the lack of it (Jordan 1998). In safety, too, safe systems defined merely by the absence of negative events are making room for the idea that safe, or resilient, systems are capable of recognizing and adapting to challenges and disruptions—even those that fall outside the system's original design base. Rather than safety as an absence of negatives, it becomes defined by the presence of capacities, capabilities and competencies (Hollnagel et al. 2006). Yet the basic commitment of human factors and safety could be seen as this: to make the world a better place, a more workable and survivable place, a more enjoyable place. And perhaps, most recently, a more sustainable place.

TRANSFORMATION, BUT A CONSTANT OF MODERNISM

There is a constant underneath the transformation halfway the 20th century, however. And it makes the shift perhaps less fundamental than we might want to believe. That constant is modernism.

HIGH MODERNISM

High modernism is a characteristic of our modern age (particularly during the 1950s and 1960s) even though its ideas were already visible in, for example, Taylorism. High-modernist thinking is inspired by the capitalist and industrial successes triggered by the Scientific Revolution and the Enlightenment. It is characterized foremost by an unfailing confidence in science and technology: we can and should use them to reorder both nature and the social world. High modernism aims to make the world a better place by rendering it more linear, more predictable, more reliable, more standardized, more symmetrical, more organized. High modernism is recognizable in the post-war period in things as varied as how buildings were designed and cities were planned; development aid was disbursed; psychiatric and medical care was provided; vast nuclear power schemes became popular; agriculture was further mechanized and made dependent on chemical fertilization; local languages and dialects were subordinated to one central national language; measurement systems were ordered, simplified, and standardized; wars were planned and fought. The world was to be turned into a machine that could be run ever more efficiently, with great results for humanity, its "beneficiaries." French architect and urban planner Le Corbusier thought of buildings as "machines for living in." The Cartesian idea of people as machines was reproduced in high modernism as well. Local, historical, spiritual, geographic, or social contexts of human development were generally disregarded as frivolities, in favor of pragmatic, broad, generic schemes with (presumed) universal applicability. High modernism relied on a strong confidence in the expertise of its new elites: its engineers, planners, scientists, software writers, bureaucrats. The earlier Taylorist separation of planning (done by those smart people) from execution (done by others) was taken as a given for high modernism. Nature, including human nature, could and should be reordered in order to master the fates that had befallen earlier generations (natural disasters, disease, famines, riots). Many elements of high modernism can still be recognized in ways that are central to human factors and safety. For example:

- Authority is vested in those who plan (rather than execute) work, as in Taylor's time. This includes managers, engineers, central planners, safety professionals, and, to a lesser extent, supervisors (Scott 1998).
- Workers, citizens, or users are basically unable to resist the imposition of high modernism on their daily lives. It tends to make them passively receptive of new regimes. Even silly procedures and rules are silently complied with, for example, like the wearing of hard hats on a flat, open field where the only thing that could fall would be the sky itself.
- An enduring belief in scientific and technical progress, as during the Scientific Revolution and the Enlightenment. The late 1980s and early 1990s, for instance, saw this optimism expressed in the far-reaching automation of civil airliner cockpits, which would improve the reliability, universal applicability, and efficiency of airline operations. In a response typical for those against impositions of high modernism, the approach would later be doubted as not "human-centered" (Billings 1997).
- The administrative ordering of nature and society, hierarchically imposed from above. Already visible in Taylorism, such control was thought to be necessary to ensure progress and efficiency. It would also reduce the chances of unpredictable events.
- Standardization is seen as a good thing—through instruments as varied as checklists, computer-based expert systems, evidence-based medicine, procedures, and standardized designs. Craftsmanship, local expertise, improvisation, and other expressions of diversity are officially frowned upon in most operational safety-critical worlds.

Despite the difficulty to resist what is seen as an imposition of high modernism, some of it is visible today. Take medicine. In general, it has a rather complicated relationship with standardization. It has always appealed to local, anatomical, physiological, and pathological variations; to the diagnostic uncertainties and partially known medical histories; and to the multiple conditions and interactions between diseases and cures (Dekker 2011c). And, of course, it has appealed to the special status of the autonomous healer, historically and institutionally, whose insight and interlocution with the metaphysical as arbiter between life and death *have* to be better than the standard procedure or checklist (Bosk 2003; Dekker 2011c). "Evidence-based" medicine, which is about as modernist as beliefs in the rationality and truth of science can become, has tried, and often failed, to broadside this special status, despite the fact that it tends to lack evidence of its own efficacy, has very narrow definitions of "evidence," and is indeed of limited use to individual patients (Murray et al. 2008).

Other worlds of practice face similar modernist pressures. Standardization and compliance with "best practice" have long been a popular intervention in occupational health and safety (OH&S), for instance. A profusion of standards and codes of practice for everything from hazardous material is now available in every industrialized nation, as well as regulations and inspections to back them up. The doctrine of compliance tends to disempower middle management and site-level supervisors. It curtails their authority in industries ranging from construction to mining, refining, catering, manufacturing, and more. They might think a site or a job is ready to go ahead, but work can often not be started without OH&S approval. This is a deeply modernist belief in the administrative ordering of nature and society, like in Taylor's time. Office-bound OH&S people, who might never do the work, sometimes get to dictate its one best method to others who will. As a result, procedures and checklists pushed down onto work sites are used, but can become empty tick-and-flick exercises that are intended more to fend off someone else's liability than to support the execution of work. They may no longer have much to do with protecting those who do the industry's dirty and dangerous work.

FROM ONE MODERNISM TO THE NEXT

Modernism fuels both approaches to safety and the human factor that span the 20th century. In the first half of the 20th century, modernism was present in people's unfailing faith in the rationality of science and technology, their belief in measurement, standardization, prediction, and calculation, to fit humans to tasks, and to weed out the unfit. In the second half of the 20th century, it was still present in people's unfailing faith in the rationality of science and technology, their belief in engineering and design, to match systems to people:

- In the first half of the 20th century, people were the problem that needed to be controlled. Modernist ordering of people, using bureaucracy and administration, as well as science and technology, was there to help. Humans were the target of safety intervention.
- In the second half of the 20th century, systems were the target of safety intervention. With better systems, safety could be assured—independent of the differences between people in them. Better systems would ensure that people were not the recipients of bad technologies, designs, organization,

or procedures. Modernist ordering of work and organization, using bureaucracy and administration, as well as science and technology, was there to help.

In a sense, though, people remained the central problem to control, even after the human factors reformation. Designers had to start thinking about making their equipment and technologies "idiot-proof," that is, able to tolerate and resist the ill-advised actions of even the most hapless users. The problematic (or erratic, or unreliable) human had to be controlled through changes in the technology, rather than changes in behavior or attitude, but basic tenets of a dystopian view of humanity could remain in place. Designers were smart; users were dumb. Designers had to make it impossible for those dumb users to mess up the technologies provided to them. This is behind the belief that we have great systems, which behave as designed, but that still get defeated in the hands of "bad apple" users. One such case, of an aircraft accident in which nothing on the airplane had broken or was malfunctioning, will be explored in the next chapter.

In many places, modernism seems to be as far as we have come in our thinking about safety and the human factor. It is a modernism of design, systems, bureaucracy, administrative rationality, and processes. And it has much in common with Tayloristic thinking of the first half of the 20th century. Managers and designers are smart; workers and users are dumb. Traces of such modernist beliefs can be seen in one of the most iconic images of safety thinking of the second half of the 20th century. It is the image of the "Swiss Cheese," as shown in Figure 1.1. Of course, it expresses how safety interventions need to target the systems that surround and precede the practitioners who work at the sharp end. This is a reflection of the

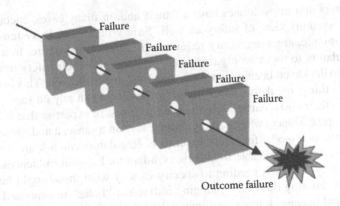

FIGURE 1.1 The so-called "Swiss Cheese" image of a system. An outcome failure is the result of many smaller, prior failures in organizational and administrative layers upstream from the sharp end where people work. These failures are related to, for example, procedures, design, supervision, and management. It conveys the modernist idea that safety at the sharp end is assured by good administrative and technological ordering of the upstream system—we need to find and fix the holes or failures there to prevent outcome failures. It also, of course, reifies centuries-old Cartesian–Newtonian ideas about cause–effect relationships, linearity, and closed systems (see Chapter 2).

reformation in safety thinking halfway the 20th century, and it has been emancipatory and empowering (Reason 2013). By pointing to the system, rather than the sharp end human, it represents "systems" thinking. But it is still modernist, which puts limits on what it can do. Let's look at that in more detail now.

Modernist "Systems" Thinking

Inspired by the Swiss Cheese image, and the modernist premises underlying it, many organizations have turned to its version of systems thinking. Better administrative and technological ordering of the organization upstream of the human at the sharp end for them has meant finding and fixing the many small failures (or holes) that they might have in their procedures, their designs, their compliance, and their systems. Organizations can prevent bigger outcome failures in the end if they fix the smaller failures that come before. This is the canon for most industries. Safety management systems, loss prevention systems, and auditing systems are just some of the parts of a bureaucratic machinery of safety. This machinery is seen as critical in the hunt for holes in upstream processes and structures, and the bureaucratic output is a paperworked assurance that outcome failures can be averted. It seems, sometimes, that organizations believe that they have a great safety culture because they have the paperwork to show it. Bureaucratic systems have burgeoned in almost every industry. They do the counting, tabulating, recording, and investigating (and actually not much hole-fixing). The growth of the health and safety profession—in numbers, cost, power, and influence—has been remarkable in many countries, to the point where the British Prime Minister recently referred to it as an albatross around the neck of business, a monster that needs to be waged war against (Anon 2012).

Regulators and inspectorates have adopted and, in many cases, encouraged the modernist systems view of safety as well. Safety management systems in aviation, for example, are a regulatory requirement almost everywhere. In a way, they allow regulators to focus on organizations' paperwork—their safety management systems—to decide on license approvals, citations, or certification. This is not necessarily a bad thing: regulators under budgetary pressure can rely on such systems to help control their costs, particularly the need to invest in expertise that is at least as good as the practitioners who need regulating. It is, in a sense, a nod toward accepting resilience as a model for safety assurance. Regulators can ask an organization to show that they will be able to absorb, or adapt to, harmful influences that may come their way—without needing to specify exactly what those might be, without the regulator hunting down and plugging individual "holes" in supposed layers of organizational defense. It has also directed the attention of regulators, investigators, insurers, investors, and other stakeholders away from the sharp end, and toward the context surrounding that work, toward the places where the conditions of possibility for success and failure are created. Data in loss prevention systems, for example, are now frequently taken as guidance for setting premium rates or awarding contracts. Showing stakeholders the numbers (in annual reports, tenders) gets precedence over showing the messy details of what it takes to make or break safety in the diversity of an organization's local practice.

Yet this has also reinvented and reified central tenets of modernism:

- People are once again, paradoxically, a problem to control. Safety is assured through strong administrative and hierarchical ordering of work and the organization around it. People in that organization need to be controlled by complying with standard procedures and other administrative require-ments; otherwise, their unsafe acts are seen to undermine the system of multiple defenses.
- Safety has become defined largely as an absence of negatives. A system free of holes (or without the kind of injury or incident evidence that points to holes) is believed to be a safe system. As a result, organizations count what they can count and try to show low numbers (typically on high-frequency, low-consequence events). This, however, may not be what counts. Their predictive value for low-frequency, high-consequence events is very low.
- Safety can become an administrative or bureaucratic accountability that needs to be managed upward (to show low numbers on negatives). It is then no longer an ethical responsibility managed downward, into the organiza-tion where people do others' dangerous and dirty work.
- Failures are seen as linear, proportional effects of other failures. Cause–effect relationships are straightforward in the Newtonian sense and similar to that seen in the domino model of accidents. Finding causes for failures is a simple matter of tracing backward in time and space.
- It takes us back to early 20th-century thinking about humans as the cause of safety trouble. All those organizational layers higher up are, after all, controlled or governed by people too (supervisors, managers, directors, and inspectors). We have simply shifted the burden to *other* people, but our logic of failure has not changed.

To be sure, there is nothing "wrong" with modernist beliefs per se. Science, technology, and human rationality have made the world a better place over the past centuries—a much better place. But modernism locks into place a host of ideas and assumptions about how that world works, and it implies that we can keep achieving successes and making progress with those same ideas. *That* is not necessarily true. The world of our Enlightenment forebears is not our world. Our world is vastly more complex; it is many times fuller with people, and it is deeply interconnected, in many places tightly coupled, and highly technological. We build higher and travel faster and farther. Ideas and capital zoom across the globe, often blind to national borders, yet local practices and diversity are prized more than they ever were during the uni-versalizing impulses of the Enlightenment. This is a world for which modernism is not always well equipped.

The biggest problem this leaves us with is not necessarily ideological. Perhaps these are all defensible and morally reasonable ideas. After all, the goals they once set out to pursue, as did modernism in general, are certainly laudable and still worth-while. The problem, however, is practical. Practical progress on safety has slowed down in many industries. This is not only true for ultra-safe industries such as sched-uled airline flying in the Western world. Some industries are hardly getting any safer as measured by negatives (incidents, accidents, and fatalities), independent of how many new rules or designs they adopt. Safety progress in construction, for example, has plateaued in many countries. Industries, like oil and gas, that show low or zero injuries, still manage to keep killing scores of workers each year. In fact, between 1990 and 2010, fatalities as a proportion of all recordable injuries have grown, not declined (perhaps because some have become better at hiding the "negatives" of injuries and incidents, which they cannot do with fatalities) (Townsend 2013). The trajectory of the safety journeys of many industries has become asymptotic. Asymptotes point to dying strategies. Asymptotes tell us that more of the same is not going to lead to something different. The sorts of solutions that created the asymptote are not going to help us break through it. They will just maintain the status quo.

SAFETY: FROM OPERATIONAL VALUE TO BUREAUCRATIC ACCOUNTABILITY

TWO SIDES OF THE ENLIGHTENMENT

The Enlightenment aimed to liberate people from the imposed truths and morals of church and crown. It was what is called emancipatory—setting people free from traditional, political, legal, or social restrictions. The Enlightenment saw people as self-contained individuals, driven by the power of reason. This position has become synonymous with scientific, rational thought in the West. In Descartes' *Meditations* (1641), human rationality was seen as the basis for systematic, true knowledge about the world. Individual human capacities and concerns started taking precedence, emphasizing the centrality of the human individual and her or his rational ways of solving problems.

But in this emancipation of the human subject, there is a tension. Instead of relying on handed-down tradition and authority, humans now turned to the application of their own reason and to the powers of their science and their technology. This has been accompanied by a growing faith in the rational, administrative ordering of both nature and society. From faith in a god and a monarchy, we went to a faith in technical rationality. Thanks to the Scientific Revolution, we started to believe in our methods and techniques and in our ability to parse up the world into its constituent bits. That was the way in which we figured out what was wrong about it (like Taylor did with his meat processing production line). That was how we reordered the world so as to make it run more reliably, more efficiently. Better control from above, better administrative control, would help us achieve safe, reliable results.

HUMAN FACTORS AS BOTH EMANCIPATORY AND RESTRICTIVE

Human factors was emancipatory relative to what had gone before. Recall from the above discussion its insights: Errors come from somewhere. They are related to problems in the design of people's tools and tasks. They are the effect of trouble deeper in the organization. This kind of thinking "liberates" individuals at the sharp end from sole responsibility for inefficiencies or failures. The emancipation lies in the realization that it is not just their behavior, or that problems can simply be adjusted with more clever incentives or sanctions oriented at the person. There is a whole organizational and engineered context that surrounds them: a system that pretty much sets the conditions for their success or failure. Taylor's insight was similar, perhaps surprisingly so. It wasn't that individual workers were lazy or deficient (although he left that possibility open, and thought it should be addressed by personnel selection and person-to-task matching). Rather, inefficiencies at the operational end were the result of a lack of control from above: a lack of supervision, coordination, engineering design, or managerial pre-work.

The Swiss Cheese image says that safety problems at the operational or sharp end are not necessarily *created* by the sharp end, but rather *inherited* by the sharp end. This has been emancipatory and hugely empowering. An accident investigation today is not complete or legitimate if it does not somehow invoke organizational levels and their contribution to things going wrong at the sharp end. But look at it from another angle–just for a moment Taylor might have said a similar thing about the efficiency and reliability problems of his meat plant workers. The production issues they had, the problems of balancing the line—those were inherited from administrative, bureaucratic, and planning levels and abetted by the design of the machinery. They were not created at the level of the line. In safety thinking, such insights, fueled by those of human factors thinkers, led to a wholesale reorientation in the 1970s. We need, said Barry Turner, to look at the administrative, the bureaucratic, and the organizational (Turner 1978). That is where we see failures brew. That is where accidents are incubated. The processes that set failure in motion are not unmotivated workers or their errors at the sharp end. Instead, the seeds of disaster need to be sought among the normal processes of bureaucratic life at the blunt end, in everyday managerial processes of information gathering, decision making, and communicating. The potential for having an accident is highly connected to the way we organize, operate, maintain, regulate, manage, and administer our safety-critical technologies. As man-made disaster theory argues (and shows): many of the things we do to prevent accidents from happening actually help bring them forth. Failure piggybacks, nonrandomly, opportunistically, onto the very structures and processes we put in place to prevent it (Pidgeon and O'Leary 2000).

Normal accident theory took a slightly different focus not much later, but it, too, was directed upward toward the organization and system rather than downward toward the worker at the operational end. We need, said Perrow, to look at the interactive complexity and coupling of the systems we ask people to maintain and operate. Errors and failures in doing so are related not to how these people do their work, as much as they are the result of how these systems are planned, conceived, built, or left to grow in size, complexity, and safety-criticality (Perrow 1984). As we just saw, looking upstream rather than downstream was further popularized decades ago in the image of defenses-in-depth (the Swiss Cheese Model). Investigation methods based on this model (e.g., ICAM [the Incident Cause Analysis Method] and HFACS [the Human Factors Analysis and Classification System] [Shappell and Wiegmann 2001]) have been inspired by it and are widely used today.

Investigations often go beyond the organization that suffered the accident, taking into account the regulations, inspections, and policy-making environments that surrounded it—layer after layer. Some have commented that this dilutes responsibility and accountability, which is an important issue. (It is not the topic of this book, but has been of others [Dekker 2012, 2013].)

On the other hand—and here is the tension of high modernism—this kind of thinking, as well as the models and methods it has spawned, suggests that better hierarchical ordering of organization and work will translate into better risk management. The Swiss Cheese image may not have looked so foreign to Taylor (each act of the worker at the sharp end needs to be preceded by several acts of a manager, a planner—someone upstream, who ensures that there are no holes in the processes reaching the worker). And it is not difficult to recognize some of the tenets of high modernity in the image either:

- Efficiency and safety at the operational sharp end (Taylor's production line workers) are largely dependent on the managers, planners, and engineers higher up in the organization (at the "blunt" rather than "sharp" end). Problems at the sharp end are largely inherited from a lack of administrative and organizational control at the blunt end.
- A belief in the administrative ordering of work to ensure that it is both safe and efficient. This involves not only several layers of line management and supervision but also extensive safety bureaucracies. These (e.g., a typical Health, Safety and Environment Department) instill practices in which accountability for safety is generated chiefly through process, paperwork, audit trails, and administrative work—all at an increasing distance from the operation. Rather than safety as a practical and ethical responsibility downward, it may become a bureaucratic accountability upward.
- A machinery for the surveillance and monitoring of human behavior is largely accepted and hard to resist from below. It happens through safety management systems, error counting and classification systems, data recording and monitoring systems, operational safety audits and other auditing systems, so-called "just culture" interventions, loss prevention systems, and similar schemes.
- Emancipation of the individual is thus constrained by hierarchical control, surveillance, and interventions from above (e.g., demands to comply with personal protective equipment rules). If such control is not effective, or compliance is not followed up, it can be characterized as a "hole" in a layer of defense and called "deficient supervision" or "poor leadership" (Reason 1990; Shappell and Wiegmann 2001).
- Better organization of the blunt end will enhance the possibility of an accident-free future. The best chance of minimizing accidents is by identifying and correcting delayed action failures (latent failures) before they combine with local triggers to breach or circumvent the system's defenses (Reason 1997).
- This thinking fits the ultimate aspiration of high modernity: a world free of harm; organizations free of accidents. The hope that accident-free organizations are possible is deeply embedded in most societies today. In fact, many

societies and their civil bureaucracies have transformed this from hope to commitment. When an accident happens, it is deemed so unnatural, so unusual, that a scapegoat for it might need to be found: somebody who, in the hierarchical, predictable order of things, did not do her or his job well. Such is our high modernist belief that the "accidental" no longer really exists. What there is, is risk that was not managed well. And behind that, a person or persons can be found culpable or at least responsible (Dekker 2012).

Even earlier assumptions (stemming from Newton and Descartes) are visible in this, too, and will become clearer after you have read Chapter 2. Organizations are modeled as static structures with an "above" and a "below" (which is a rather Cartesian idea). Accidents are the result of linear cause–effect trajectories through these organizational layers, from top to bottom. Failures can be understood by looking for other failures, they are each other's linear cause and effect. And ultimately, a disaster in a system can be reduced to the failure of components or layers, including those higher up in the organization. Such Cartesian–Newtonian assumptions, often unseen or taken for granted in how we think about accidents, have significant consequences for what we believe we can achieve to understand and prevent them. More will be said about this in the chapter on accidents. For now, what matters is the extent to which this perpetuates the tension of high modernity. It liberates, and it limits. Current thinking in human factors and safety (like in any high modernist endeavor) tells us simultaneously that

- People are the solution. They are smart; they can think rationally and solve great problems with their reasoning, their science, and their technologies.
- People are the problem. They need to be controlled, monitored, contained, instructed, and restricted; otherwise, their errors will leak through the defenses and create problems.

The belief that safety is generated chiefly or in part through planning, process, paperwork, audit trails, and administrative work—all at an increasing distance from the operation—has become entrenched in many industries. Originally, the ideas that gave rise to this were empowering and emancipatory, like human factors intended. Safety was not seen as a problem of only the sharp or operational end of practice. Rather, it had everything to do with how work was organized, resourced, supervised, planned, designed, and managed. Increasingly, however, this might become constraining. A focus on safety systems and procedural compliance, on surveillance and monitoring, has put new limits on the people who do work at the sharp end. The emancipation promised by the Enlightenment, and implemented by human factors thinking, has become perverted into a new kind of captivity. Safety is now sometimes held hostage by deference to liability concerns and to protocol, insurance, and fear of regulation. Rules are put in place and held in place not necessarily because they help create safety, but because they help manage or deflect liability for any bad outcomes. Indeed, you'd almost suspect that some industries no longer have safety management systems, but liability management systems (yet they still call them safety management systems). As a result, some of those concerned with safety in many organizations are in a place that is organizationally, physically, culturally, and psychologically different and distant from where the safety-critical work goes on. Safety has morphed in part from an operational value into a bureaucratic accountability.

SAFETY BUREAUCRACIES

Expansion of a bureaucracy of safety creates various tensions with what most organizations would like to achieve:

- There is an element of self-perpetuation in safety bureaucracies and the government regulations that keep them in place (often known as red or green tape). Bureaucratic accountability is demanded because of bureaucratic accountability; paperwork begets paperwork; nonoperational positions grow more nonoperational positions.
- This is known as bureaucratic entrepreneurialism. Fear of the consequences of curtailing a safety function is combined with a promise of future useful work and reminders of past successes. This is what most bureaucracies do, as it helps perpetuate their existence.
- Safety bureaucracies tend to institutionalize and further legitimate the counting and tabulation of negatives (incidents, violations). They are organized around vocabularies of deficit and control. Incentive structures around the absence of negatives get people to push bad news out of view and fudge injury or incident numbers. This hampers learning and honesty.
- Safety bureaucracies are mostly organized around lagging indicators: measuring that which has already been. The predictive value of most lagging indicators is known to be poor.
- Systems of bureaucratic accountability tend to value technical expertise less, disempower middle management, and delegitimize operational experience. People may no longer feel as able or empowered to think for themselves. This can stifle innovation, choke off initiative, and erode problem ownership. Rather than inspiring people to take responsibility, such systems can encourage its shirking.
- The gap between how an organization or operation works and how bureaucratic systems believe it works can paradoxically be left to grow. Innovation that is noncompliant may be driven underground and unlikely to see widespread adoption, thus hampering possible sources of future efficiency or competitiveness.
- Time and opportunity for richer communication with operational staff by supervisors and managers can get compromised by the daily demands of bureaucratic accountability. Managers might typically report that their ability to interact with the workforce is reduced because of meetings, paperwork demands, and email.

Expanding our safety bureaucracies may become self-defeating. Ever-expanding systems of compliance can eventually lead to less compliance, because the job still needs to get done. This can actually harm safety. Man-made disaster theory, as well as Normal Accident Theory, confirms how the potential for accidents brews at the core of the very processes and structures that are set up to enable production and prevent failure (Pidgeon and O'Leary 2000). The more these processes and structures are developed or enforced bureaucratically by those who are at a distance from the operation, the greater the risk becomes that they produce "fantasy documents" that bear little relation to actual work or operational expertise (Clarke and Perrow 1996). In 2008, for example, two years before the Macondo well blowout, BP had identified what it called priority gaps in its Gulf operations. The first of these was that there were "too many risk processes" going on, but they had collectively become "too complicated and cumbersome to effectively manage" (Elkind and Whitford 2011, p. 9). Such overbureaucratization of safety helps create "structural secrecy": a by-product of the cultural, organizational, physical, and psychological separation between operations on the one hand, and safety regulators and bureaucracies on the other. Under such conditions, critical information may not cross organizational boundaries, and mechanisms for constructive interplay are lacking (Vaughan 1996). In the meantime, bureaucracies do what they do best—classifying, counting, tabulating, monitoring, storing, and calculating. With limited potential for progress on safety from what we have now, there is a need to start thinking differently about safety. This spells a new era for human factors—an era adept at recognizing complexity and the epistemological and ontological limits of our reach.

STUDY QUESTIONS

1. Explain the transformation in our thinking about the relationship between safety and the human factor that occurred during the 20th century. What was the typical target for safety interventions before and after this transformation?
2. Why was the Enlightenment a necessary innovation in the Western world that enabled us to think about safety, risk, and their control in the first place?
3. Which characteristics of Scientific Management do you recognize in safety work even today?
4. What were the developments responsible for changing craftsmen into workers during the Industrial Revolution, and what did that mean for the control and organization of work? Please come up with a few examples of your own.
5. What does "accident-proneness" mean, and when was this a popular explanation for workplace accidents? What is a major flaw in the argument of accident-proneness?
6. Explain the events that gave rise to the emergence of "human factors" as a field of scientific and practical inquiry. What did this mean for the role of psychology in understanding and influencing safety?
7. How can you recognize high-modernist assumptions in safety work during the 20th century (both before and after the transformation) and even today?

2 It Was Human Error

CONTENTS

KEY POINTS

- The absence of mechanical failure in an accident is often taken as automatic evidence that human error is the cause. This is driven in large part by the Cartesian–Newtonian models and ideas we use in safety when trying to understand the human factor.
- This language and worldview has us make particular assumptions (often without knowing it) about the relationship between cause and effect, about a clear separation between the mind and the world outside of it, and suggests that we can understand complexity by reducing it to simple components.
- This means we often try to understand the human contribution to failure in terms of a Newtonian social physics, where knowledge available in the world did not make it into somebody's mind, and where actions or omissions and their consequences are clearly linked.
- To do justice to the complexity of failure and its human contribution, we need to challenge this realist view of human factors and understand how the hindsight bias, counterfactual reasoning, and judging instead of explaining give us a mere illusion that we can explain what happened.

NO EVIDENCE OF MECHANICAL FAILURE...

On July 6, 2013, an Asiana Airlines Boeing 777 crashed on final approach to San Francisco. Federal investigators' evidence on Asiana Airlines Flight 214 appears to point to pilot error. But the National Transportation Safety Board Chairwoman pointedly refused to endorse that conclusion, even as she related initial findings that included nothing indicating a mechanical malfunction just before the fatal crash. Briefing the press in San Francisco for the first time Sunday, she repeatedly refused to finger the pilots as the culprits. She admonished reporters that the agency is nowhere near assigning blame for the crash, which killed two people and injured as many as 180. "Everything is on the table right now," she said.

The data she communicated, gleaned from the plane's voice and data recorders, suggest serious problems with the plane's approach to the runway—primarily that it was traveling far too slowly, and that the crew tried just 1.5 seconds before impact to abort the landing and try again. She said the plane had been cleared to land at 137 knots, or 158 mph, but was going much more slowly as it approached the runway. She wouldn't reveal a specific speed, but said that "we're not talking about a few knots here or there, we're talking about a significant amount of speed below 137 knots."

The plane's speed had bled off so much and other factors, including its pitch, were such that it triggered the plane's "stick shaker," a kind of warning that shakes the plane's control yoke to alert pilots that a stall is imminent. In aviation parlance, a stall happens when a plane no longer has enough "lift" to maintain its altitude. According to cockpit voice data from the last two hours of flight, there was "no discussion of any aircraft anomalies or concerns with the approach." But seven seconds before the crash, a crew member called for an increase in speed, followed three seconds later by the sound of the stick shaker. A second and a half before impact, the voice data show a crew member asking for a "go around." The flight data recorders also show that the throttles were at idle and airspeed was "slowed below the target" as the plane was on landing approach. The crew pushed the throttles open "a few seconds prior to impact, and the engines appeared to respond normally"—which, if those data hold, would appear to rule out a mechanical problem in which the engines did not perform as asked.

No mechanical failure? Then it must be human error. But is that distinction so obvious? And where does our confidence in making that crisp distinction come from? Human factors pioneers showed how human error was systematically connected to features of people's tools and tasks. Errors are deeply intertwined with the mechanical, with the organizational.

Take the electronic glideslope toward the runway that was out of service at San Francisco that day, or the low numbers of approaches and landings that long-haul pilots make yearly, particularly (hand-flown) visual approaches without the aid of an electronic signal, or the forest of modes that the various systems of this particular aircraft can go into during a visual approach. A visual approach, or one that just follows the localizer (or electronic runway centerline), pilots say, could be flown in vertical speed mode or flight path angle mode. Both these modes put the autothrottle in so-called speed mode, in which speed protections are not as extensive as when the aircraft were to fly down the electronic beams of the instrument landing system. The approach can also be flown using GPS data, but that requires the aircraft to capture the right vertical path down to the runway and fly in an automation mode that helps it do just that. If that path is somehow lost (and approaches to big airports can put pilots in positions where they are both high and fast, then the aircraft's vertical guidance reverts to an off-path mode, in which, again, some speed and other protections are lost. Another possibility in aircraft like this is to fly the approach in a level change mode, where it will fly down to a different altitude, selected by the pilot. To do that, the autothrottles will usually stay where they are (or go to idle, as necessary), but the throttles remain available for pilots to move manually if needed. On the instruments in front of the pilots, a so-called flight director will show them what pitch to fly so that they get to their selected airspeed on the mode control panel. If those commands are followed, it works well. In manual flight, however, it is of course possible to raise the nose above those commands (in order to slow down the aircraft, or correct for an under shoot by pulling up the nose). The autothrottles, however, will not do anything in this configuration of modes. The speed will bleed away, and the throttles will not respond at all.

So this was just pilot error? There is nothing "mechanical" about this sort of crash at all? Let us approach that question with a bit more of an open mind.

LOOKING FOR FAILURES TO EXPLAIN FAILURES

Our most entrenched beliefs and assumptions often lie locked up in the simplest of questions. The question about mechanical failure or human error is one of them. Was the accident caused by mechanical failure or by human error? It is a stock question in the immediate aftermath of a mishap. Indeed, it seems such a simple, innocent question. To many, it is a normal question to ask. If you have had an accident, it makes sense to find out what broke. The question, however, embodies a particular understanding of how accidents occur, and it risks confining our causal analysis to that understanding. It lodges us into a fixed interpretative repertoire. Escaping from this repertoire may be difficult. It sets out the questions we ask, provides the leads we pursue and the clues we examine, and determines the conclusions we will eventually draw. Which components were broken? Was it something engineered, or some human? How long had the component been bent or otherwise deficient? Why did it eventually break? What were the latent factors that conspired against it? Which defenses had eroded?

These are the types of questions that dominate inquiries in human factors and system safety today. We organize accident reports and our discourse about mishaps around the struggle for answers to them. Investigations turn up broken mechanical components (e.g., perforated heat tiles and wing on the Columbia Space Shuttle), underperforming human components (e.g., breakdowns in crew resource management, a pilot who has a checkered training record), and cracks in the organizations responsible for running the system (e.g., weak organizational decision chains, deficient maintenance, failures in regulatory oversight). Looking for failures—human, mechanical, or organizational—in order to explain failures is so commonsensical that most investigations never stop to think whether these are indeed the right clues to pursue. That failure is caused by failure is prerational—we do not consciously consider it any longer as a question in the decisions we make about where to look and what to conclude.

DECONSTRUCTION, DUALISM, AND STRUCTURALISM

What is that language, then, and the perhaps increasingly obsolete technical world-view it represents? Its defining characteristics are deconstruction, dualism, and structuralism. *Deconstruction* means that a system's functioning can be understood exhaustively by studying the arrangement and interaction of its constituent parts. Scientists and engineers typically look at the world this way. Accident investigations deconstruct too. In order to rule out mechanical failure, or to locate the offending parts, accident investigators speak of "reverse engineering." They recover parts from the rubble and reconstruct them into a whole again, often quite literally. Think of the TWA800 Boeing 747 that exploded in midair after takeoff from New York's Kennedy airport in 1998. It was recovered from the Atlantic Ocean floor and painstakingly pieced back together—if heavily scaffolded—in a hangar. With the puzzle as complete as possible, the broken part(s) should eventually get exposed, allowing investigators to pinpoint the source of the explosion. Accidents are puzzling wholes. But it continues to defy sense; it continues to be puzzling only when the functioning (or nonfunctioning) of its parts fails to explain the whole. The part that caused the explosion, that ignited it, was never actually pinpointed. This is what makes the TWA800 investigation scary. Despite one of the most expensive reconstructions in history, the reconstructed parts refused to account for the behavior of the whole. In such a case, a frightening, uncertain realization creeps into the investigator corps and into industry. A whole failed without a failed part. An accident happened without a cause; no cause—nothing to fix, nothing to fix—it could happen again tomorrow, or today.

The second defining characteristic is dualism. *Dualism* means that there is a distinct separation between material and human cause—between human error and mechanical failure. In order to be a good dualist, you of course have to deconstruct: You have to disconnect human contributions from mechanical contributions. The rules of the International Civil Aviation Organization that govern aircraft accident investigators prescribe exactly that. They force accident investigators to separate human contributions from mechanical ones. Specific paragraphs in accident reports are reserved for tracing the potentially broken human components. Investigators explore the anteceding 24- and 72-hour histories of the humans who would later be involved in a mishap. Was there alcohol? Was there stress? Was there fatigue? Was there a lack of proficiency or experience? Were there previous problems in the training or operational record of these people? How many flight hours did the pilot really have? Were there other distractions or problems? This investigative requirement reflects a primeval interpretation of human factors, an aeromedical tradition where human error is reduced to the notion of "fitness for duty." This notion has long been overtaken by developments in human factors toward the study of normal people doing normal work in normal workplaces (rather than physiologically or mentally deficient miscreants), but the overextended aeromedical model is retained as a kind of comforting positivist, dualist, deconstructive practice. In the fitness-for-duty paradigm, sources of human error must be sought in the hours, days, or years before the accident, when the human component was already bent and weakened and ready to break. Find the part of the human that was missing or deficient, the "unfit part," and the human part will carry the interpretative load of the accident. Dig into recent history, find the deficient pieces, and put the puzzle together: deconstruction, reconstruction, and dualism.

The third defining characteristic of the technical worldview that still governs our understanding of success and failure in complex systems is structuralism. The language we use to describe the inner workings of successful and failed systems is a language of structures. We speak of layers of defense, of holes in those layers. We identify the "blunt ends" and "sharp ends" of organizations and try to capture how one has effects on the other. Even safety culture gets treated as a structure consisting of other building blocks. How much of a safety culture an organization has depends on the routines and components it has in place for incident reporting (this is measurable), to what extent it is just in treating erring operators (this is more difficult to measure, but still possible), and what linkages it has between its safety functions and other institutional structures. A deeply complex social reality is thus reduced to a limited number of measurable components. For example, does the safety department have a direct route to highest management? What is the reporting rate compared to other companies?

Our language of failures is also a language of mechanics. We describe accident trajectories, and we seek causes and effects, and interactions. We look for initiating failures, or triggering events, and trace the successive domino-like collapse of the system that follows it. This worldview sees sociotechnical systems as machines with parts in a particular arrangement (blunt vs. sharp ends, defenses layered throughout), with particular interactions (trajectories, domino effects, triggers, initiators), and a mix of independent or intervening variables (blame culture vs. safety culture). This is the worldview inherited from Descartes and Newton, the worldview that has successfully driven technological development since the Scientific Revolution half a millennium ago. The worldview, and the language it produces, is based on particular notions of natural science and exercises a subtle but very powerful influence on our understanding of sociotechnical success and failure today. As it does with most of Western science and thinking, it pervades and directs the orientation of human factors and system safety.

Yet language, if used unreflectively, easily becomes imprisoning. Language expresses but also determines what we can see and how we see it. Language constrains how we construct reality. If our metaphors encourage us to model accident chains, then we will start our investigation by looking for events that fit in that chain. But which events should go in? Where should we start? As Nancy Leveson (2002) pointed out, the choice of which events to put in is arbitrary, as are the length, the starting point, and level of detail of the chain of events. What, she asked, justifies assuming that initiating events are mutually exclusive, except that it simplifies the mathematics of the failure model? These aspects of technology, and of operating it, raise questions about the appropriateness of the dualist, deconstructed, structuralist model that dominates human factors and system safety. In its place, we may seek a true systems view, which not only maps the structural deficiencies behind individual human errors (if indeed it does that at all) but also appreciates the organic, ecological adaptability of complex sociotechnical systems.

HUMAN ERROR AS THE ABSENCE OF MECHANICAL FAILURE

Here is an example. A twin-engine Douglas DC-9-82 landed at a regional airport in the Southern Highlands of Sweden in the summer of 1999. Rainshowers had passed through the area earlier, and the runway was still wet. While on approach to the runway, the aircraft got a slight tailwind, and after touchdown, the crew had trouble slowing down. Despite increasing crew efforts to brake, the jet overran the runway and ended up in a field a few hundred feet from the threshold. The 119 passengers and crew onboard were unhurt. After coming to a standstill, one of the pilots made his way out of the aircraft to check the brakes. They were stone cold. No wheel braking had occurred at all. How could this have happened? Investigators found no mechanical failures on the aircraft. The braking systems were fine. Instead, as the sequence of events was rolled back in time, investigators realized that the crew had not armed the aircraft's ground spoilers before landing. Ground spoilers help a jet aircraft brake during roll out, but they need to be armed before they can do their work. Arming them is the job of the pilots, and it is a before-landing checklist item and part of the procedures that both crewmembers are involved in. In this case, the pilots forgot to arm the spoilers. "Pilot error," the investigation concluded. Or actually, they called it "Breakdowns in CRM (Crew Resource Management)" (Tillbud vid landning med flygplanet LN-RLF den 23/6 på Växjö/Kronoberg flygplats, G län (Rapport RL 2000:38). [Incident during landing with aircraft LN-RLF on June 23 at Växjö/Kronoberg airport] 2000, p. 12), a more modern, more euphemistic way of saying "pilot error." The pilots did not coordinate what they should have; for some reason, they failed to communicate the required configuration of their aircraft. Also, after landing, one of the crew members had not called "Spoilers!" as the procedures dictated. This could, or should, have alerted the crew to the situation, but it did not happen. Human errors had been found. The investigation was concluded.

"Human error" is our default when we find no mechanical failures. It is a forced, inevitable choice that fits nicely into an equation, where human error is the inverse of the amount of mechanical failure. Equation 2.1 shows how we determine the ratio of causal responsibility:

$$\text{human error} = f(1 - \text{mechanical failure}) \qquad (2.1)$$

If there is no mechanical failure, then we know what to start looking for instead. In this case, there was no mechanical failure. Equation 2.1 came out as a function of 1 minus 0. The human contribution was 1. It was human error, a breakdown of CRM. Investigators found that the two pilots onboard the MD-80 were actually both captains, and not a captain and a copilot as is usual. It was a simple and not altogether uncommon scheduling fluke, a stochastic fit flying onboard the aircraft since that morning. With two captains on a ship, responsibilities risk getting divided unstably and incoherently. Division of responsibility easily leads to its abdication. If it is the role of the copilot to check that the spoilers are armed, and there is no copilot, the risk is obvious. The crew was in some sense "unfit," or at least prone to breaking down. It did (there was a "breakdown of CRM"). But what does this explain? These are processes that themselves require an explanation. Perhaps there is a different meaning beneath the prima facie particulars of such an incident, a reality where machinistic and human cause are much more deeply intertwined than our formulaic approaches to investigations allow us to grasp. In order to get better glimpses of this reality, we first have to turn to dualism. It is dualism that lies at the heart of the choice between human error and mechanical failure. We take a brief peek at its past and confront it with the unstable, uncertain empirical encounter of an unarmed spoilers case.

The urge to separate human cause from machinistic cause is something that must have puzzled even the early nascent human factors tinkerers. Think of the fiddling with World War II cockpits that had identical control switches for a variety of functions. Would a flap-like wedge on the flap handle and a wheel-shaped cog on the gear lever avoid the typical confusion between the two? Both common sense and experience said "yes." By changing something in the world, human factors engineers (to the extent that they existed already) changed something in the human. By toying with the hardware that people worked with, they shifted the potential for correct versus incorrect action, but only the potential. For even with functionally shaped control levers, some pilots, in some cases, still got them mixed up. At the same time, pilots did not always get identical switches mixed up. Similarly, not all crews consisting of two captains fail to arm the spoilers before landing. Human error, in other words, is suspended, unstably, somewhere between the human and the engineered interfaces. The error is neither fully human, nor fully engineered. At the same time, mechanical "failures" (providing identical switches located next to one another) get to express themselves in human action. So, if a confusion between flaps and gear occurs, then what is the cause? Human error or mechanical failure? You need both to succeed; you need both to fail. Where one ends and the other begins is no longer so clear. One insight of early human factors work was that machinistic feature and human action are intertwined in ways that resist the neat, dualist, deconstructed disentanglement still favored by investigations (and their consumers) today.

The choice between human cause and material cause is not just a product of recent human factors engineering or accident investigations. The choice is firmly rooted in the Cartesian–Newtonian worldview that governs much of our thinking to this day, particularly in technologically dominated professions such as human factors engineering, safety and risk management, and accident investigation.

THE CARTESIAN–NEWTONIAN WORLDVIEW

This book calls for an innovation from modernism in our thinking about safety and about the role of humans in our systems. Modernism, as has been pointed out, has its roots in the Scientific Revolution and the Enlightenment. Key ideas introduced by Descartes and Newton during the Scientific Revolution are fundamental to modernism. These ideas are in a sense preconditions. For modernism to work, for it to be believable, we need a set of concepts and ideas about the world that were first formulated by Descartes and Newton. Indeed, we buy into those ideas more than we know—most are so commonsensical to our own experiences with the world that they are all but transparent; invisible. They are part of the condition we live in. And the condition we live in is hard to see if we do not exert ourselves, with the help of others, to make it explicit. So let us first look back at those ideas here, and see how they influence how we think about failure and the role of humans even today. Then we cast our glance way ahead—beyond modernism, to a conception of the world as complex and holistic. This sketches a kind of ideal type that we might want to strive for if we want to move from modernism.

Isaac Newton and René Descartes were two towering figures in the Scientific Revolution between 1500 and 1700 CE, which produced a dramatic shift in worldview, as well as profound changes in knowledge and in ideas on how to acquire and test knowledge. Descartes proposed a sharp distinction between what he called res cogitans, the realm of mind and thought, and res extensa, the realm of matter. Although Descartes admitted to some interaction between the two, he insisted that mental and physical phenomena cannot be understood by reference to each other. Problems that occur in either realm require entirely separate approaches and different concepts to solve them. The notion of separate mental and material worlds became known as dualism and its implications can be recognized in much of what we think and do today. According to Descartes, the mind is outside of the physical order of matter and is in no way derived from it. The choice between human error and mechanical failure is such a dualist choice: According to Cartesian logic, human error cannot be derived from material things. As we will see, this logic does not hold up well—in fact, on closer inspection, the entire field of human factors is based on its abrogation.

Separating the body from the soul, and subordinating the body to the soul, not only kept Descartes' out of trouble with the Church. His dualism, his division between mind and matter, addressed an important philosophical problem that had the potential of holding up scientific, technological, and societal progress: What is the link between mind and matter, between the soul and the material world? How could we, as humans, take control of and remake our physical world as long as it was indivisibly allied to or even synonymous with an irreducible, eternal soul? A major aim during the 16th- and 17th-century Scientific Revolution was to see and understand (and become able to manipulate) the material world as a controllable, predictable, programmable machine. This required it to be seen as nothing but a machine: No life, no spirit, no soul, no eternity, no immaterialism, no unpredictability. Descartes' res extensa, or material world, answered to just that concern. The res extensa was described as working like a machine, following mechanical rules and allowing explanations in terms of the arrangement and movement of its constituent parts. Scientific progress became easier because of what it excluded. What the Scientific Revolution required, Descartes' disjunction provided. Nature became a perfect machine, governed by mathematical laws that were increasingly within the grasp of human understanding and control, and away from things humans cannot control. Newton, of course, is the father of many of the laws that still govern our understanding of the universe today. His third law of motion, for example, lies at the basis of our presumptions about cause and effect, and causes of accidents: For every action, there is an equal and opposite reaction. In other words, for each cause, there is an equal effect, or rather, for each effect, there must be an equal cause. Such a law, though applicable to the release and transfer of energy in mechanical systems, is misguiding and disorienting when applied to sociotechnical failures, where the small banalities and subtleties of normal work done by normal people in normal organizations can slowly degenerate into enormous disasters, into disproportionately huge releases of energy. The cause–consequence equivalence dictated by Newton's third law of motion is quite inappropriate as a model for organizational accidents.

Attaining control over a material world was critically important for people 500 years ago. The inspiration and fertile ground for the ideas of Descartes and Newton can be understood against the background of their time. Europe was emerging from the Middle Ages—popularly understood to have been fearful and fateful times, where life spans were cut short by wars, disease, and epidemics. We should probably not underestimate anxiety and apprehension about humanity's ability to make it at all against apocryphal odds. The population of Newton's native England, for example, took until 1650 to recover to the level of 1300. People were at the mercy of ill-understood and barely controllable forces. In the preceding millennium, piety, prayer, and penitence were among the chief mechanisms through which people could hope to attain some kind of sway over ailment and disaster.

The growth of insight produced by the Scientific Revolution slowly began to provide an alternative, with measurable empirical success. The Scientific Revolution provided new means for really controlling the natural world. Telescopes and microscopes gave people new ways of studying components that had thus far been too small or too far away to see, cracking open a new view on the universe and for the first time revealing causes of phenomena hitherto ill understood. Nature was not a monolithic, inescapable bully, and people were no longer just on the receiving, victimized end of its vagaries. By studying it in new ways, with new instruments, nature could be decomposed, broken into smaller bits, measured, and, through all of that, better understood and eventually controlled. Advances in mathematics (geometry, algebra, calculus) generated models that could account for and begin to predict newly discovered phenomena in, for example, medicine and astronomy. By discovering some of the building blocks of life and the universe, and by developing mathematical imitates of their functioning, the Scientific Revolution reintroduced a sense of predictability and control that had long been dormant during the Middle Ages. Humans could achieve dominance and preeminence over the vicissitudes and unpredictabilities of nature. The route to such progress would come from measuring, breaking down (known variously today as reducing, decomposing, or deconstructing) and mathematically modeling the world around us—to subsequently rebuild it on our terms.

Measurability and control are themes that animated the Scientific Revolution, and they resonate strongly today. Even the notions of dualism (material and mental worlds are separate) and deconstruction (larger wholes can be explained by the arrangement and interaction of their constituent lower-level parts) have long outlived their initiators. The influence of Descartes is judged so great in part because he wrote in his native tongue, rather than in Latin, thereby presumably widening access and popular exposure to his thoughts. The mechanization of nature spurred by his dualism, and Newton's and others' enormous mathematical advances, heralded centuries of unprecedented scientific progress, economic growth, and engineering success. As Fritjof Capra (1982) put it, NASA would not have been able to put a man on the moon without René Descartes.

FROM DESCARTES AND NEWTON TO HOLISM AND COMPLEXITY

If we are to depart, where appropriate and possible, from Cartesian–Newtonian thinking in safety, then we need to revise or replace the following:

- Our beliefs in the simple relationship between causes and effects
- The idea that we can understand system complexity by reducing it to the behavior of constituent components
- Assumptions about the foreseeability of harm (and the resulting outcome and hindsight biases)
- Time-reversibility (which suggests we can "reconstruct" the events that led up to an incident or accident)

- The notion that knowledge is accurate when it corresponds to some objective outside world
- The ability to come up with one best method for how to do a task, and with one "true story" of what happened if things did not go well

Descartes and Newton favor linear thinking: a process that follows a chain of causal reasoning from a premise to a single outcome. In contrast, complexity thinking regards an outcome as emerging from a complex network of causal interactions and, therefore, not necessarily the result of a single factor. The remainder of this chapter lays out how a Newtonian analysis of our systems makes particular assumptions about the relationship between cause and effect, foreseeability of harm, time-reversibility, and the ability to come up with the "true story" of an incident. Recognize how the Cartesian–Newtonian worldview is fundamental to high modernism: without that worldview, neither the premises nor the promises of high modernism would carry much weight at all. The emerging contrast to this worldview acknowledges the complex, systemic nature of the functioning of our systems. It calls for a different approach, and this entire book is an installment in helping to build it.

The logic behind Newtonian science is easy to formulate, although its implications for how we think about failure and success in our systems are subtle and pervasive. Classical mechanics, as formulated by Newton and further developed by Laplace and others, encourages a reductionist, mechanistic methodology and worldview. Many still equate "scientific thinking" with "Newtonian thinking." The mechanistic paradigm is compelling in its simplicity, coherence, and apparent completeness and largely consistent with intuition and common sense. The remainder of this chapter will consider the most important aspects of Cartesian–Newtonian thinking in summary form. They will be unpacked throughout the rest of the book, as we trace their impact in case studies as well as theorizing around safety and the human factor.

Reductionism and the Eureka Part

The best known principle of Newtonian science, formulated well before Newton by the philosopher-scientist Descartes, is that of analysis or reductionism. The functioning or nonfunctioning of the whole can be explained by the functioning or nonfunctioning of constituent components. Attempts to understand the failure of a complex system in terms of failures or breakages of individual components in it—whether those components are human or machine—is very common (Galison 2000). The investigators of the Trans World Airlines 800 crash off New York called it their search for the "eureka part": the part that would have everybody in the investigation declare that the broken component, the trigger, the original culprit, had been located and could carry the explanatory load of the loss of the entire Boeing 747. But for this crash, the so-called "eureka part" was never found (Langewiesche 1998). The defenses-in-depth metaphor (Hollnagel 2004; Reason 1990) relies on the linear parsing-up of a system to locate broken layers or parts. This was recently used by BP in its own analysis of the Deepwater Horizon accident and subsequent Gulf of Mexico oil spill (BP 2010), seen by others as an effort to divert liability onto multiple participants in the construction of the well and the platform (Levin 2010).

The philosophy of Newtonian science is one of simplicity: the complexity of the world is only apparent, and to deal with it, we need to analyze phenomena into their basic components. This is applied in the search for psychological sources of failure, for example. Methods that subdivide "human error" into further component categories, such as perceptual failure, attention failure, memory failure, or inaction, are in use in, for example, air traffic control (Hollnagel and Amalberti 2001). It is also applied in legal reasoning in the wake of accidents, by separating out one or a few actions (or inactions) on the part of individual people.

Causes for Effects Can Be Found

In the Newtonian vision of the world, everything that happens has a definitive, identifiable cause and a definitive effect. There is symmetry between cause and effect (they are equal but opposite). The determination of the "cause" or "causes" is of course seen as the most important function of accident investigation but assumes that physical effects can be traced back to physical causes (or a chain of causes–effects) (Leveson 2002). The assumption that effects cannot occur without specific causes influences legal reasoning in the wake of accidents too. For example, to raise a question of negligence in an accident, harm must be caused by the negligent action (GAIN 2004). Assumptions about cause–effect symmetry can be seen in what is known as the outcome bias (Fischhoff 1975). The worse the consequences, the more any preceding acts are seen as blameworthy (Hugh and Dekker 2009).

Newtonian ontology is materialistic: all phenomena, whether physical, psychological, or social, can be reduced to (or understood in terms of) matter, that is, the movement of physical components inside three-dimensional Euclidean space. The only property that distinguishes particles is where they are in that space. Change, evolution, and indeed accidents can be reduced to the geometrical arrangement (or misalignment) of fundamentally equivalent pieces of matter, whose interactive movements are governed exhaustively by linear laws of motion, of cause and effect. The Newtonian model may have become so pervasive and coincident with "scientific" thinking that, if analytic reduction to determinate cause–effect relationships cannot be achieved, then the accident analysis method or agency isn't entirely worthy. The Chairman of the NTSB at the time, Jim Hall, raised the specter of his agency not being able to find the eureka part in TWA800, which would challenge its entire reputation (p. 119): "What you're dealing with here is much more than an aviation accident… What you have at stake here is the credibility of this agency and the credibility of the government to run an investigation" (Dekker 2011b).

Foreseeability of Harm

According to Newton's image of the universe, the future of any part of it can be predicted with absolute certainty if its state at any time was known in all details. With enough knowledge of the initial conditions of the particles and the laws that govern their motion, all subsequent events can be foreseen. In other words, if somebody can be shown to have known (or should have known) the initial positions and momentum of the components constituting a system, as well as the forces acting on those components (which are not only external forces but also those determined by the positions of these and other particles), then this person could, in principle, have predicted the further evolution of the system with complete certainty and accuracy.

If such knowledge is in principle attainable, then harmful outcomes are foreseeable too. Where people have a duty of care to apply such knowledge in the prediction of the effects of their interventions, it is consistent with the Newtonian model to ask how they failed to foresee the effects. Did they not know the laws governing their part of the universe (i.e., were they incompetent, unknowledgeable)? Did they fail to plot out the possible effects of their actions? Indeed, legal rationality in the determination of negligence follows this feature of the Newtonian model (p. 6): "Where there is a duty to exercise care, reasonable care must be taken to avoid acts or omissions which can reasonably be foreseen to be likely to cause harm. If, as a result of a failure to act in this reasonably skilful way, harm is caused, the person whose action caused the harm, is negligent" (GAIN 2004).

In other words, people can be seen as negligent if the person did not avoid actions that could be foreseen to lead to effects—effects that would have been predictable and thereby avoidable if the person had sunk more effort into understanding the starting conditions and the laws governing the subsequent motions of the elements in that Newtonian sub-universe. Most road traffic legislation is based on this Newtonian commitment to foreseeability too. For example, a road traffic law in a typical Western country might specify how a motorist should adjust speed so as to be able stop the vehicle before colliding with some obstacle, at the same time remaining aware of the circumstances that could influence such selection of speed. Both the foreseeability of all possible hindrances and the awareness of circumstances (initial conditions) critical for determining speed are steeped in Newtonian epistemology. Both are also heavily subject to outcome bias: if an accident suggests that an obstacle or particular circumstance was not foreseen, then speed was surely too high. The system's user, as a consequence, is always wrong (Tingvall and Lie 2010).

TIME-REVERSIBILITY

The trajectory of a Newtonian system is determined not only toward the future but also toward the past. Given its present state, we can in principle reverse the evolution to reconstruct any earlier state that it has gone through. Such assumptions give accident investigators the confidence that an event sequence can be reconstructed by starting with the outcome and then tracing its causal chain back into time. The notion of reconstruction reaffirms and instantiates Newtonian physics: knowledge about past events is not original, but merely the result of uncovering a preexisting order. The only thing between an investigator and a good reconstruction is the limits on the accuracy of the representation of what happened. It follows that accuracy can be improved by "better" methods of investigation (Shappell and Wiegmann 2001).

COMPLETENESS OF KNOWLEDGE

Newton argued that the laws of the world are discoverable and ultimately completely knowable. God created the natural order (though kept the rulebook hidden from man) and it was the task of the investigator to discover this hidden order underneath the apparent disorder (Feyerabend 1993). It follows that the more facts an analyst or investigator collects, the more it leads, inevitably, to a better investigation: a better representation of "what happened." In the limit, this can lead to a perfect, objective representation of the world outside (Heylighen 1999), or one final (true) story, the one in which there is no gap between external events and their internal representation. Those equipped with better methods, and particularly those who enjoy greater "objectivity" (i.e., those who have no bias, which distorts their perception of the world, and who will consider all the facts), are better positioned to construct such a true story. Formal, government-sponsored accident investigations can sometimes enjoy this idea of objectivity and truth—if not in the substance of the story they produce, then at least in the institutional arrangements surrounding its production. Putative objectivity can be deliberately engineered into the investigation as a sum of subjectivities: all interested parties (e.g., vendors, the industry, operator, unions, and professional associations) can officially contribute (though dissenting voices can be silenced or sidelined [Byrne 2002; Perrow 1984]). Other parties often wait until a formal report is produced before publicly taking either position or action, legitimizing the accident investigation as original arbitrator between fact and fiction.

SOCIOTECHNICAL SYSTEMS AND DESCARTES AND NEWTON

Together, taken-for-granted assumptions about decomposition, cause–effect symmetry, foreseeability of harm, time-reversibility, and completeness of knowledge give rise to a Newtonian analysis. It can be summed up as follows:

- To understand the functioning of our systems, we need to search for clear connections between the outcome (whether failure or success) and the functioning or malfunctioning of its constituent components. The relationship between component behavior and system behavior is analytically nonproblematic.
- Causes for effects can always be found, because there are no effects without causes. In fact, the larger the effect, the larger (e.g., the more egregious) the cause must have been.
- If they put in more effort, people can more reliably foresee outcomes. After all, they would have a better understanding of the starting conditions and they are already supposed to know the laws by which the system behaves (otherwise, they wouldn't be allowed to work in it). With those two in hand, all future system states can be predicted and harmful states can be foreseen and avoided.
- An event sequence can be reconstructed by starting with the outcome and tracing its causal chain back into time. Knowledge thus produced about past events is the result of uncovering a preexisting order.
- One official account of what happened is possible and desirable. Not just because there is only one preexisting order to be discovered, but also because knowledge (or the story) is the mental representation or mirror of that order. The truest story is the one in which the gap between external events and internal representation is the smallest. The true story is the one in which there is no gap.

These assumptions can remain largely transparent and closed to critique in safety and human factors precisely because they are so self-evident and commonsensical. The way they get retained and reproduced is perhaps akin to what Althusser (1984) called "interpellation," by the confluences of shared relationships, shared discourses, institutions, and knowledge. Foucault called the practices that produce knowledge and keep knowledge in circulation an episteme: a set of rules and conceptual tools for what counts as factual. Such practices are exclusionary. They function in part to establish distinctions between those statements that will be considered true and those that will be considered false (Foucault 1980). A sociotechnical Newtonian physics is thus read into events that could yield much more complexly patterned interpretations. People involved in accident analysis, for example, may be expected to explain themselves in terms of the dominant assumptions; they will make sense of events using those assumptions; they will then reproduce the existing order in their words and actions. Organizational, institutional, and technological arrangements surrounding their work don't leave plausible alternatives (in fact, they implicitly silence them). For instance, investigators are mandated to find the probable cause(s) and turn out enumerations of broken components as their findings. Technological–analytical support (incident databases, error analysis tools) emphasizes linear reasoning and the identification of malfunctioning components (Shappell and Wiegmann 2001). Also, organizations and those held accountable for internal failures need something to "fix," which further valorizes condensed accounts. If these processes fail to satisfy societal accountability requirements, then courts can decide to pursue individuals criminally, which also could be said to represent a hunt for a broken component (Dekker 2009; Thomas 2007).

VIEW FROM SYSTEMS AND COMPLEXITY

Analytic reduction cannot tell how a number of different things and processes act together when exposed to a number of different influences at the same time. This is complexity, a characteristic of a system. Complex behavior arises because of the interaction between the components of a system. It asks us to focus not on individual components but on their relationships. The properties of the system emerge as a result of these interactions; they are not contained within individual components. Complex systems generate new structures internally; they are not reliant on an external designer. In reaction to changing conditions in the environment, the system has to adjust some of its internal structure.

COMPLICATED VERSUS COMPLEX

There is an important distinction between "complex" and "complicated" systems:

- Complicated systems may be quite intricate and consist of a huge number of parts (e.g., a jet airliner). Nevertheless, it can be taken apart and put together again. Even if such a system cannot be understood completely by a single person, it is understandable and describable in principle. This makes them complicated.

- Complex systems, on the other hand, come to be in the interaction of the components. Jet airliners become complex systems when they are deployed in a nominally regulated world with cultural diversity, receiver-oriented versus transmitter-oriented communication expectations, different hierarchical gradients in a cockpit and multiple levels of politeness differentiation, effects of fatigue, procedural drift, and varied training and language standards (Hutchins et al. 2002), as well as cross-cultural differences in risk perceptions, attitudes, and behavior (Lund and Rundmo 2009).

This is where complicated systems become complex: open to influences that lie way beyond engineering reliability predictions. In a complex system, each component is ignorant of the behavior of the system as a whole and cannot know the full influences of its actions. Components respond locally to information presented to them, and complexity arises from the huge, multiplied webs of relationships and interactions that result from these local actions. The boundaries of what constitutes the system become fuzzy; interdependencies and interactions multiply and mushroom. Their nonlinearity offers opportunities not only for dampening and modulating but also for amplifying any risky influences that get into the system.

Complexity is a feature of the system, not of components inside of it. The knowledge of each component is limited and local, and there is no component that possesses enough capacity to represent the complexity of the entire system in that component itself. The behavior of the system cannot be reduced to the behavior of the constituent components. If we wish to study such systems, we have to investigate the system as such. It is exactly at this point that reductionist methods fail.

Complex systems have a history, a path dependence, which also spills over those fuzzy boundaries. Their past, and the past of events around them, is co-responsible for their present behavior, and descriptions of complexity should take history into account. Take as an example the takeoff accident of SQ006 at Taipei where deficient runway and taxiway lighting/signage was implicated in the crew's selection of a runway that was actually closed for traffic (ASW 2002). A critical taxiway light had burnt out, and the story could stop there. But as only one of two countries in the world (the Vatican being the other), Taiwan is not bound by airport design rules from the International Civil Aviation Organization, a United Nations body. Cross-strait relationships between Taiwan (formerly Formosa) and China, and their tortured post-WWII history (Mao's communists vs. Chiang Kai-Shek's Kuomintang Nationalists) and their place in a larger global context are responsible for keeping Taiwan out of the UN and thus out of formal reach for international safety regulations that apply to even the smallest markings and lights and their maintenance on an airport. A broken light and exceptionally bad visibility on the night of the SQ006 takeoff constituted the small changes in starting conditions that led to a large event in a system that had been living far from equilibrium for a long time—all consistent with complexity.

In describing the macro-behavior (or emergent behavior) of the system, not all its micro-features can be taken into account. The description on the macro-level is thus a reduction or compression of complexity and cannot be an exact description of what the system actually does. Moreover, the emergent properties on the macro-level can influence the micro-activities, a phenomenon sometimes referred to as "top-down

TABLE 2.1

Contrasting Cartesian–Newtonian and Complexity Visions of the World

Cartesian–Newtonian	Complexity
Reductionism: to understand the system, you decompose it into its parts, as their functioning linearly accounts for behavior at the system level. You go *down and in*.	Holism and synthesis: to understand the system, you also have to go *up and out* and see how it interacts with, and is configured among, other systems.
Effects *result* from causes; these can be found in the behavior of individual components.	Effects *emerge* from the complex interaction between components.
Outcomes are foreseeable if you know the initial conditions *and* the rules by which the system operates.	We can only estimate the probabilities of outcomes, not know them with certainty.
The system is time-reversible. From any moment, you can reconstruct how it looked any time before and predict how it will look at any time in the future.	Time is not reversible. Systems continually change as relationships and connections evolve internally and adapt to their changing environment.
Complete knowledge is attainable. It requires a fully accurate correspondence to the outside world, for which we are developing ever-better methods.	Complete knowledge is unattainable. Different descriptions of a complex system decompose it in different ways that cannot be reduced to each other.

causation." Nevertheless, macro-behavior is not the result of anything else but the micro-activities of the system, keeping in mind that these are influenced not only by their mutual interaction and by top-down effects but also by the interaction of the system with its environment. These insights have important implications for the knowledge claims made when analyzing accidents with complex systems. Since we do not have direct access to the complexity itself, our knowledge of such systems is in principle limited. Let's turn to an outline of the implications of these insights for understanding safety and the human factor, revisiting the topics above (see also Table 2.1).

SYNTHESIS AND HOLISM

Perrow pointed out decades ago that a Newtonian focus on parts as the cause of accidents may drive a false belief that redundancy is the best way to protect against hazard (Perrow 1984; Sagan 1993). The downside is that barriers, as well as professional specialization, policies, procedures, protocols, redundant mechanisms, and structures, all add to a system's complexity.

MORE BARRIERS, MORE RISK: THE FALLACY OF SOCIAL REDUNDANCY

Recall from the first chapter how the Newtonian belief that failure of parts is the cause of system failure is represented in the Swiss Cheese Model as well. One hospital decided to take its cues for harm prevention from this model and decided to put in place a double barrier in order to prevent medication misadministration to their patients. The double barrier consisted of a double check by nurses before the drug was allowed to go into the patient. After a while, results counterintuitively showed that the risk of medication misadministration had gone up, not down, with the introduction of the double barrier procedure.

Scott Sagan has referred to this as the fallacy of social redundancy (Sagan 1993), a phenomenon well documented in Janis' work on groupthink too (Janis 1982). Social *duplication*, that is, having two different units, teams, or individuals do the same thing in a single production sequence, does not reliably generate the effects of technical redundancy or duplication. The reason is that the "components" in social redundancy are not independent. People know each other, they talk, they are influenced by each other's ideas and views (even of them), and they are both impacted by common factors such as production pressures, fatigue, knowledge of the patient in question, and more. Building in social redundancy, in other words, is a fallacy. It is one of those places where following a particular model of safety (more barriers) can be actively harmful to safety, because it does not fit the circumstances and specifics of the risk to be managed.

The introduction of more barriers can also entail an explosion of new relationships (between parts and layers and components) that spread through the system. System accidents result from the relationships between components, not from the workings or dysfunctioning of any component part (Leveson 2002). Failures involve the unanticipated interaction of a multitude of events in a complex system—events and interactions, often very normal, whose combinatorial explosion can quickly outwit people's best efforts at predicting and mitigating trouble. In order to understand where problems may come from in a complex system, the idea is to think "up and out" rather than "down and in" (Dekker 2011b). Consider not how one component may be reliable or erratic, but rather how it is configured in relationships with many other components, and how the system of which it is part relates to many other systems. For example, the nurses doing the medication double-checking in the example above are configured in social and organizational relationships. For their work, they are dependent on many things and factors over which they have no or little control, and some over which they do. The packaging of the medication they have to double-check, for example, is out of their control but can influence the success of their interaction with it. Scheduling (and whom they are scheduled to work with), fatigue, and other factors may also largely be outside of their control. The location of the medication may be partly under their control but involve relationships with pharmacists, and they may not be familiar with it. Prescribing the medication largely depends on physicians, whom nurses are expected to trust, yet speak up against if they don't. Rather than analysis (down and in), synthesis (up and out) can reveal relationships and constraints that affect the probability of a failure. Holistic thinking encompasses the systems in which humans and their technologies are configured, not just the components inside.

EMERGENCE

Safety has been characterized as an emergent property, something that cannot be predicted on the basis of the components that make up the system (Leveson 2002). Accidents have similarly been characterized as emergent properties of complex systems (Hollnagel 2004). They cannot be predicted on the basis of the constituent parts. Rather, they are one emergent feature of constituent components doing their (normal) work. A systems accident is possible in an organization where people themselves suffer no noteworthy incidents, in which everything looks normal, and everybody is abiding by their local rules, common solutions, or habits (Vaughan 2005). This means that the behavior of the whole cannot be explained by, and is not mirrored in, the behavior of constituent components. Snook (2000) expressed the realization that bad effects can happen with no causes in his study of the shooting down of two US Black Hawk helicopters by two US fighter jets in the no-fly zone over Northern Iraq in 1993:

> This journey played with my emotions. When I first examined the data, I went in puzzled, angry, and disappointed—puzzled how two highly trained Air Force pilots could make such a deadly mistake; angry at how an entire crew of AWACS controllers could sit by and watch a tragedy develop without taking action; and disappointed at how dysfunctional Task Force OPC must have been to have not better integrated helicopters into its air operations. Each time I went in hot and suspicious. Each time I came out sympathetic and unnerved... If no one did anything wrong; if there were no unexplainable surprises at any level of analysis; if nothing was abnormal from a behavioral and organizational perspective; then what...?

Snook's impulse to hunt down the broken components (deadly pilot error, controllers sitting by, a dysfunctional Task Force) led to nothing. There was no "eureka part." The accident defied Newtonian logic.

Asymmetry or nonlinearity means that an infinitesimal change in starting conditions can lead to huge differences later on. This sensitive dependence on initial conditions removes proportionality from the relationships between system inputs and outputs. The evaluation of damage caused by debris falling off the external tank prior to the fatal 2003 Space Shuttle Columbia flight can serve as an example (CAIB 2003; Starbuck and Farjoun 2005). Always under pressure to accommodate tight launch schedules and budget cuts (in part because of a diversion of funds to the International Space Station), certain problems became seen as maintenance issues rather than flight safety risks. Maintenance issues could be cleared through a nominally simpler bureaucratic process, which allowed quicker shuttle vehicle turnarounds. In the mass of assessments to be made between flights, the effect of foam debris strikes was one. Gradually converting this issue from safety to maintenance was not different from a lot of other risk assessments and decisions that NASA had to do as one shuttle landed and the next was prepared for flight—one more decision, just like tens of thousands of other decisions. While any such decision can be quite rational given the local circumstances and the goals, knowledge, and attention of the decision makers, interactive complexity of the system can take it onto unpredictable pathways to hard-to-foresee system outcomes.

This complexity has implications for the ethical load distribution in the aftermath of complex system failure. Consequences cannot form the basis for an assessment of the gravity of the cause (or the quality of the decision leading up to it), something that has been argued in the safety and human factors literature (Orasanu and Martin 1998). It suggests that everyday organizational decisions, embedded in masses of similar decisions and only subject to special consideration with the wisdom of hindsight, cannot be fairly singled out for purposes of exacting accountability (e.g., through criminalization) because their relationship to the eventual outcome is complex and nonlinear and was probably impossible to foresee (Jensen 1996).

FORESEEABILITY OF PROBABILITIES, NOT CERTAINTIES

Decision makers in complex systems are capable of assessing the probabilities, but not the certainties of particular outcomes. Knowledge of initial conditions and total knowledge of the laws governing a system (the two Newtonian conditions for assessing foreseeability of harm) are unobtainable in complex systems. Often, in retrospect, we nevertheless assume that other people had such knowledge. And then we base our judgment of those people's decisions on that assumption. That we do this is not just the irresistibility of well-documented psychological biases of outcome and hindsight, but also consistent with Newtonian thinking: Complete knowledge of the world is possible, if only we apply ourselves diligently enough. And that goes, in retrospect, for other people too. Thus, with an outcome in hand, its (presumed) foreseeability becomes quite obvious, and it may appear as if a decision in fact determined an outcome, an outcome that the decision inevitably led up to (Fischhoff and Beyth 1975).

TIME-IRREVERSIBILITY

The conditions of a complex system are irreversible. The precise set of conditions that gave rise to the emergence of a particular outcome (e.g., an accident) is something that can never be exhaustively reconstructed. Complex systems continually experience change as relationships and connections evolve internally and adapt to their changing environment. Given the open, adaptive nature of complex systems, the system after the accident is not the same as the system before the accident—many things will have changed, not only as a result of the outcome, but as a result of the passage of time. This also means that the predictive power of retrospective analysis of failure is severely limited (Leveson 2002). Decisions in organizations, for example, to the extent that they can be excised and described separate from context at all, were not the single beads strung along some linear cause–effect sequence that they may seem afterward. Complexity argues that they are spawned and suspended in the messy interior of organizational life that influences and buffets and shapes them in a multitude of ways. Many of these ways are hard to trace retrospectively as they do not follow documented organizational protocol but rather depend on unwritten routines, implicit expectations, professional judgments, and subtle oral influences on what people deem rational or doable in any given situation (Vaughan 1999).

Reconstructing events in a complex system, then, is impossible, primarily as a result of the characteristics of complexity. Psychological characteristics of retrospective investigation make it so too. As soon as an outcome has happened, whatever past events can be said to have led up to it undergo a whole range of transformations (Fischhoff and Beyth 1975; Hugh and Dekker 2009). Take the idea that it is a sequence of events that precedes an accident. Who makes the selection of the "events" and on the basis of what? The very act of separating important or contributory events from unimportant ones is an act of construction, of the creation of a story, not the reconstruction of a story that was already there, ready to be uncovered. Any sequence of events or list of contributory or causal factors already smuggles a whole array of selection mechanisms and criteria into the supposed "re"-construction. There is no objective way of doing this—all these choices are affected, more or less explicitly, by the analyst's background, preferences, experiences, biases, beliefs, and purposes. "Events" are themselves defined and delimited by the stories with which the analyst configures them and are impossible to imagine outside this selective, exclusionary, narrative fore-structure (Cronon 1992).

PERPETUAL INCOMPLETENESS AND UNCERTAINTY OF KNOWLEDGE

The Newtonian belief that is both instantiated and reproduced in official accident investigations is that there is a world that is objectively available and apprehensible. This epistemological stance represents a kind of aperspectival objectivity. It assumes that investigators are able to take a "view from nowhere" (Nagel 1992), a value-free, background-free, position-free view that is true. This reaffirms the classical or Newtonian view of nature (an independent world exists to which investigators, with proper methods, can have objective access). It rests on the belief that observer and the observed are separable. Knowledge is nothing more than a mapping from object to subject. Investigation is not a creative process: it is merely an "uncovering" of distinctions that were already there and simply waiting to be observed (Heylighen et al. 2006).

Complexity, in contrast, suggests that the observer is not just the contributor to, but in many cases the creator of, the observed (Wallerstein 1996). Cybernetics introduced this idea to complexity and systems thinking: knowledge is intrinsically subjective, an imperfect tool used by an intelligent agent to help it achieve its personal goals. Not only does the agent not need an objective reflection of reality, it can actually never achieve one. Indeed, the agent does not have access to any "external reality": it can merely sense its inputs, note its outputs (actions), and from the correlations between them induce certain rules or regularities that seem to hold within its environment. Different agents, experiencing different inputs and outputs, will in general induce different correlations and therefore develop a different knowledge of the environment in which they live. There is no objective way to determine whose view is right and whose is wrong, since the agents effectively live in different environments (Heylighen et al. 2006).

Different descriptions of a complex system, then (from the point of view of different agents), decompose the system in different ways. It follows that the knowledge gained by any description is always relative to the perspective from which the description was made. This does not imply that any description is as good as any other. It is merely the result of the fact that only a limited number of characteristics of the system can be taken into account by any specific description. Although there is no a priori procedure for deciding which description is correct, some descriptions will deliver more interesting results than others. It is not that some complex readings are "truer" in the sense of corresponding more closely to some objective state of affairs (as that would be a Newtonian commitment). Rather, the acknowledgement of complexity in accident analysis can lead to a richer understanding and thus it holds the potential to improve safety and help to expand the ethical response in the aftermath of failure.

SOCIOTECHNICAL SYSTEMS AND COMPLEXITY

When sociotechnical systems are seen as complex, there is no longer an obvious relationship between the behavior of parts in the system (or their dysfunctioning, e.g., "human errors") and system-level outcomes. Instead, system-level behaviors emerge from the multitude of relationship and interconnections deeper inside the system, and cannot be reduced to those relationships or interconnections. The selection of "causes" (or "events" or "contributory factors") is always an act of construction by the investigation. There is no objective way of doing this—all analytical choices are affected, more or less explicitly, by the investigation's own position in a complex system, by its background, preferences, language, experiences, biases, beliefs, and purposes. It can never construct one true story of what happened. Truth, then, lies in diversity of explanations and narratives, not in singularity (Cilliers 2010). Studies in safety and human factors that embrace complexity, then, might stop looking for the "causes" of failure or success. Instead, they gather multiple narratives from different perspectives inside of the complex system, which give partially overlapping and partially contradictory accounts of how emergent outcomes come about. The complexity perspective dispenses with the notion that there are easy answers to a complex systems event—supposedly within reach of the one with the best method or most objective investigative viewpoint. It allows us to invite more voices into the conversation and to celebrate their diversity and contributions.

REDUCTION AND COMPONENTS: THE CARTESIAN–NEWTONIAN WAY TO THINK ABOUT FAILURE

The Cartesian–Newtonian heritage is a mixed blessing. Human factors and systems safety are stuck with a language, with metaphors and images that emphasize structure, components, mechanics, parts and interactions, cause and effect. While giving us initial direction for building safe systems and for finding out what went wrong when it turns out we have not, there are limits to the usefulness of this inherited vocabulary.

BACK TO SPOILERS

Let us go back to that summer day of 1999 and the MD-80 runway overrun. In good Cartesian–Newtonian tradition, we can begin by opening up the aircraft a bit more, picking apart the various components and procedures to see how they interact, second by second. Initially, we will be met with resounding empirical success—as indeed Descartes and Newton frequently were. But when we want to recreate the whole on the basis of the parts we find, a more troubling reality swims into view: It does not go well together anymore. The neat, mathematically pleasing separation between human and mechanical cause, between social and structural issues, has blurred. The whole no longer seems a linear function of the sum of the parts. As Scott Snook (2000) explained it, the two classical Western scientific steps of analytic reduction (the whole into parts) and inductive synthesis (the parts back into a whole again) may seem to work, but simply putting the parts we found back together does not capture the rich complexity hiding inside and around the incident. What is needed is a holistic, organic integration. What is perhaps needed is a new form of analysis and synthesis, sensitive to the total situation of organized sociotechnical activity. But first let us examine the analytical, componential story.

Spoilers are those flaps that come up into the airstream on the topside of the wings after an aircraft has touched down. Not only do they help brake the aircraft by obstructing the airflow, they also cause the wing to lose the ability to create lift, forcing the aircraft's weight onto the wheels. Extension of the ground spoilers also triggers the automatic braking system on the wheels: The more weight the wheels carry, the more effective their braking becomes. Before landing, pilots select the setting they wish on the automatic wheel-braking system (minimum, medium, or maximum), depending on runway length and conditions. After landing, the automatic wheel-braking system will slow down the aircraft without the pilot having to do anything, and without letting the wheels skid or lose traction. As a third mechanism for slowing down, most jet aircraft have thrust reversers, which redirect the outflow of the jet engines into the oncoming air, instead of out toward the back.

In this case, no spoilers came out, and no automatic wheel braking was triggered as a result. While rolling down the runway, the pilots checked the setting of the automatic braking system multiple times to ensure it was armed and even changed its setting to maximum as they saw the end of the runway coming up. But it would never engage. The only remaining mechanism for slowing down the aircraft was the thrust reversers. Thrust reversers, however, are most effective at high speeds. By the time the pilots noticed that they were not going to make it before the end of the runway, the speed was already quite low (they ended up going into the field at 10–20 knots) and thrust reversers no longer had much immediate effect. As the jet was going over the edge of the runway, the captain closed the reversers and steered somewhat to the right in order to avoid obstacles.

How are spoilers armed? On the center pedestal, between the two pilots, are a number of levers. Some are for the engines and thrust reversers, one is for the flaps, and one is for the spoilers. In order to arm the ground spoilers, one of the pilots needs to pull the lever upward. The lever goes up by about one inch and sits there, armed until touchdown. When the system senses that the aircraft is on the ground (which it does in part through switches in the landing gear), the lever will come back automatically and the spoilers come out. Asaf Degani, who studied such procedural problems extensively, has called the spoiler issue not one of human error, but one of timing (Degani et al. 1999). On this aircraft, as on many others, the spoilers should not be armed before the landing gear has been selected down and is entirely in place. This has to do with the switches that can tell when the aircraft is on the ground. These are switches that compress as the aircraft's weight settles onto the wheels, but not only then. There is a risk in this type of aircraft that the switch in the nose gear will even compress as the landing gear is coming out of its bays. This can happen because the nose gear folds out into the oncoming airstream. As the nose gear is coming out and the aircraft is slicing through the air at 180 knots, the sheer wind force can compress the nose gear, activate the switch, and subsequently risk extending the ground spoilers (if they had been armed). This is not a good idea: The aircraft would have trouble flying with ground spoilers out. Hence the requirement: The landing gear needs to be all the way out, pointing down. Only when there is no more risk of aerodynamic switch compression can the spoilers be armed. This is the order of the before-landing procedures:

> Gear down and locked.
> Spoilers armed.
> Flaps FULL.

On a typical approach, pilots select the landing gear handle down when the so-called glide slope comes alive: when the aircraft has come within range of the electronic signal that will guide it down to the runway. Once the landing gear is out, spoilers must be armed. Then, once the aircraft captures that glide slope (i.e., it is exactly on the electronic beam) and starts to descend down the approach to the runway, flaps need to be set to FULL (typically 40°). Flaps are other devices that extend from the wing, changing the wing size and shape. They allow aircraft to fly more slowly for a landing. This makes the procedures conditional on context. It now looks like this:

> Gear down and locked (when glide slope live).
> Spoilers armed (when gear down and locked).
> Flaps FULL (when glide slope captured).

But how long does it take to go from "glide slope live" to "glide slope capture"? On a typical approach (given the airspeed) this takes about 15 seconds. On a simulator, where training takes place, this does not create a problem. The whole gear cycle (from gear lever down to the "gear down and locked" indication in the cockpit) takes about 10 seconds. That leaves 5 seconds for arming the spoilers, before the crew needs to select flaps FULL (the next item in the procedures). In the simulator, then, things look like this:

> At $t = 0$ Gear down and locked (when glide slope live).
> At $t + 10$ Spoilers armed (when gear down and locked).
> At $t + 15$ Flaps FULL (when glide slope captured).

But in real aircraft, the hydraulic system (which, among other things, extends the landing gear) is not as effective as it is on a simulator. The simulator, of course, only has simulated hydraulic aircraft systems, modeled on how the aircraft is when it has flown zero hours, when it is sparkling new, straight out of the factory. On older aircraft, it can take up to half a minute for the gear to cycle out and lock into place. This makes the procedures look like this:

At $t = 0$ Gear down and locked (when glide slope live).
At $t + 30$ Spoilers armed (when gear down and locked).
BUT! at $t + 15$ Flaps FULL (when glide slope captured).

In effect, then, the "flaps" item in the procedures intrudes before the "spoilers" item. Once the "Flaps" item is completed and the aircraft is descending toward the runway, it is easy to go down the procedures from there, taking the following items. Spoilers in this case never get armed. Their arming has tumbled through the cracks of a time warp. An exclusive claim to human error (or CRM breakdown) becomes more difficult to sustain against this background. How much human error was there, actually? Let us remain dualist for now and revisit Equation 2.1. Now apply a more liberal definition of *mechanical failure*. The nose gear of the actual aircraft, fitted with a compression switch, is designed so that it folds out into the wind while still airborne. This introduces a systematic mechanical vulnerability that is tolerated solely through procedural timing (a known leaky mechanism against failure): first the gear, then the spoilers. In other words, "gear down and locked" is a mechanical prerequisite for spoiler arming, but the whole gear cycle can take longer than there is room in the procedures and the timing of events driving their application. The hydraulic system of the old jet does not pressurize as well: It can take up to 30 seconds for a landing gear to cycle out. The aircraft simulator, in contrast, does the same job inside of 10 seconds, leaving a subtle but substantive mechanical mismatch. One work sequence is introduced and rehearsed in training, whereas a delicately different one is necessary for actual operations. Moreover, this aircraft has a system that warns if the spoilers are not armed on takeoff, but it does not have a system for warning that the spoilers are not armed on approach. Then there is the mechanical arrangement in the cockpit. The armed spoiler handle looks different from the unarmed one by only one inch and a small red square at the bottom. From the position of the right-seat pilot (who needs to confirm their arming), this red patch is obscured behind the power levers as these sit in the typical approach position. With so much mechanical contribution going around (landing gear design, eroded hydraulic system, difference between simulator and real aircraft, cockpit lever arrangement, lack of spoiler warning system on approach, procedure timing) and a helping of scheduling stochastics (two captains on this flight), a whole lot more mechanical failure could be plugged into the equation to rebalance the human contribution.

But that is still dualist. When reassembling the parts that we found among procedures, timing, mechanical erosion, and design trade-offs, we can begin to wonder where mechanical contributions actually end and where human contributions begin. The border is no longer so clear. The load imposed by a wind of 180 knots on the nose wheel is transferred onto a flimsy procedure: first the gear, then the spoilers. The nose wheel, folding out into the wind and equipped with a compression switch, is incapable of carrying that load and guaranteeing that spoilers will not extend, so a procedure gets to carry the load instead. The spoiler lever is placed in a way that makes verification difficult, and a warning system for unarmed spoilers is not installed. Again, the error is suspended, uneasily and unstably, between human intention and engineered hardware—it belongs to both and to neither uniquely. And then there is this: The gradual wear of a hydraulic system is not something that was taken into account during the certification of the jet. An MD-80 with an anemic hydraulic system that takes more than half a minute to get the whole gear out, down, and locked, violating the original design requirement by a factor of three, is still considered airworthy. The worn hydraulic system cannot be considered a mechanical failure. It does not ground the jet. Neither does the hard-to-verify spoiler handle or lack of warning system on approach. The jet was once certified as airworthy with or without all of that. That there is no mechanical failure, in other words, is not because there are no mechanical issues. There is no mechanical failure because social systems, made up of manufacturers, regulators, and prospective operators—undoubtedly shaped by practical concerns and expressed through situated engineering judgment with uncertainty about future wear—decided that there would not be any (at least not related to the issues now identified in an MD-80 overrun). Where does mechanical failure end and human error begin? Dig just deeply enough and the question becomes impossible to answer.

RES EXTENSA AND RES COGITANS, OLD AND NEW

IMPOSING A WORLDVIEW

Separating *res extensa* away from *res cogitans*, like Descartes did, is artificial. It is not the result of natural processes or conditions, but rather an imposition of a worldview. This worldview, though initially accelerating scientific progress, is now beginning to seriously hamper our understanding. In modern accidents, machinistic and human causes are blurred. The disjunction between material and mental worlds and the requirement to describe them differently and separately are debilitating our efforts to understand sociotechnical success and failure.

The distinction between the old and new views of human error, which was discussed earlier in *Field Guide to Human Error Investigations* (Dekker 2002), actually rides roughshod over these subtleties. Recall how the investigation into the runway overrun incident found "breakdowns in CRM" as a causal factor. This is old-view thinking. Somebody, in this case a pilot, or rather a crew of two pilots, forgot to arm the spoilers. This was a human error, an omission. If they had not forgotten to arm the spoilers, the accident would not have happened, end of story. But such an analysis of failure does not probe below the immediately visible surface variables of a sequence of events. As Perrow (1984) puts it, it judges only where people should have zigged instead of zagged. The old view of human error is surprisingly common. In the old view, error—by any other name (e.g., complacency, omission, breakdown of CRM)—is accepted as a satisfactory explanation. This is what the new view of human error tries to avoid. It sees human error as a consequence, as a result of failings and problems deeper inside the systems in which people work. It resists seeing human error as the cause. Rather than judging people for not doing what they should have done, the new view presents tools for explaining why people did what they did. Human error becomes a starting point, not a conclusion. In the spoiler case, the error is a result of design trade-offs, mechanical erosion, procedural vulnerabilities, and operational stochastics. Granted, the commitment of the new view is to resist neat, condensed versions in which a human choice or a failed mechanical part led the whole structure onto the road to perdition. The distinction between the old and new views is important and needed. Yet even in the new view the error is still an effect, and effects are the language of Newton. The new view implicitly acknowledges the existence, the reality of error. It sees error as something that is out there, in the world, and caused by something else, also out there in the world. As the next chapters show, such a (naively) realist position is perhaps untenable.

Recall how the Newtonian–Cartesian universe consists of wholes that can be explained and controlled by breaking them down into constituent parts and their interconnections (e.g., humans and machines, blunt ends and sharp ends, safety cultures and blame cultures). Systems are made up of components and of mechanical-like linkages between those components. This lies at the source of the choice between human and material cause (is it human error or mechanical failure?). It is Newtonian in that it seeks a cause for any observed effect, and Cartesian in its dualism. In fact, it expresses both Descartes' dualism (either mental or material: you cannot blend the two) and the notion of decomposition, where lower-order properties and interactions completely determine all phenomena. They are enough; you need no more. Analyzing which building blocks go into the problem, and how they add up, is necessary and sufficient for understanding why the problem occurs. Equation 2.1 is a reflection of the assumed explanatory sufficiency of lower-order properties. Throw in the individual contributions, and the answer to why the problem occurred rolls out. An aircraft runway overrun today can be understood by breaking the contributions down into human and machine causes, analyzing the properties and interactions of each, and then reassembling it back into a whole. "Human error" turns up as the answer. If there are no material contributions, the human contribution is expected to carry the full explanatory load.

As long as progress is made using this worldview, there is no reason to question it. In various corners of science, including human factors, many people still see no reason to do so. Indeed, there is no reason that structuralist models cannot be imposed on the messy interior of sociotechnical systems. That these systems, however, reveal machine-like properties (components and interconnections, layers and holes) when we open them up postmortem does not mean that they are machines, or that they, in life, grew and behaved like machines. As Leveson (2012) has pointed out, analytic reduction assumes the following:

- The separation of a whole into constituent parts is feasible.
- Subsystems operate independently.
- Analysis results are not distorted by taking the whole apart.
- This in turn implies that the components are not subject to feedback loops and other nonlinear interactions and that they are essentially the same when examined singly as when they are playing their part in the whole.

- Moreover, it assumes that the principles governing the assembly of the components into the whole are straightforward; the interactions among components are simple enough that they can be considered separate from the behavior of the whole.

NEWTONIAN SOCIAL PHYSICS

HUMAN FACTORS AND ITS REALIST WORLDVIEW

Human factors as a discipline takes a rather realist view. It lives in a world of real things, of facts and concrete observations. It presumes the existence of an external world in which phenomena occur that can be captured and described objectively. In this world, there are "errors" and "violations," and these errors and violations are quite real. They form part of a kind of Newtonian social or psychological physics that can be described and understood using the same approaches as those of the natural sciences. The flight deck observer from the last chapter, for example, would see that pilots do not arm the spoilers before landing and marks this up as an error or a procedural violation. The observer considers his observation quite "true," and the error quite "real." Upon discovering that the spoilers had not been armed, the pilots themselves too may see their omission as an "error," as something that they missed but should not have missed. But just as it did for the flight deck observer, the error becomes real only because it is visible from outside the stream of experience. From the inside of this stream, while things are going on and work is being accomplished, there is no "error." In this case, there are only procedures that get inadvertently mangled through the timing and sequence of various tasks. And not even this gets noticed by those applying the procedures.

Paul Feyerabend (1993) has pointed out that all observations are ideational. "Facts" do not exist without an observer wielding a particular theory that tells him what to look for. Observers are not passive recipients but active creators of the empirical reality they encounter. There is no clear separation between observer and observed. As said in the previous chapter, none of this makes the "error" any less real to those who observe it. But it does not mean that the error exists "out there," in some independent empirical universe. This was the whole point of ontological relativism: what it means to be in a particular situation and make certain observations is quite flexible and connected systematically to the observer. None of the possible worldviews can be judged superior or privileged uniquely by empirical data about the world, since objective, impartial access to that world is impossible. Yet in the pragmatic and optimistically realist spirit of human factors, error counting methods have gained popularity by selling the belief that such impartial access is possible. The claim to privileged access lies (as modernism and Newtonian science would dictate) in method. The method is strong enough to discover errors that the pilots themselves had not seen.

Errors appear so "real" when we step or set ourselves outside the stream of experience in which they occur. They appear so real to an observer sitting behind the pilots. They appear so real to even the pilot himself after the fact. But why? It cannot be because the errors are real, since the autonomy principle has been proven false. As an observed "fact," the error only exists by virtue of the observer and his position on the outside of the stream of experience. The error does not exist because of some objective empirical reality in which it putatively takes place, since there is no such thing and if there was, we could not know it. Recall the air traffic control test of the previous chapter: actions, omissions, and postponements related to air traffic clearances carry entirely different meanings for those on the inside and on the outside of the work experience. Even different observers on the outside cannot agree on a common denominator because they have diverging backgrounds and conceptual looking glasses. The autonomy principle is false: "facts" do not exist without an observer.

Errors Are Active, Corrective Interventions in History

To paraphrase Giddens, errors are an active, corrective intervention in (immediate) history. It is impossible for us to give a mere chronicle of our experiences: our assumptions, past experiences, and future aspirations make that we impress a certain organization on that which we just went through or saw. "Errors" are a powerful way to impose structure onto past events. "Errors" are a particular way in which we as observers (or even participants) reconstruct the reality we just experienced. Such reconstruction, however, inserts a severe discontinuity between past and present. The present was once an uncertain, perhaps vanishingly improbable future. Now we see it as the only plausible outcome of a pretty deterministic past. Being able to stand outside an unfolding sequence of events (either as participants from hindsight or as observers from outside the setting) makes it exceedingly difficult to see how unsure we once were (or could have been if we had been in that situation) of what was going to happen. History as seen through the eyes of a retrospective outsider (even if the same observer was a participant in that history not long ago) is substantially different from the world as it appeared to the decision makers of the day. This endows history, even immediate history, with a determinism it lacked when it was still unfolding.

"Errors," then, are ex post facto constructs. The research base on the hindsight bias contains some of the strongest evidence on this. "Errors" are not empirical facts. As Philip Tetlock has pointed out, they are the result of outside observers squeezing now-known events into the most plausible, or convenient deterministic scheme. In the research base on hindsight, it is not difficult to see how such retrospective restructuring embraces a liberal take on the history it aims to recount. The distance between reality as portrayed by a retrospective observer and as experienced by those who were there (even if these were once the same people) grows substantially with the rhetoric and discourse employed and the investigative practices used. We shall shortly see a lot of this.

We will also take a look at developments in psychology that have (not so long ago) tried to get away from the normativist bias in our understanding of human performance and decision making. This intermezzo is necessary because "errors" and "violations" do not exist without some norm, even if implied. Hindsight of course has a powerful way of importing criteria or norms from outside people's situated contexts and highlighting where actual performance at the time fell short. To see errors as ex post constructs rather than as objective, observed facts, we have to understand the influence of implicit norms on our judgments of past performance. Doing without errors means doing without normativism. It means that we cannot question the accuracy of insider accounts (something human factors consistently does, for example, when it asserts a "loss of situation awareness"), as there is no objective, normative reality to hold such accounts up to, and relative to which we can deem them accurate or inaccurate. Reality as experienced by people at the time was reality as it was experienced by them at the time. It was that experienced world that drove their assessments and decisions, not our (or even their) retrospective, outsider rendering of that experience. We have to use local norms of competent performance to understand why what people did made sense to them at the time.

MAKING TANGLED HISTORIES LINEAR

The hindsight bias is one of the most consistent "biases" in psychology. One effect is that "people who know the outcome of a complex prior history of tangled, indeterminate events, remember that history as being much more determinant, leading 'inevitably' to the outcome they already knew" (Weick 1995, p. 28). Hindsight allows us to change past indeterminacy and complexity into order, structure, and oversimplified causality.

A TURN FOR THE WORSE

As an example, take the turn toward the mountains that a Boeing 757 made just before an accident near Cali, Colombia in 1995. According to the investigation, the crew did not notice the turn, at least not in time (Aeronautica Civil 1996). What should the crew have seen in order to know about the turn? They had plenty of indications, according to the manufacturer of their aircraft:

> "Indications that the airplane was in a left turn would have included the following: the EHSI (Electronic Horizontal Situation Indicator) Map Display (if selected) with a curved path leading away from the intended direction of flight; the EHSI VOR display, with the CDI (Course Deviation Indicator) displaced to the right, indicating the airplane was left of the direct Cali VOR course, the EaDI indicating approximately 16 degrees of bank, and all heading indicators moving to the right. Additionally the crew may have tuned Rozo in the ADF and may have had bearing pointer information to Rozo NDB on the RMDI" (Boeing 1996, p. 13).

This is a standard response after mishaps: point to the data that would have revealed the true nature of the situation. In hindsight, there is an overwhelming array of evidence that did point to the real nature of the situation, and if only people had paid attention to even some of it, the outcome would have been different. Confronted with a litany of indications that could have prevented the accident, we wonder how people at the time could not have known all of this. We wonder how this "epiphany" was missed, why this bloated shopping bag full of revelations was never opened by the people who most needed it.

But knowledge of the "critical" data comes only with the omniscience of hindsight. We can only know what really was critical or highly relevant once we know the outcome. Yet if data can be shown to have been physically available, we often assume that it should have been picked up by the practitioners in the situation. The problem is that pointing out that something should have been noticed does not explain why it was not noticed, or why it was interpreted differently back then. This confusion has to do with us, not with the people we are investigating. What we, in our reaction to failure, fail to appreciate is that there is a dissociation between data availability and data observability—between what can be shown to have been physically available and what would have been observable given people's multiple interleaving tasks, goals, attentional focus, expectations, and interests. Data, such as the litany of indications in the example above, do not reveal themselves to practitioners in one big monolithic moment of truth. In situations where people do real work, data can get drip-fed into the operation: a little bit here, a little bit there. Data emerge over time. Data may be uncertain. Data may be ambiguous. People have other things to do too. Sometimes, the successive or multiple data bits are contradictory; often, they are unremarkable. It is one thing to say how we find some of these data important in hindsight. It is quite another to understand what the data meant, if anything, to the people in question at the time.

The same kind of confusion occurs when we, in hindsight, get an impression that certain assessments and actions point to a common condition. This may be true at first sight. In trying to make sense of past performance, it is always tempting to group individual fragments of human performance that seem to share something, that seem to be connected in some way, and connected to the eventual outcome. For example, "hurry" to land was such a leitmotif extracted from the evidence in the Cali investigation. Haste in turn is enlisted to explain the errors that were made:

IN A RUSH?

"Investigators were able to identify a series of errors that initiated with the flightcrew's acceptance of the controller's offer to land on runway 19... The CVR indicates that the decision to accept the offer to land on runway 19 was made jointly by the captain and the first officer in a 4-second exchange that began at 2136:38. The captain asked: 'would you like to shoot the one nine straight in?' The first officer responded, 'Yeah, we'll have to scramble to get down. We can do it.' This interchange followed an earlier discussion in which the captain indicated to the first officer his desire to hurry the arrival into Cali, following the delay on departure from Miami, in an apparent attempt to minimize the effect of the delay on the flight attendants' rest requirements. For example, at 2126:01, he asked the first officer to 'keep the speed up in the descent'... (This is) evidence of the hurried nature of the tasks performed" (Aeronautica Civil 1996, p. 29).

But in the case above, the fragments used to build the argument of haste come from over half an hour of extended performance. Outside observers have treated the record as if it were a public quarry to pick stones from, and the accident explanation the building he needs to erect. The problem is that each fragment is meaningless outside the context that produced it: each fragment has its own story, background, and reasons for being, and when it was produced, it may have had nothing to do with the other fragments it is now grouped with. Also, behavior takes place in between the fragments. These intermediary episodes contain changes and evolutions in perceptions and assessments that separate the excised fragments not only in time but also in meaning. Thus, the condition, and the constructed linearity in the story that binds these performance fragments, arises not from the circumstances that brought each of the fragments forth; it is not a feature of those circumstances. It is an artifact of the outside observer. In the case described above, "hurry" is a condition identified in hindsight, one that plausibly couples the start of the flight (almost two hours behind schedule) with its fatal ending (on a mountainside rather than an airport). "Hurry" is a retrospectively invoked leitmotif that guides the search for evidence about itself. It leaves the investigator with a story that is admittedly more linear and plausible and less messy and complex than the actual events. Yet it is not a set of findings, but of tautologies, of objects constructed and subsequently found.

COUNTERFACTUAL REASONING

Tracing the sequence of events back from the outcome—that we as outside observers already know about—we invariably come across joints where people had opportunities to revise their assessment of the situation but failed to do so; where people were given the option to recover from their route to trouble, but did not take it. These are counterfactuals—quite common in accident analysis. For example, "The airplane could have overcome the windshear encounter if the pitch attitude of 15 degrees nose-up had been maintained, the thrust had been set to 1.93 EPR (Engine Pressure Ratio) and the landing gear had been retracted on schedule" (NTSB 1995, p. 119). Counterfactuals prove what could have happened if certain minute and often utopian conditions had been met. Counterfactual reasoning may be a fruitful exercise when trying to uncover potential countermeasures against such failures in the future. But saying what people could have done in order to prevent a particular outcome does not explain why they did what they did. This is the problem with counterfactuals. When they are enlisted as explanatory proxy, they help circumvent the hard problem of investigations: finding out why people did what they did. Stressing what was not done (but if it had been done, the accident would not have happened) explains nothing about what actually happened, or why. In addition, counterfactuals are a powerful tributary to the hindsight bias. They help us impose structure and linearity on tangled prior histories. Counterfactuals can convert a mass of indeterminate actions and events, themselves overlapping and interacting, into a linear series of straightforward bifurcations. For example, people could have perfectly executed the go-around maneuver but did not; they could have denied the runway change but did not. As the sequence of events rolls back into time, away from its outcome, the story builds. We notice that people chose the wrong prong at each fork, time and again—ferrying them along inevitably to the outcome that formed the starting point of our investigation (for without it, there would have been no investigation).

But human work in complex, dynamic worlds is seldom about simple dichotomous choices (as in: to err or not to err). Bifurcations are extremely rare—especially those that yield clear previews of the respective outcomes at each end. Choice moments (such as there are) typically reveal multiple possible pathways that stretch out, like cracks in a window, into the ever denser fog of futures not yet known. Their outcomes are indeterminate; hidden in what is still to come. Actions need to be taken under uncertainty and under the pressure of limited time and other resources. What from the retrospective outside may look like a discrete, leisurely two-choice opportunity to not fail, is from the inside one fragment caught up in a stream of surrounding actions and assessments. From the inside, it may not look like a choice at all. These are often choices only in hindsight. To the people caught up in the sequence of events, there was perhaps not any compelling reason to reassess their situation or decide against anything (or else they probably would have) at the point the investigator has now found significant or controversial. They were likely doing what they were doing because they thought they were right, given their understanding of the situation, their pressures. The challenge for an investigator becomes to understand how this may not have been a discrete event to the people whose actions are under investigation. The investigator needs to see how other people's "decisions" to continue were likely nothing more than continuous behavior—reinforced by their current understanding of the situation, confirmed by the cues they were focusing on, and reaffirmed by their expectations of how things would develop.

JUDGING INSTEAD OF EXPLAINING

When outside observers use counterfactuals, even as explanatory proxy, they themselves often require explanations as well. After all, if an exit from the route to trouble stands out so clearly to outside observers, how was it possible for other people to miss it? If there was an opportunity to recover, to not crash, then failing to grab it demands an explanation. The place where observers often look for clarification is the set of rules, professional standards, and available data that surrounded people's operation at the time, and how people did not see or meet that which they should have seen or met. Recognizing that there is a mismatch between what was done or seen and what should have been done or seen—as per those standards—we easily judge people for not doing what they should have done. Where fragments of behavior are contrasted with written guidance that can be found to have been applicable in hindsight, actual performance is often found wanting; it does not live up to procedures or regulations. For example, "One of the pilots... executed (a computer entry) without having verified that it was the correct selection and without having first obtained approval of the other pilot, contrary to procedures" (Aeronautica Civil 1996, p. 31). Investigations invest considerably in organizational archeology so that they can construct the regulatory or procedural framework within which the operations took place or should have taken place. Inconsistencies between existing procedures or regulations and actual behavior are easy to expose when organizational records are excavated after-the-fact and rules are uncovered that would have fit this or that particular situation.

This is not, however, very informative. There is virtually always a mismatch between actual behavior and written guidance. Pointing that there is a mismatch sheds little light on the why of the behavior in question. And for that matter, mismatches between procedures and practice are not unique to mishaps. There are also less obvious or undocumented standards. These are often invoked when a controversial fragment (e.g., a decision to accept a runway change [Aeronautica Civil 1996], or the decision to go around or not [NTSB 1995]) knows no clear preordained guidance but relies on local, situated judgment. For these cases, there are always "standards of good practice" that are based on convention and putatively practiced across an entire industry. One such standard in aviation is "good airmanship," which, if nothing else can, will explain the variance in behavior that had not yet been accounted for.

While micromatching, the observer frames people's past assessments and actions inside a world that she or he has invoked retrospectively. Looking at the frame as overlay on the sequence of events, she or he sees that pieces of behavior stick out in various places and at various angles: a rule not followed here; available data not observed there; professional standards not met over there. But rather than explaining controversial fragments in relation to the circumstances that brought them forth, and in relation to the stream of preceding as well as succeeding behaviors that surrounded them, the frame merely boxes performance fragments inside a world the observer now knows to be true. The problem is this after-the-fact world may have very little relevance to the actual world that produced the behavior under study. The behavior is contrasted against the observer's reality, not the reality surrounding the behavior at the time. Judging people for what they did not do relative to some rule or standard does not explain why they did what they did. Saying that people failed to take this or that pathway—only in hindsight the right one—judges other people from a position of broader insight and outcome knowledge that they themselves did not have. It does not explain a thing yet; it does not shed any light on why people did what they did given their surrounding circumstances. The outside observer has become caught in what William James called the psychologist's fallacy a century ago: he has substituted his own reality for the one of his object of study.

INVERTING PERSPECTIVES

Knowing about and guarding against the psychologist's fallacy, this mixing of realities is critical to understanding "error." When looked at from the position of retrospective outsider, the "error" can look so very real, so compelling. They failed to notice, they did not know, they should have done this or that. But from the point of view of people inside the situation, as well as potential other observers, this same "error" is often nothing more than normal work. If we want to begin to understand why it made sense for people to do what they did, we have to reconstruct their local rationality. What did they know? What was their understanding of the situation? What were their multiple goals, resource constraints, and pressures? Behavior is rational within situational contexts: people do not come to work to do a bad job. As historian Barbara Tuchman puts it:

Every scripture is entitled to be read in the light of the circumstances that brought it forth. To understand the choices open to people of another time, one must limit oneself to what they knew; see the past in its own clothes, as it were, not in ours (Tuchman 1981, p. 75).

This position turns the social and operational context into the only legitimate interpretive device. This context becomes the constraint on what meaning we, who were not there when it happened, can now give to past controversial assessments and actions. Historians are not the only ones to encourage this switch, this inversion of perspectives, this persuasion to put ourselves in the shoes of other people. In hermeneutics, it is known as the difference between exegesis (reading out of the text) and eisegesis (reading into the text). The point is to read out of the text what it has to offer about its time and place, not to read into the text what we want it to say or reveal now. Jens Rasmussen pointed out that if we cannot find a satisfactory answer to questions such as "how could they not have known?," then this is not because these people were behaving bizarrely. It is because we have chosen the wrong frame of reference for understanding their behavior. The frame of reference for understanding people's behavior is their own normal, individual work context, the context they are embedded in and from which point of view the decisions and assessments made are mostly normal, daily, unremarkable, perhaps even unnoticeable. A challenge is to understand how assessments and actions that from the outside look like "errors" become neutralized or normalized so that from the inside they appear nonremarkable, routine, and normal.

If we want to understand why people did what they did, then the adequacy of the insider's representation of the situation cannot be called into question. The reason is that there are no objective features in the domain on which we can base such a judgment. In fact, as soon as we make such a judgment, we have imported criteria from the outside—from another time and place, from another rationality. Ethnographers have always championed the point of view of the person on the inside. Emerson, as did Rasmussen, advised that instead of using criteria from outside the setting to examine mistake and error, we should investigate and apply local notions of competent performance that are honored and used in particular social settings (Vaughan 1999). This excludes generic rules and motherhoods (e.g., "pilots should be immune to commercial pressures"). Such "criteria" ignore the subtle dynamics of localized skills and priority setting; they run roughshod over what would be considered "good" or "competent" or "normal" from inside actual situations. Indeed, such criteria impose a rationality from the outside, impressing a frame of context-insensitive, idealized concepts of practice upon a setting where locally tailored and subtly adjusted criteria rule instead.

The ethnographic distinction between etic and emic perspectives was coined in the 1950s to capture the difference between how insiders view a setting and how outsiders view it. Emic originally referred to the language and categories used by people in the culture studied, while etic language and categories were those of the outsider (e.g., the ethnographer) based on their analysis of important distinctions. Today, emic is often understood to be the view of the world from the inside-out, that is, how the world looks from the eyes of the person studied. The point of ethnography is to develop an insider's view of what is happening, an inside-out view. Etic is contrasted as the perspective from the outside-in, where researchers or observers attempt to gain access to some portions of an insider's knowledge through psychological methods such as surveys or laboratory studies.

Emic research considers meaning-making activities. It studies the multiple realities that people construct from their experiences. It assumes that there is no direct access to a single, stable, and fully knowable external reality. Nobody has this access. Instead, all understanding of reality is contextually embedded and limited by the local rationality of the observer. Emic research points at the unique experience of each human, suggesting that any observer's way of making sense of the world is as valid as any other, and that there are no objective criteria by which this sensemaking can be judged correct or incorrect. Emic researchers resist distinguishing between "objective" features of a situation, and "subjective" ones. Such a distinction distracts the observer from the situation as it looked to the person on the inside, and in fact distorts this insider perspective.

A fundamental concern is to capture and describe the point of view of people inside a system or situation; to make explicit that which insiders take for granted, see as common sense, find unremarkable or normal. When we want to understand error, we have to embrace ontological relativity not out of philosophical intransigence or philanthropy, but for trying to get the inside-out view. We have to do this for the sake of learning what makes a system safe or brittle. As we will see in Chapter 5, for example, the notion of what constitutes an "incident" (i.e., what is worthy of reporting as a safety threat) is socially constructed; shaped by history, institutional constraints, cultural and linguistic notions; and negotiated among insiders in the system. None of the structural measures an organization takes to put an incident reporting system in place will have effect if insiders do not see safety threats as "incidents" that are worth sending into the reporting system. Nor will the organization ever really improve reporting rates if it does not understand the notion of "incident" (and conversely, the notion of "normal practice") from the point of view of the people who do it every day.

To succeed at this, outsiders need to take the inside-out look; they need to embrace ontological relativity, as only this can crack the code to system safety and brittleness. All the processes that set complex systems onto their drifting paths toward failure— the conversion of signals of danger into normal, expected problems; the incremental-ist borrowing from safety; the assumption that past operational success is a guarantee of future safety—are sustained through implicit social–organizational consensus, driven by insider language and rationalizations. The internal workings of these pro-cesses are simply impervious to outside inspection and thereby numb to external pressure for change. Outside observers cannot attain an emic perspective or study the multiple rationalities created by people on the inside if they keep seeing "errors" and "violations." Outsiders can perhaps get some short-term leverage by (re-)impos-ing context-insensitive rules, regulations, or exhortations and making moral appeals for people to follow them, but the effects are generally short lived. Such measures cannot be supported by operational ecologies. There, practice is under pressure to adapt in an open system, exposed to pressures of scarcity and competition. It will once again inevitably drift into niches that generate greater operational returns at no apparent cost to safety.

ERROR AND (IR)RATIONALITY

Understanding error against the background of local rationality, or rationality for that matter, has not been an automatic by-product of studying the psychology of error. In fact, research into human error had a very rationalist bias up into the 1970s (Reason 1990), and in some quarters in psychology and human factors, such rationalist partiality has never quite disappeared. Rationalist means that mental processes can be understood with reference to normative theories that describe optimal strategies. Strategies may be optimal when the decision maker has perfect, exhaustive access to all relevant information, takes time enough to consider it all, and applies clearly defined goals and preferences to making the final choice. In such cases, errors are explained by reference to deviations from this "rational" norm, this ideal. If the decision turns out wrong, it may be because the decision maker did not take enough time to consider all information or that she or he did not generate an exhaustive set of choice alternatives to pick from. Errors, in other words, are deviant. They are departures from a standard. Errors are irrational in the sense that they require a motivational (as opposed to cognitive) component in their explanation. If people did not take enough time to consider all information, it is because they could not be bothered too. They did not try hard enough, and they should try harder next time, perhaps with the help of some training or procedural guidance. Investigative practice in human factors is still rife with such rationalist reflexes.

It did not take long for cognitive psychologists to find out how humans could not or should not even behave like perfectly rational decision makers, or perfectly rational anythings. While economists clung to the normative assumptions of decision making (decision makers have perfect and exhaustive access to information for their decisions, as well as clearly defined preferences and goals about what they want to achieve), psychology, with the help of artificial intelligence, posited that there is no such thing as perfect rationality (i.e., full knowledge of all relevant information, possible outcomes, and relevant goals), because there is not a single cognitive system in the world (neither human nor machine) that has sufficient computational capacity to deal with that all. Rationality is bounded. Psychology subsequently started to chart people's imperfect, or bounded, or local rationality. Reasoning, it discovered, is governed by people's local understanding, by their focus of attention, goals, and knowledge, rather than some global ideal. Human performance is embedded in, and systematically connected to, the situation in which it takes place: it can be understood (i.e., makes sense) with reference to that situational context, not by reference to some universal standard. Human actions and assessments can be described meaningfully only in reference to the localized setting in which they are made; they can be understood by intimately linking them to details of the context that produced and accompanied them. Such research has given "rationality" an interpretive flexibility: what is locally rational does not need to be globally rational. If a decision is locally rational, it makes sense from the point of view of the decision maker—which is what matters if we want to learn about the underlying reasons for what from the outside looks like "error." The notion of local rationality removes the need to rely on irrational explanations of error. "Errors" make sense: they are rational, if only locally so, when seen from the inside of the situation in which they were made.

But psychologists themselves often have trouble with this. They discover biases and aberrations in decision making (e.g., groupthink, confirmation bias, and routine violations) that seem hardly rational, even from inside a situational context. These deviant phenomena rather require motivational explanations. They call for motivational solutions. People should be motivated to do the right thing, to pay attention, to double check. If they do not, then they should be reminded that it is their duty, their job. Notice how easily we slip back into prehistoric behaviorism: through a modernist system of rewards and punishments (job incentives, bonuses, and threats of retribution), we hope to mold human performance after supposedly fixed features of the world.

That psychologists and others would call such actions irrational, referring it back to some motivational component, may be due to the limits of the conceptual language of the discipline. Putatively motivational issues (such as deliberately breaking rules) must themselves be put back into context, to see how human goals (getting the job done fast by not following all the rules to the letter) are made congruent with system goals through a collective of subtle pressures, subliminal messages about organizational preferences, and empirical success of operating outside existing rules. The system wants fast turnaround times, maximization of capacity utilization, and efficiency. Given those system goals (which are often kept implicit), rule-breaking is not a motivational shortcoming, but rather an indication of a well-motivated human operator: personal goals and system goals are harmonized, which in turn can lead to total system goal displacement: efficiency is traded off against safety. But psychology often keeps seeing motivational shortcomings. And human factors keeps suggesting soporific countermeasures (injunctions to follow the rules, better training, more top-down task analysis). Human factors has trouble incorporating the more subtle but powerful influences of organizational environments, structures, processes, and tasks into accounts of individual cognitive practices. In this regard, the discipline has space for further conceptual development. Indeed, how unstated cultural norms and values travel from the institutional, organizational level to express themselves in individual assessments and actions (and vice versa) is a concern central to sociology, not human factors. Bridging this macro–micro connection in the systematic production of rule violations means understanding the dynamic interrelationships between issues as wide ranging as organizational characteristics and preferences, its environment and history, incrementalism in trading safety off against production, unintentional structural secrecy that fragments problem-solving activities across different groups and departments, patterns and representations of safety-related information that are used as imperfect input to organizational decision making, the influence of hierarchies and bureaucratic accountability on people's choice, and more (e.g., Vaughan 1996, 1999). The structuralist lexicon of human factors and system safety today has no words for many of these concepts, let alone models for how they go together.

From Decision Making to Sensemaking

In another move away from rationalism and toward the inversion of perspectives (i.e., trying to understand the world the way it looked to the decision maker at the time), large swathes of human factors have embraced the ideas of naturalistic decision making (known as NDM) over the last decade. By importing cyclical ideas about cognition (situation assessment informs action, which changes the situation, which in turn updates assessment) (Neisser 1976) into a structuralist, normativist psychological lexicon, NDM virtually reinvented decision making (Orasanu and Connolly 1993). The focus shifted from the actual decision moment, back into the preceding realm of situation assessment. This shift was accompanied by a methodological reorientation, where decision making and decision makers were increasingly studied in their complex, natural environments. Real decision problems, it quickly turned out, resist the rationalistic format dictated for so long by economics: options are not enumerated exhaustively, access to information is incomplete at best, and people spend more time assessing and measuring up situations than making decisions—if that is indeed what they do at all (Klein 1998). In contrast to the prescriptions of the normative model, decision makers tend not to generate and evaluate several courses of action concurrently, in order to then determine the best choice. People do not typically have clear or stable sets of preferences along which they can even rank the enumerated courses of action, picking the best one. Nor do most complex decision problems actually have a single correct answer. Rather, decision makers in action tend to generate single options at the time, mentally simulate whether this option would work in practice, and then either act on it or move on to a new line of thought. Naturalistic decision making also takes the role of expertise more seriously than previous decision making paradigms: what distinguishes good decision makers from bad decision makers most is their ability to make sense of situations by using a highly organized experience base of relevant knowledge. Once again neatly folding into ideas presented by Neisser (1976), such reasoning about situations is more schema driven, heuristic, and recognitional than it is computational. The typical naturalistic decision setting does not allow the decision maker enough time or information to generate perfect solutions with perfectly rational calculations. Naturalistic decision making calls for judgments made under uncertainty, ambiguity, and time pressure, and options that seem to work are better than perfect options that never get computed.

The same reconstructive, corrective intervention into history that produces our clear perceptions of "errors" also generates discrete "decisions." What we see as "decision making" from the outside is action embedded in larger streams of practice, something that flows naturally and continually from situation assessment and reassessment.

A "DECISION" TO GO AROUND

In hindsight, for a number of aviation accidents, we can ask why the crew didn't make a go-around when they realized that they would not be able to complete a landing in a safe manner. The Flight Safety Foundation sponsored a study in the late 1990s to analyze the factors that play into approach and landing accidents (Khatwa and Helmreich 1998) and concluded in part, predictably, that executing a missed approach is one of the best investments that pilots can make in safety to prevent approach and landing accidents. Such advice should not be confused with an explanation for why many crews do not do so. In fact, getting crews to execute go-arounds, particularly in cases of unstabilized approaches, remains one of the most vexing problems facing most chief pilots and managers of flight operations across the world. Characterizations such as "press-on-itis" do little to explain why crews press on; such words really only label a very difficult problem differently without offering any deeper understanding.

Concomitant recommendations directed at flight crews are, as a result, difficult to implement. For example, one recommendation is that "flight crews should be fully aware of situations demanding a timely go-around" (Khatwa and Helmreich 1998, p. 53). Even if crews can be shown to possess such knowledge in theory, which most actually can (e.g., they are able to recite the criteria for a stabilized approach from the airline's Operations Manual), becoming aware that a timely go-around is demanded hinges on a particular awareness of the situation itself. The data from continued approaches suggest that crews do not primarily interpret situations in terms of stabilized approach criteria, but in terms of their ability to continue the approach. Soft and hard gates (e.g., 1000 feet, 500 feet), when set in context of the end of a flight at a busy, major airport on a scheduled flight, become norms against which to plan and negotiate the actual approach vis-à-vis the air traffic and weather situation, not iron-fisted stop rules for that approach.

Thus, there seems little mileage in just reminding crews of the situations that demand a go-around, as almost all crews are able to talk about such situations and offer advice about how to do the right thing—until they are players in a rapidly unfolding situation themselves. When on approach themselves, it is not primarily generic criteria that crews see against which they can make some cognitive calculus of the legality and wisdom of continuing their approach. When on an approach themselves, crews see a situation that still looks doable, a situation that looks like they can make it, a situation that may hold a useful lesson for their less experienced colleague in the right seat, a situation that suggests things will be all right before passing over the threshold.

Discontent with discrete decisions as explanation for these sorts of events has been growing in human factors research. Operational decisions are no longer believed to be based on a "rational" analysis of all parameters that are relevant to the decision. Instead, the decision, or rather a continual series of assessments of the situation, is focused on elements in the situation that allow the decision maker to distinguish between reasonable options. The psychology of decision making is such that a situation is not assessed in terms of all applicable criteria (certainly not quantitative ones), but in terms of the options the situation appears to present (Klein 1993; Orasanu and Martin 1998).

One promising countermeasure based on insights from this work on naturalistic decision making seems to be not to remind crews of the criteria for a stabilized approach, but to offer generic rewards for all the cases in which crews execute a missed approach. Chief pilots or managers of flight operations who offer a no-questions-asked reward (e.g., a bottle of wine) to crews who make a go-around generally report modest success in reducing the number of unstabilized approaches. It is crucial, in setting up such policies, to communicate to crews that breaking off an approach is not only entirely legitimate but actually desired: getting pilots to buy into the idea that "each landing is a failed go-around." The handing out of such rewards should be advertised broadly to everybody else. Such encouragement is of course difficult to uphold in the face of production and economic pressures and incredibly easy to undermine by sending subliminal messages (or more overt ones) to crews that on-time performance or cost/fuel savings are important. The paying of bonuses for on-time performance, which has been, and still is, custom in some airlines, is an obvious way to increase the likelihood of unstabilized approaches that are not broken off.

PLAN CONTINUATION

An interesting line of research has come from NASA Ames Research Center (Orasanu and Martin 1998). A phenomenon that Judith Orasanu and her colleagues called "plan continuation" captures a considerable amount of the data available from cases where an approach was continued despite cues that, in hindsight or in written guidance, pointed to the wisdom of a go-around. In fact, three out of four cases in which crews made tactical decisions that turned out erroneous in hindsight fit the plan-continuation pattern. The NASA research takes as its starting point the psychology of decision making consistent with the last decades of research into it. Decision making in complex, dynamic settings such as an approach is not an activity that involves a weighty comparison of options against prespecified criteria. Rather, such decision making is "front-loaded": this means that most human cognitive resources are spent on assessing the situation and then reassessing it for its continued doability. In other words, decision making on an approach is hardly about making decisions, but rather about continually sizing up the situation. The "decision" is often simply the outcome, the automatic by-product of the situation assessment. This is what turns a go-around decision into a continually (re-)negotiable issue: even if the decision to go around is not made on the basis of an assessment of the situation now, it can be pushed ahead and be made a few or more seconds later when new assessments of the situation have come in.

Even more important than the cognitive processes involved in decision making are the contextual factors that surround a crew at the time. The order in which cues about the developing situation come in and their relative persuasiveness are two key determinants for plan continuation. Conditions often deteriorate gradually and ambiguously, not precipitously and unequivocally. In such a gradual deterioration, there are almost always strong initial cues that suggest that the situation is under control and can be continued without increased risk. This sets a crew on the path to plan continuation. Weaker and later cues that suggest that another course of action could be safer then have a hard time dislodging the plan as it is being continued, and as evidenced from the situation as it has so far been handled.

Note how plan continuation is different from a characterization of "confirmation bias." Confirmation bias suggests that crews seek out the evidence that supports (or confirms) their hypothesis of the situation, at the expense of other evidence. In most cases, there is little that suggests operators are actively avoiding evidence that speaks against the plan as it was being continued. Indeed, the "bias" in confirmation bias seems to be produced more in the mind of the retrospective observer, the one calling it a "confirmation bias," rather than in the mind of the person observed. Once again, it is hindsight that endows certain indications with a particular salience over others—a hindsight interpretation against which observed performance can be judged to have been "biased." This, of course, is hardly a meaningful conclusion: only hindsight would have shown which cues were more important than others, and people inside the situation didn't have hindsight and so cannot meaningfully be judged to have been "biased" relative to it. Hindsight should not be the source of our explanation for why people did what they did. Instead, their own unfolding context should be that source.

Contextual dynamics are a joint product of how problems in the world are developing and the actions taken to do something about it. Decision and action are interleaved rather than temporally segregated. The decision maker is thus seen as in step with the continuously unfolding environment, simultaneously influenced by it and influencing it through his or her next steps. Understanding "decision making," then, requires an understanding of the dynamics that lead up to those supposed "decision moments," because by the time we get there, the interesting stuff has evaporated, gotten lost in the noise of action. Naturalistic decision making research is frontloaded: it studies the front end of decision making, rather than the back end. It is interested, indeed, in sensemaking more than in decision making.

Removing "decision making" from the vocabulary of human factors investigations is the logical next step, suggested by Snook (2000). It would be an additional way to avoid counterfactual reasoning and judgmentalism, as decisions that eventually led up to a bad outcome all too quickly become "bad" decisions (p. 206):

> Framing such tragedies as decisions immediately focuses our attention on an individual making choices…, such a framing puts us squarely on a path that leads straight back to the individual decision maker, away from the potentially powerful contextual features and right back into the jaws of the fundamental attribution error. 'Why did they decide…?' quickly becomes 'Why did they make the wrong decision?'. Hence, the attribution falls squarely onto the shoulders of the decision maker and away from potent situational factors that influence action. Framing the… puzzle as a question of meaning rather than deciding shifts the emphasis away from individual decision makers toward a point somewhere 'out there' where context and individual action overlap.

Yet sensemaking is not immune to counterfactual pressure either. If what made sense to the person inside the situation still makes no sense given the outcome, then human factors hastens to point that out (see Chapter 4). Even in sensemaking, the normativism is an ever-present risk.

STUDY QUESTIONS

1. Explain how deconstruction, dualism, and structuralism easily creep into our investigations of incidents and accidents. Try to find an accident report that bears this out.
2. Summarize the Cartesian–Newtonian worldview and explain how, even to this day, it directs how we conduct investigations.

3. Why does it make sense, in a Newtonian universe, to aim to reconstruct an accident, whereas in a complex world doing that is impossible?

4. If we want to understand why it made sense for people to do what they did in their situation, what steps do we need to take as investigators? And what should we avoid?

5. What does bounded, or local, rationality mean? What are some of the bases for it both in practical accident investigation and in cognitive science?

6. What would be the characteristics of an investigation in the new era, which takes complexity seriously? Would it find "causes," for example? How would it write recommendations which acknowledge that control over the future of a complex system is impossible, but influencing it might well be possible?

3 People as a Problem to Control

CONTENTS

KEY POINTS

- Modernism typically brings us back to square one: people are a problem to control, particularly when they do not comply with protocol, procedures, rules, and checklists that have been built up around them—often by other people who themselves do not do the safety-critical work every day.
- The observation and measurement of errors are an example of the machinery of a safety bureaucracy. It assumes that it is possible to tabulate and

compare errors from different observations as objective data rather than context-dependent interpretations and attributions. Error counting also tends to see safety as an absence of negatives and people as a problem to control.

- Procedures seen as algorithmic if–then rules, rather than resources for action whose application depends on situated insight, also exemplifies the push–pull between seeing people as a problem to control versus a solution to harness. Applying and adapting procedures successfully are a substantive cognitive skill.

- This raises questions about the role of a regulator (or any safety monitor) in complex systems. Compliance may make sense in simple, or Newtonian, systems (or aspects of systems), which have one best method of working. But in complex systems, safety monitoring, inspection, and regulation represent one more way of decomposing that system, producing one more narrative.

- Complex operational systems are characterized by goal conflicts and resource constraints. Multiple irreconcilable aims need to be pursued simultaneously, safety among them. What is seen as "noncompliance" from the outside is, for people at work, compliance with a whole complex of expectations, demands, and pressures, many of them unstated and inexplicit.

If the belief is that people are a problem to control, and that we need to intervene at the level of their attitudes and behavior, then what do we typically do? This chapter runs through some of what we do. We count errors, we proceduralize, we regulate and inspect for compliance. This would not be a chapter in this book, however, if it does not also stop to look critically at what exactly we are doing, and what we hope to achieve. Error counting, for example, is set against the epistemological and ontological assumptions that hold it up. A critical examination of the categories we use to parse up our observations of performance shows how little observation, and how much our own construction they are. When we put faith in such efforts, we engage in an act of epistemological alchemy. We turn our own social constructions, observations, and psychological attributions into countable facts that can be hardened and stored in worldwide databases and considered "real" by others. Proceduralization can be considered as one more Tayloristic intervention in complex operating worlds, where safety is not the result of following all the rules, but linked to knowing when and how to adapt. This needs to be understood against the backdrop of goal conflicts that are typical of work at the sharp end (and elsewhere), where reconciling fundamentally irreconcilable expectations is part of daily business.

COUNTING AND CONTROLLING ERRORS

As organizations and other stakeholders (e.g., trade and industry groups, regulators) try to assess and control the "safety health" of their operations, counting and tabulating errors appear to be a meaningful measure. Not only does it provide an immediate, numeric estimate of the probability of accidental death, injury, or any other undesirable event, it also allows the bureaucratic comparison of systems and components in it (this hospital vs. that hospital, this airline vs. that one, this aircraft fleet or pilot vs. that one, this site vs. that site, this country vs. that country). Keeping track of adverse events is thought to provide relatively easy, quick, and accurate access to the internal safety workings of a system. Adverse events and errors, after all, supposedly point to the holes in the layers of defense. And as it was with Taylor, adverse events can be seen as the start of—or reason for—probing higher up, in order to search for the unfavorable organizational conditions that could be changed to prevent recurrence. Many industries have endeavored to quantify safety problems and find potential sources of vulnerability and failure. This has spawned a number of error-classification systems. Some classify decision errors together with the conditions that helped produce them; some have a specific goal, for example, to categorize information transfer problems (e.g., instructions, errors during watch changeover briefings, coordination failures); others try to divide error causes into cognitive, social, and situational (physical, environmental, ergonomic) factors; yet others attempt to classify error causes along the lines of a linear information-processing or decision-making model, and some apply the Swiss Cheese metaphor (i.e., systems have multiple layers of defense, but all of them have holes) in the identification of errors and vulnerabilities up the causal chain. Error-classification systems are used both after an event (e.g., during incident investigations) or for observations of current human performance.

THE MORE WE MEASURE, THE LESS WE KNOW

In pursuing categorization and tabulation of errors, human factors makes a number of assumptions and takes certain philosophical positions. Little of this is made explicit in the description of these methods, yet it carries consequences for the utility and quality of the error count as a measure of safety health and as a tool for directing resources for improvement. Here is an example. In one of the methods, the observer is asked to distinguish between "procedure errors" and "proficiency errors." Proficiency errors are related to a lack of skills, experience, or (recent) practice, whereas procedure errors are those that occur while carrying out prescribed or normative sequences of action (e.g., checklists). This seems straightforward. Yet, as Croft reported, the following problem confronts the observer while using one error counting and classification system popular in aviation. One type of error (a pilot entering a wrong altitude in a flight computer) can legitimately end up in either of two categories of the error-counting method (a procedural error or proficiency error):

> For example, entering the wrong flight altitude in the flight management system is considered a procedural error... Not knowing how to use certain automated features in an aircraft's flight computer is considered a proficiency error (Croft 2001, p. 77).

If a pilot enters the wrong flight altitude in the flight-management system, is that a procedural or a proficiency issue, or both? How should it be categorized? Thomas Kuhn encouraged science to turn to creative philosophy when confronted with the inklings of problems in relating theory to observations (as in the problem of categorizing an observation into theoretical classes). It can be an effective way to elucidate and, if necessary, weaken the grip of a tradition on the collective mind, and suggest the basis for a new one. This is certainly appropriate when epistemological questions arise: questions on how we go about knowing what we (think we) know. To understand error classification and some of the problems associated with it, let us link it once again to the philosophical tradition that governs much of human factors and safety research, and the worldview in which it takes place.

REALISM AND POSITIVISM

The position human factors takes when it uses observational tools to measure "errors" is a realist one: It presumes that there is a real, objective world with verifiable patterns that can be observed, categorized, and predicted. Errors, in this sense, are a kind of Durkheimian fact. Emile Durkheim, a founding father of sociology, believed that social reality is objectively "out there," available for neutral, impartial empirical scrutiny. Reality exists, truth is worth striving for. Of course, there are obstacles to getting to the truth, and reality can be hard to pin down. Yet, pursuing a close mapping or correspondence to that reality is a valid, legitimate goal of theory development. It is that goal, of achieving a close mapping to reality, that governs error-counting methods. If there are difficulties in getting that correspondence, then these difficulties are merely methodological in nature. The difficulties call for refinement of the observational instruments or additional training of the observers.

These presumptions are modernist; inherited from the ideas of the Scientific Revolution. In the finest of scientific spirits, method is called on to direct the searchlight across empirical reality, more method is called on to correct ambiguities in the observations, and even more method is called on to break open new portions of hitherto unexplored empirical reality, or to bring into focus those portions that so far were vague and elusive. Other labels that fit such an approach to empirical reality could include positivism, which holds that the only type of knowledge worth bothering with is that which is based directly on experience. Positivism is associated with the doctrine of Auguste Comte: the highest, purest (and perhaps only true) form of knowledge is a simple description of sensory phenomena. In other words, if an observer sees an error, then there was an error. For example, the pilot failed to arm the spoilers. This error can then be written up and categorized as such.

But positivist has obtained a negative connotation, really meaning "bad" when it comes to social science research. Instead, a neutral way of describing the position of error-counting methods is realist, if naively so. Operating from a realist stance, researchers are concerned with validity (a measure of that correspondence they seek) and reliability. If there is a reality that can be captured and described objectively by outside observers, then it is also possible to generate converging evidence with multiple observers, and consequently achieve agreement about the nature of that reality. This means reliability: Reliable contact has been made with empirical reality, generating equal access and returns across observations and observers.

Error-counting methods rely on realism and positivism. It is possible to tabulate errors from different observers and different observations (e.g., different flights or airlines) and build a common database that can be used as some kind of aggregate norm against which new and existing entrants can be measured. But such absolute objectivity is impossible to obtain. The world is too messy for that, phenomena that occur in the empirical world too confounded, and methods forever imperfect. It comes as no surprise, then, that error-counting methods have different definitions, and different levels of definitions, for *error*, because error itself is a messy and confounded phenomenon:

- Error as the cause of failure: For example, the pilot's failure to arm the spoilers led to the runway overrun.
- Error as the failure itself: Classifications rely on this definition when categorizing the kinds of observable errors operators can make (e.g., decision errors, perceptual errors, skill-based errors) and probing for the causes of this failure in processing or performance. According to Helmreich, "Errors result from physiological and psychological limitations of humans. Causes of error include fatigue, workload, and fear, as well as cognitive overload, poor interpersonal communications, imperfect information processing, and flawed decision making" (Helmreich 2000a, p. 781).
- Error as a process or, more specifically, as a departure from some kind of standard: This standard may consist of operating procedures. Violations, whether exceptional or routine, or intentional or unintentional, are one example of error according to the process definition. Depending on what they use as standard, observers of course come to different conclusions about what is an error.

Not differentiating among these different possible definitions of error is a well-known problem. Is error a cause, or is it a consequence? To the error-counting methods, such causal confounds and messiness are neither really surprising nor really problematic. Truth, after all, can be elusive. What matters is getting the method right. More method may solve problems of method. That is, of course, if these really are problems of method. The modernist would say "yes." "Yes" would be the stock answer from the Scientific Revolution onward. Methodological wrestling with empirical reality, where empirical reality plays hard to catch and proves pretty good at the game, is just that: methodological. Find a better method, and the problems go away. Empirical reality will swim into view, unadulterated.

DOCTORS ARE MORE DANGEROUS THAN GUN OWNERS

There are about 700,000 physicians in the United States. The US Institute of Medicine estimates that each year between 44,000 and 98,000 people die as a result of medical errors (Kohn et al. 2000). This makes for a yearly accidental death rate per doctor of between 0.063 and 0.14. In other words, up to one in seven doctors will kill a patient each year by mistake. Take gun owners in contrast. There are 80,000,000 gun owners in the United States. Yet their errors lead to "only" 1500 accidental gun deaths per year. This means that the accidental death rate, caused by gun-owner error, is 0.000019 per gun owner per year. Only about 1 in 53,000 gun owners will kill somebody by mistake. Doctors then, are 7500 times more likely to kill somebody by mistake. While not everybody has a gun, almost everybody has a doctor (or several doctors), and is thus severely exposed to the human error problem.

DID YOU REALLY SEE THE ERROR HAPPEN?

But does a single, stable reality that can be approached by the best of methods, and described in terms of correspondence with that reality, exist? And if it does, could we know that reality? If we describe reality in a particular way (e.g., this was a "procedure error"), then that may not imply any type of mapping onto an objectively attainable external reality—close or remote, good or bad. Perhaps such language, or constituting such an object, does not describe phenomena as though they reflect or represent something stable, objective, something "out there." Rather, capturing and describing a phenomenon are the result of a collective generation and agreement of meaning that, in this case, human factors researchers and their industrial counterparts have reached. The reality of a *procedure error*, in other words, is socially constructed. It is a piece of discourse that is shaped by, and dependent on, models and paradigms of knowledge that have evolved through group consensus. This meaning is enforced and handed down through systems of observer training, labeling and communication of the results, and industry acceptance and promotion. As philosophers like Kuhn have pointed out, these paradigms of language and thought at some point adopt a kind of self-sustaining energy, or "consensus authority" (Angell and Straub 1999). It is sustained by the following circularity:

- If human factors auditors count errors for managers, they, as (putatively scientific) measurers, have to presume that errors exist.
- But in order to prove that errors exist, auditors have to measure them. In other words, measuring errors becomes the proof of their existence.
- And in turn, this existence of errors was preordained by their measurement.

In the end, everyone agrees that counting errors is a good step forward on safety because almost everyone seems to agree that it is a good step forward. The practice is not questioned because few seem to question it. The procedural error becomes true (or appears to people as a close correspondence to some objective reality) only because a community of specialists has contributed to the development of the tools that make it appear so and has agreed on the language that makes it visible. There is nothing inherently true about the error at all. In accepting the utility of error counting, it is likely that industry accepts its theory (and thereby the reality and validity of the observations it generates) on the authority of authors, teachers, and their texts, not because of evidence. In his headline, Croft (2001) announced that researchers have now perfected ways to monitor pilot performance in the cockpit. "Researchers" have "perfected." There is little that an industry can do other than to accept such authoritarian high modernism. Not to mention the pilots who are thus monitored. What alternatives have they, asks Kuhn, what power?

Realism, that product and accompaniment of the Scientific Revolution and high modernism, assumes that a common denominator can be found for all systems of belief and value, and that we should strive to converge on those common denominators through our (scientific) methods. There is a truth, and it is worth looking for through method. As it was with the "ground truth" in situation awareness measurements, it takes courage to start coping without such common denominators. This challenges the high modernist culture of realism and empiricism, of which error-counting methods are but an instance. In the words of Varela et al. (1991), it is hard to give up this "Cartesian anxiety." We seem to need the idea of a fixed, stable reality that surrounds us, independent of who looks at it. To give up that idea would be to descend into uncertainty, into idealism, into subjectivism. There would be no more groundedness, no longer a set of predetermined norms or standards, only a constantly shifting chaos of individual impressions, leading to relativism and, ultimately, nihilism.

Was This an Error? It Depends on Whom You Ask

Although people might live in the same empirical world, they typically arrive at rather different, yet often equally valid, conclusions about what is going on inside of it. They propose different vocabularies and models to capture those phenomena and activities. Philosophers sometimes use the example of a tree. Though at first sight an objective, stable entity in some external reality, separate from us as observers, the tree can mean entirely different things to someone in the logging industry as compared to, say, a wanderer in the Sahara. Both interpretations can be valid because validity is measured in terms of local relevance, situational applicability, and social acceptability—not in terms of correspondence with a real, external world. Among different characterizations of the world, there is no *more real* or *more true*. Validity is a function of how the interpretation conforms to the worldview of those to whom the observer makes his appeal. A *procedure error* is a legitimate, acceptable form of capturing an empirical encounter only because there is a consensual system of like-minded coders and consumers who together have agreed on the linguistic label. The appeal falls onto fertile ground.

But the validity of an observation is negotiable. It depends on where the appeal goes, on who does the looking and who does the listening. This is known as *onto logical relativism*: there is flexibility and uncertainty in what it means to be in the world or in a particular situation. The meaning of observing a particular situation depends entirely on what the observer brings to it. The tree is not just a tree. It is a source of shade, sustenance, survival. Following Kant's ideas, social scientists argue that the act of observing and perceiving objects (including humans) is not a passive, receiving process, but an active one that engages the observer as much as it changes or affects the observed. This creates the epistemological uncertainty that creeps into all error-counting methods, incident classification systems, and other high-modernist schemes, which, after all, attempt to shoehorn observations into numerical objectivity. Most social observers, incident coders, or error counters will have felt this uncertainty at one time or another. Was this a procedure error, or a proficiency error, or both? Or was it perhaps no error at all? Was this the cause, or was it the consequence? If it were up to Kant, not having felt this uncertainty would serve as an indication of being a particularly obtuse observer. It would certainly not be proof of the epistemological astuteness of either method or error counter. The uncertainty suffered by them is epistemological because it is realized that certainty about what we know, or even about how to know whether we know it or not, seems out of reach. Yet those within the ruling paradigm have their stock answer to this challenge, just as they have it whenever confronted with problems of bringing observations and theories in closer correspondence. More methodological agreement and refinement, including observer training and standardization, may close the uncertainty. Better-trained observers will be able to distinguish between a procedure error and proficiency error, and an improvement to the coding categories may also do the job. Similar modernist approaches have had remarkable success for five centuries, so there is no reason to doubt that they may offer routes to some progress even here. Or is there?

CAN MORE METHOD SOLVE PROBLEMS OF METHOD?

Perhaps more method may not solve problems seemingly linked to method. Consider a study reported by Hollnagel and Amalberti (2001), whose purpose was to test a new error-measurement instrument. This instrument was designed to help collect data on, and get a better understanding of, air-traffic controller errors, and to identify areas of weakness and find possibilities for improvement. The method asked observers to count errors (primarily error rates per hour) and categorize the types of errors using a taxonomy proposed by the developers. The tool had already been used to pick apart and categorize errors from past incidents but would now be put to test in a real-time field setting—applied by pairs of psychologists and air-traffic controllers who would study air-traffic control work going on in real time. The observing air-traffic controllers and psychologists, both trained in the error taxonomy, were instructed to take note of all the errors they could see.

Despite common indoctrination, there were substantial differences between the numbers of errors each of the two groups of observers noted, and only a very small number of errors were actually observed by both. People watching the same performance, using the same tool to classify behavior, came up with totally different error counts. Closer inspection of the score sheets revealed that the air-traffic controllers and psychologists tended to use different subsets of the error types available in the tool, indicating just how negotiable the notion of error is. The same fragment of performance means entirely different things to two different (but similarly trained and standardized) groups of observers. Air-traffic controllers relied on external working conditions (e.g., interfaces, personnel, and time resources) to refer to and categorize errors, whereas psychologists preferred to locate the error somewhere in presumed quarters of the mind (e.g., working memory) or in some mental state (e.g., attentional lapses). Moreover, air-traffic controllers who actually did the work could tell both groups of error coders that they both had it wrong. Debriefing sessions exposed how many observed errors were not errors at all to those said to have committed them, but rather normal work, expressions of deliberate strategies intended to manage problems or foreseen situations that the error counters had either not seen, or not understood if they had. Croft (2001) reported the same result in observations of cockpit errors: More than half the errors seen by error counters were never seen as such by the flight crews themselves. Some realists may argue that the ability to discover errors not seen by people themselves is a good thing: It confirms the strength or superiority of the method. But in Hollnagel and Amalberti's (2001) case, error coders were forced to disavow such claims to epistemological privilege. They reclassified the errors as normal actions, rendering the score sheets virtually devoid of error counts. Early transfers of aircraft were not an error, for example, but turned out to correspond to a deliberate strategy connected to a controller's foresight, planning ahead, and workload management. Rather than an expression of weakness, such strategies uncovered sources of resilience that would never have come out, or would even have been misrepresented and mischaracterized, with just the data in the classification tool. Such normalization of actions, which at first appear deviant from the outside, is a critical aspect to understanding the human factor and its strengths and weaknesses (see Vaughan 1996). Without understanding such processes of normalization, it is impossible to penetrate the situated meaning of errors or violations. The realist idea is that errors are "out there," that they exist and can be observed, captured, and documented independently of the observer. This would mean that it makes no difference who does the observing (which it empirically does). Such presumed realism is naive because all observations are ideational—influenced (or made possible in the first place) to a greater or lesser extent by who is doing the observing and by the worldview governing those observations. Realism does not work well in human factors and safety research, because it is ultimately impossible to separate the observer from the observed—something we will revisit in the chapter on methods and models.

PRESUMED REALITY OF ERRORS

The test of the air-traffic control error-counting method revealed how "an action should not be classified as an 'error' only based on how it appears to an observer" (Hollnagel and Amalberti 2001, p. 13). The test confirms ontological relativism. Yet sometimes the observed "error" should be entirely noncontroversial, should it not? Take the spoiler example from earlier in the book. The flight crew forgot to arm the spoilers. They made a mistake. It was an error. You can apply the new view to human error, and explain all about context and situation and mitigating factors. Explain why they did not arm the spoilers, but that they did not arm the spoilers is a fact. The error occurred. Even multiple different observers would agree on that. The flight crew failed to arm the spoilers. How can one not acknowledge the existence of that error? It is there, it is a fact, staring us in the face.

THE AUTONOMY PRINCIPLE

But what is a fact? Facts always privilege the ruling paradigm. Facts always favor current interpretations, as they fold into existing constructed renderings of what is going on. Facts actually exist by virtue of the current paradigm. They can neither be discovered nor given meaning without it. There is no such thing as observations without a paradigm; research in the absence of a particular worldview is impossible. In the words of Paul Feyerabend (1993, p. 11): "On closer analysis, we even find that science knows no 'bare facts' at all, but that the 'facts' that enter our knowledge are already viewed in a certain way and are, therefore, essentially ideational." Feyerabend called the idea that facts are available independently and can thereby objectively favor one theory over another, the autonomy principle (Feyerabend 1993, p. 26).

The autonomy principle asserts that the facts that are available as empirical content of one theory (e.g., procedural errors as facts that fit the threat and error model) are objectively available to alternative theories too. But this does not work. As the spoiler example showed, errors occur against and because of a background, in this case a background so systemic, so structural, that the original human error pales against it. The error almost becomes transparent, it is normalized, it becomes invisible. Against this backdrop, this context of procedures, timing, engineering trade-offs, and weakened hydraulic systems, the omission to arm the spoilers all but dissolves. Figure and ground trade places: no longer is it the error that is really observable or even at all interesting. With deeper investigation, ground becomes figure. The backdrop begins to take precedence as the actual story, subsuming, swallowing the original error. No longer can the error be distinguished as a singular, failed decision moment. Somebody who applies a theory of naturalistic decision making will not see a procedure error. What will be seen instead is a continuous flow of actions and assessments, coupled and mutually cued, a flow with nonlinear feedback loops and interactions, inextricably embedded in a multilayered evolving context. Human interaction with a system, in other words, is seen as a continuous control task. Such a characterization is hostile to the digitization necessary to fish out individual human errors.

Whether individual errors can be seen depends on the theory used. Observers in error counting are themselves participants, participating in the very creation of the observed fact, and not just because they are there, looking at how other people are working. Of course, through their sheer presence, error counters probably distort people's normal practice, perhaps turning situated performance into a mere window-dressed posture. More fundamentally, however, observers in error counting are participants, because the facts they see would not exist without them. They are created through the method. Observers are participants because it is impossible to separate observer and object. Errors are not "read out" of the observations. Rather, they are "read into" the observations—remember: errors are active interventions in history (in this case, someone else's history).

None of this, by the way, makes the procedure error less real to those who observe it. This is the whole point of ontological relativism. But it does mean that the autonomy principle is false. Facts are not stable aspects of an independent reality, revealed to scientists who wield the right instruments and methods. The discovery and description of every fact are dependent on a particular theory. In the words of Einstein, it is the theory that determines what can be seen. Facts are not available "out there," independent of theory. To suppose that a better theory should come along to account for procedure errors in a way that more closely matches reality is to stick with a model of scientific progress that was disproved long ago. It follows the idea that theories should not be dismissed until there are compelling reasons to do so, and compelling reasons arise only because there is an overwhelming number of facts that disagree with the theory. Scientific work, in this idea, is the clean confrontation of observed fact with theory. The problem is, those facts do not exist without the theory.

RESISTING CHANGE: THE THEORY IS RIGHT. OR IS IT?

A common assumption, not uncommon in human factors and safety research, is that theoretical progress happens through the accumulation of observed facts (Parasuraman et al. 2008). And that disagreeing facts will ultimately manage to topple a theory. The issue, however, is that counterinstances (i.e., facts that disagree with the theory) are not always seen as such. Instead, if observations reveal counterinstances (such as errors that resist unique classification in any of the categories of the error-counting method), then researchers tend to see these as further puzzles in the match between observation and theory (Kuhn 1962)—puzzles that can be addressed by further refinement of their method. Counterinstances, in other words, are not seen as speaking against the theory. According to Kuhn, one of the defining responses to paradigmatic crisis is that scientists do not treat anomalies as counterinstances, even though that is what they are. It is extremely difficult for people to renounce the paradigm that has led them into a crisis. Instead, the epistemological difficulties suffered by error-counting methods (Was this a cause or a consequence? Was this a procedural or a proficiency error?) are dismissed as minor irritants and reasons to engage in yet more methodological refinement consonant with the current paradigm. Neither researchers nor their supporting communities in industry are willing to forego a paradigm until and unless there is a viable alternative ready to take its place. This is among the most sustained arguments surrounding the continuation of error counting: Researchers engaging in error classification are willing to acknowledge that what they do is not perfect, but vow to keep going until shown something better. And industry concurs. As Kuhn pointed out, the decision to accept one paradigm necessarily coincides with the acceptance of another. Proposing a viable alternative theory that can deal with its own facts, however, is exceedingly difficult, and has proven to be so even historically. Facts, after all, privilege the status quo.

GALILEO AND HIS TELESCOPE

Galileo's telescopic observations of the sky generated observations that motivated an alternative explanation about the place of the earth in the universe. His observations favored the Copernican heliocentric interpretation (where the earth goes around the sun) over the Ptolemaic geocentric one (where the sun goes around the earth). The Copernican interpretation, however, was a worldview away from what was currently accepted, and many doubted Galileo's data as a valid empirical window on that heliocentric reality. People were highly suspicious of the new instrument: Some asked Galileo to open up his telescope to prove that there was no little moon hiding inside of it. How, otherwise, could the moon or any other celestial body be seen so closely if it was not itself hiding in the telescope? One problem was that Galileo did not offer a theoretical explanation for why this could be so, and why the telescope was supposed to offer a better picture of the sky than the naked eye. He could not, because relevant theories (optica) were not yet well developed. Generating better data (like Galileo did), and developing entirely new methods for better access to these data (such as a telescope), does in itself little to dislodge an established theory that allows people to see the phenomenon with their naked eye and explain it with their common sense. Similarly, people see the error happen with their naked eye, even without the help of an error-classification method: The pilot fails to arm the spoilers. Even their common sense confirms that this is an error. The sun goes around the earth. The earth is fixed. The Church was right, and Galileo was wrong. None of his observed facts could prove him right, because there was no coherent set of theories ready to accommodate his facts and give them meaning. The Church was right, as it had all the facts. And it had the theory to accommodate them comprehensively and consistently.

Interestingly, the Church kept closer to reason as it was defined at the time. It considered the social, political, and ethical implications of Galileo's alternatives and deemed them too risky to accept—certainly on the grounds of tentative, rickety evidence. Disavowing the geocentric idea would be disavowing Creation itself, removing the common ontological denominator of the past millennium and severely undermining the authority and political power the Church derived from it.

Error-classification methods, too, guard a piece of rationality that many people in industry and research would not want to see disintegrate. Errors occur, they can be distinguished objectively. Errors can be an indication of unsafe performance. There is good performance and bad performance; there are identifiable causes for why people perform well or less well and for why failures happen. Without such a supposedly factual basis, without such hopes of an objective rationality, traditional and well-established ways for dealing with threats to safety and trying to create progress could collapse. It would be a kind of Cartesian anxiety. How can we hold people accountable for mistakes if there are no "errors"? How can we report safety occurrences and maintain expensive incident-reporting schemes if there are no errors? What can we fix if there are no causes for adverse events? Such questions fit a broader class of appeals against relativism. Relativism, the charge goes, easily leads to moral ambiguity, nihilism, and a lack of structural progress. Holding on to the realist status quo, is more desirable. And after all, most observed facts still seem to privilege it. Errors exist. They have to. The argument that errors exist is not only natural and necessary, it is also quite impeccable, quite forceful. The idea that errors do not exist, in contrast, is unnatural, even absurd. Those within the established paradigm will challenge the sheer legitimacy of questions raised about the existence of errors, and by implication even the legitimacy of those who raise the questions: "Indeed, there are some psychologists who would deny the existence of errors altogether. We will not pursue that doubtful line of argument here" (Reason and Hobbs 2003, p. 39). Because the current paradigm judges it absurd and unnatural, the question about whether errors exist is seen as not worth pursuing: It is doubtful and unscientific— and in the strictest sense (when scientific pursuits are measured and defined within the ruling paradigm), that is precisely what it is. If some scientists do not succeed in bringing statement and fact into closer agreement (they do not see a procedure error where others would), then this discredits the scientist rather than the theory.

GALILEO AND ERROR COUNTING

Galileo suffered from this too. It was the scientist who was discredited (for a while at least), not the prevailing paradigm. So what does he do? How does Galileo proceed once he introduces an interpretation so unnatural, so absurd, so countercultural, so revolutionary? What does he do when he notices that even the facts are not (interpreted to be) on his side? As Feyerabend (1993) masterfully described it, Galileo engaged in propaganda and psychological trickery. Through imaginary conversations between Sagredo, Salviati, and Simplicio, written in his native tongue rather than in Latin, he put the ontological uncertainty and epistemological difficulty of the geocentric interpretation on full display. The sheer logic of the geocentric interpretation fell apart whereas that of the heliocentric interpretation triumphed. Where the appeal to empirical facts failed (because those facts will still be forced to fit the prevailing paradigm rather than its alternative), an appeal to logic may still succeed. The same is true for error counting and classification. Just imagine this conversation:

Simplicio: Errors result from physiological and psychological limitations of humans. Causes of error include fatigue, workload, and fear, as well as cognitive overload, poor interpersonal communications, imperfect information processing, and flawed decision making.

Sagredo: But are errors in this case not simply the result of other errors? Flawed decision making would be an error. But in your logic, it causes an error. What is the error then? And how can we categorize it?

Simplicio: Well, but errors are caused by poor decisions, failures to adhere to brief, failures to prioritize attention, improper procedure, and so forth.

Sagredo: This appears to be not causal explanation, but simply relabeling. Whether you say error, or poor decision, or failure to prioritize attention, it all still sounds like error, at least when interpreted in your worldview. And how can one be the cause of the other to the exclusion of the other way around? Can errors cause poor decisions just like poor decisions cause errors? There is nothing in your logic that rules this out, but then we end up with a tautology, not an explanation.

And yet, such arguments may not help either. The appeal to logic may fail in the face of overwhelming support for a ruling paradigm—support that derives from consensus authority, from political, social, and organizational imperatives rather than a logical or empirical basis (which is, after all, pretty porous). Even Einstein expressed amazement at the common reflex to rely on measurements (e.g., error counts) rather than logic and argument: "Is it not really strange," Albert Einstein asked in a letter to Max Born, "that human beings are normally deaf to the strongest of argument while they are always inclined to overestimate measuring accuracies?" (Feyerabend 1993, p. 239). Numbers are strong. Arguments are weak. Error counting is good because it generates numbers, it relies on accurate measurements (recall Croft [2001] who announced that "Researchers" have "perfected" ways to monitor pilot performance), rather than on argument. In the end, no argument, none of this propaganda or psychological trickery can serve as a substitute for the development of alternative theory, nor did it in Galileo's case. Without a paradigm, without a worldview, there are no facts. People will reject no theory on the basis of argument or logic alone. They need another to take its place. A paradigmatic interregnum would produce paralysis. Suspended in a theoretical vacuum, researchers would no longer be able to see facts or do anything meaningful with them.

SAFETY AS THE ABSENCE OF NEGATIVES?

If we start looking for an alternative paradigm, let's first ask a question that takes us back to Chapter 1. Why do people bother with error counts in the first place? What goals do they hope these empirical measures help them accomplish, and are there better ways to achieve those goals? A final aim of error counting is to help make progress on safety, but this puts the link between errors and safety on trial:

- Can the counting of negatives (e.g., these errors) say anything useful about safety?
- What does the quantity measured (errors) have to do with the quality managed (safety)?

Error-classification methods assume a close mapping between these two, and assume that an absence or reduction of errors is synonymous with progress on safety. By treating safety as positivistically measurable, error counting may be breathing the scientific spirit of a bygone era. Human performance in the laboratory was once gauged by counting errors, and this is still done when researchers test limited, contrived task behavior in spartan settings. But how well does this export to natural settings where people carry out actual complex, dynamic, and interactive work, where determinants of good and bad outcomes are deeply confounded?

It may not matter. The idea of a realist count is compelling to industry and to many researchers for the same reasons that any numerical performance measurement is. Managers get easily infatuated with "balanced scorecards," key performance indicators, or other figures of performance. Entire business models depend on quantifying performance results, so why not quantify safety? Error counting becomes yet another quantitative basis for managerial interventions. Pieces of data from the operation that have been excised and formalized away from their origin can be converted into graphs and bar charts that subsequently form the inspiration for interventions. This allows managers, and their airlines, to elaborate their idea of control over operational practice and its outcomes. Managerial control, however, exists only in the sense of purposefully formulating and trying to influence the intentions and actions of operational people (Angell and Straub 1999). It is not the same as being in control of the consequences (by which safety ultimately gets measured industry-wide), because for that the real world is too complex and operational environments are too stochastic (e.g., Snook 2000).

There is another tricky aspect of trying to create progress on safety through error counting and classification. This has to do with not taking context into regard when counting errors. Errors, according to realist interpretations, represent a kind of equivalent category of bad performance (e.g., a failure to meet one's objective or intention), no matter who commits the error or in what situation. Such an assumption has to exist; otherwise, tabulation becomes untenable. One cannot (or should not) add apples and oranges, after all. If both apples and oranges are entered into the method (and, given that the autonomy principle is false, error-counting methods do add apples and oranges), then it can produce statistical tabulations that claim doctors are 7500 times more dangerous than gun owners. As Hollnagel and Amalberti (2001) showed, attempts to map situated human capabilities such as decision making, proficiency, or deliberation onto discrete categories are doomed to be misleading. They cannot cope with the complexity of actual practice without serious degeneration (Angell and Straub 1999). Error classification disembodies data. It removes the context that helped produce the behavior in its particular manifestation. Such disembodiment may actually retard understanding. The local rationality principle (people's behavior is rational when viewed from the inside of their situations) is impossible to maintain when context is removed from the controversial action. And error categorization does just that: it removes context. Once the observation of some kind of error is tidily locked away into some category, it has been objectified, formalized away from the situation that brought it forth. Without context, there is no way to re-establish local rationality. And without local rationality, there is no way to understand human error. And without understanding human error, there may be no way to learn how to create progress on safety.

SAFETY AS REFLEXIVE PROJECT

Safety is more than the measurement and management of negatives (errors), if it is that at all. Just as errors are epistemologically elusive (How do you know what you know? Did you really see a procedure error? Or was it a proficiency error?) and onto-logically relativist (what it means "to be" and to perform well or badly inside a particular situation is different from person to person), the notion of safety may similarly lack an objective, common denominator. The idea behind measuring safety through error counts is that safety is some kind of objective, stable (and perhaps ideal) reality, a reality that can be measured and reflected, or represented, through method. But does this idea hold? Rochlin (1999, p. 1550), for example, proposed that safety is a "constructed human concept" and others in human factors have begun to probe how individual practitioners construct safety, by assessing what they understand risk to be, and how they perceive their ability of managing challenging situations. A substantial part of practitioners' construction of safety turns out to be reflexive, assessing the person's own competence or skill in maintaining safety across different situations. Interestingly, there may be a mismatch between risk salience (how critical a particular threat to safety was perceived to be by the practitioner) and frequency of encounters (how often these threats to safety are in fact met in practice). The safety threats deemed most salient were the ones least frequently dealt with (Orasanu and Martin, 1998). Safety is more akin to a reflexive project, sustained through a revisable narrative of self-identity that develops in the face of frequently and less frequently encountered risks. It is not something referential, not something that is objectively "out there" as a common denominator, open to any type of approximation by those with the best methods. Rather, safety may be reflexive: something that people relate to themselves.

The numbers produced by error counts are a logical endpoint of a structural analysis that focuses on (supposed) causes and consequences, an analysis that defines risk and safety instrumentally, in terms of minimizing errors and presumably measurable consequences. A second, more recent approach is more socially and politically oriented, and places emphasis on representation, perception, and interpretation rather than on structural features (Rochlin 1999). The managerially appealing numbers generated by error counts do not carry any of this reflexivity, none of the nuances of what it is to "be there," doing the work, creating safety on the line. What it means to be there, however, ultimately determines safety (as outcome): People's local actions and assessments are shaped by their own perspectives. These in turn are embedded in histories, rituals, interactions, beliefs and myths, both of people's organization and organizational subculture and of them as individuals. This would explain why good, objective, empirical indicators of social and organizational definitions of safety are difficult to obtain. Operators of reliable systems "were expressing their evaluation of a positive state mediated by human action, and that evaluation reflexively became part of the state of safety they were describing" (Rochlin 1999, p. 1550). In other words, the description itself of what safety means to an individual operator is a part of that very safety, dynamic and subjective. "Safety is in some sense a story a group or organization tells about itself and its relation to its task environment" (Rochlin 1999, p. 1555).

CAN WE EVEN MEASURE "SAFETY"?

But how does an organization capture what groups tell about themselves; how does it pin down these stories? How can management measure a mediated, reflexive idea? If not through error counts, what can an organization look for in order to get some measure of how safe it is? Large recent accidents provide some clues of where to start looking (see Chapter 5). A main source of residual risk in otherwise safe transportation systems is the drift into failure described in Chapter 2. Pressures of scarcity and competition narrow an organization's focus on goals associated with production. With an accumulating base of empirical success (i.e., no accidents, even if safety is increasingly traded off against other goals such as maximizing profit or capacity utilization), the organization, through its members' multiple little and larger daily decisions, will begin to believe that past success is a guarantee of future safety, that historical success is a reason for confidence that the same behavior will lead to the same (successful) outcome the next time around. The absence of failure, in other words, is taken as evidence that hazards are not present, that countermeasures already in place are effective. Such a model of risk is embedded deeply in the reflexive stories of safety that Rochlin (1999) talked about, and it can be made explicit only through qualitative investigations that probe the interpretative aspect of situated human assessments and actions. Error counts do little to elucidate any of this. More qualitative studies could reveal how currently traded models of risk may increasingly be at odds with the actual nature and proximity of hazard, though it may of course be difficult to establish the objective, or ontologically absolutist, presence of hazard.

Particular aspects of how organization members tell or evaluate safety stories, however, can serve as markers. Woods and Hollnagel (2006), for example, has called one of these markers "distancing through differencing." In this process, organizational members look at other failures and other organizations as not relevant to them and their situation. They discard other events because they appear at the surface to be dissimilar or distant. Discovering this through qualitative inquiry can help specify how people and organizations reflexively create their idea, their story of safety. Just because the organization or section has different technical problems, different managers, different histories, or can claim to already have addressed a particular safety concern revealed by the event, does not mean that they are immune to the problem. Seemingly divergent events can represent similar underlying patterns in the drift toward hazard. High-reliability organizations characterize themselves through their preoccupation with failure: continually asking themselves how things can go wrong and could have gone wrong, rather than congratulating themselves on the fact that things went right. Distancing through differencing means underplaying this preoccupation. It is one way to prevent learning from events elsewhere, one way to throw up obstacles in the flow of safety-related information.

Additional processes that can be discovered include to what extent an organization resists oversimplifying interpretations of operational data, whether it defers to expertise and expert judgment rather than managerial imperatives. Also, it could be interesting to probe to what extent problem-solving processes are fragmented across organizational departments, sections, or subcontractors. The 1996 ValuJet accident, where flammable oxygen generators were placed in an aircraft cargo hold without shipping caps, subsequently burning down the aircraft, was related to a web of subcontractors that together made up the virtual airline of ValuJet. Hundreds of people within even one subcontractor logged work against the particular ValuJet aircraft, and this subcontractor was only one of many players in a network of organizations and companies tasked with different aspects of running (even constituting) the airline. Relevant maintenance parts (among them the shipping caps) were not available at the subcontractor, ideas of what to do with expired oxygen canisters were generated *ad hoc* in the absence of central guidance, and local understandings for why shipping caps may have been necessary were foggy at best. With work and responsibility for it distributed among so many participants, nobody may have been able anymore to see the big picture, including the regulator. Nobody may have been able to recognize the gradual erosion of safety constraints on the design and operation of the original system.

If safety is a reflexive project rather than an objective datum, we must develop new probes for measuring the safety health of an organization. Error counts do not suffice. They uphold an illusion of rationality and control, but may offer neither real insight nor productive routes for progress on safety. It is, of course, a matter of debate whether the vaguely defined organizational processes that could be part of new safety probes (e.g., distancing through differencing, deference to expertise, fragmentation of problem-solving, incremental judgments into disaster) are any more real than the errors from the counting methods they seek to replace or augment. But then, the reality of these phenomena is in the eye of the beholder: observer and observed cannot be separated; object and subject are largely indistinguishable. The processes and phenomena are real enough to those who look for them and who wield the theories to accommodate the results. Criteria for success may lie elsewhere, for example, in how well the measure maps onto past evidence of precursors to failure. Yet even such mappings are subject to paradigmatic interpretations of the evidence base. Indeed, consonant with the ontological relativity of the age human factors has now entered, the debate can probably never be closed. Are doctors more dangerous than gun owners? Do errors exist? It depends on who you ask.

The real issue, therefore, lies a step away from the fray, a level up, if you will. Whether we count errors as Durkheimian fact on the one hand or see safety as a reflexive project on the other, competing premises and practices reflect particular models of risk. These models of risk are interesting not because of their differential abilities to access empirical truth (because that may all be relative), but because of what they say about us, about human factors and system safety. It is not the monitoring of safety that we should simply pursue, but the monitoring of that monitoring. If we want to make progress on safety, one important step is to engage in such meta-monitoring, to become better aware of the models of risk embodied in our assumptions and approaches to safety.

PROCEDURES: PRESCRIPTIONS OR RESOURCES FOR ACTION?

PROCEDURES AS BASIS FOR CONTROL OR RESILIENCE

A good place to begin this metamonitoring is with procedures. Procedures are a nexus between the idea that people are a problem to control and the idea that people are a solution to harness. People do not always follow procedures. We can easily observe this when watching people at work, and managers, supervisors, and regulators (or anybody else responsible for safe outcomes of work) often consider it to be a large practical problem. In hindsight, after a mishap, rule violations seem to play such a dominant causal role. If only they had followed the procedure! Studies keep returning the basic finding that procedure violations precede accidents. For example, an analysis carried out for an aircraft manufacturer identified "pilot deviation from basic operational procedure" as a primary factor in almost 100 accidents (Lautman and Gallimore 1987, p. 2). In the optimistic or eerie terms (depending on how you look at this) of authoritarian high modernism, the title of their piece was "Control of the crew-caused accident." Crews, or humans, in other words cause accidents. Control over crews is possible through better compliance with procedures and better hierarchical order and imposition of rules and obedience. One methodological problem with such work as a piece of empirical evidence is that it selects its cases on the dependent variable (the accident). This generates tautologies rather than findings. Indeed, finding procedure violations as overriding part of the "causal" picture is not informative in itself. High-modernist thinking that is organized around vocabularies of control and hierarchical ordering. Performance variations, especially those at odds with explicit written guidance, are an anathema to high-modernist order. In such vocabularies and in that kind of thinking, of course they easily get overestimated for their role in the sequence of events:

> The interpretation of what happened may be distorted by naturalistic biases to overestimate the possible causal role of unofficial action or procedural violation... While it is possible to show that violations of procedures are involved in many safety events, many violations of procedures are not, and indeed some violations (strictly interpreted) appear to represent more effective ways of working (McDonald et al. 2002, p. 3).

Hindsight always turns complex, tangled histories laced with uncertainty and pressure into neat, linear anecdotes with obvious choices. What look like "violations" from the outside and hindsight are often actions that make sense given the pressures and trade-offs that exist on this inside of real work. Finding procedure violations as causes or contributors to mishaps, in other words, says more about us and the biases we introduce when looking back onto a sequence of events, than it does about people who were doing actual work at the time. Yet if procedure violations are construed to be such a large ingredient of mishaps, then it can be tempting, in the wake of failure, to introduce even more procedures. Or to change existing ones, or to enforce stricter compliance. For example, shortly after a fatal shootdown of two US Black Hawk helicopters over Northern Iraq by US fighter jets, "higher headquarters in Europe dispatched a sweeping set of rules in documents several inches thick to 'absolutely guarantee' that whatever caused this tragedy would never happen again" (Snook 2000, p. 201). It is a common, but not typically satisfactory reaction. Introducing more procedures does not necessarily avoid the next incident, nor do exhortations to follow rules more carefully necessarily increase compliance or enhance safety. In the end, a mismatch between procedures and practice is not unique to accident sequences. Not following procedures does not necessarily lead to trouble, and safe outcomes may be preceded by just as (relatively) many procedural deviations as accidents are.

PROCEDURE APPLICATION AS RULE-FOLLOWING

When rules are "violated," are these bad people ignoring the rules? Or are these bad rules, ill-matched to the demands of real work? To be sure, procedures, with the aim of standardization, can play an important role in shaping safe practice. Commercial aviation is often held up as prime example of the powerful effect of standardization on safety. But there is a deeper, more complex dynamic where real practice is continually adrift from official written guidance, settling at times, unsettled and shifting at others. There is a deeper, more complex interplay whereby practice sometimes precedes and defines the rules rather than being defined by them. In those cases, is a violation an expression of defiance, or an expression of compliance—people following "practical rules" rather than official, impractical ones?

These possibilities lie between two opposing models of what procedures mean, and what they in turn mean for safety. These models of procedures guide how organizations think about making progress on safety. The first model is based on the notion that not following procedures can lead to unsafe situations. These are its premises:

- Procedures represent the best thought-out, and thus the safest way to carry out a job.
- Procedure-following is mostly simple IF–THEN rule-based mental activity: IF this situation occurs, THEN this algorithm (e.g., checklist) applies.
- Safety results from people following procedures.
- For progress on safety, organizations must invest in people's knowledge of procedures and ensure that procedures are followed.

In this idea of procedures, those who violate them are often depicted as putting themselves above the law. These people may think that rules and procedures are made for others, but not for them, as they know how to really do the job. This idea of rules and procedures suggests that there is something exceptionalist or misguidedly elitist about those who choose not to follow the rules. After a maintenance-related mishap, for example, investigators for the regulator found that "the engineers who carried out the flap change demonstrated a willingness to work around difficulties without reference to the design authority, including situation where compliance with the maintenance manual could not be achieved" (JAA 2001, p. 21). The engineers demonstrated a "willingness." Such terminology embodies notions of volition (the engineers had a free choice either to comply or not to) and full rationality (they knew what they were doing). They violated willingly. Violators are wrong, because rules and procedures prescribe the best, safest way to do a job, independent of who does that job. Rules and procedures are for everyone. Such characterizations are naïve at best, and always misleading. If you know where to look, daily practice is testimony to the ambiguity of procedures; and evidence that procedures are a rather problematic category of human work.

THE "JOB PERCEPTION GAP"

Aviation line maintenance is emblematic: A "job perception gap" exists (McDonald et al. 2002) where supervisors are convinced that safety and success result from mechanics following procedures—a sign-off means that applicable procedures were followed. But mechanics may encounter problems for which the right tools or parts are not at hand; the aircraft may be parked far away from base. Or there may be too little time: aircraft with a considerable number of problems may have to be turned around for the next flight within half an hour. Mechanics consequently see success as the result of their evolved skills at adapting, inventing, compromising, and improvising in the face of local pressures and challenges on the line—a sign-off means the job was accomplished despite resource limitations, organizational dilemmas, and pressures. Those mechanics who are most adept are valued for their productive capacity even by higher organizational levels. Unacknowledged by those levels, though, are the vast informal work systems that develop so mechanics can get work done, advance their skills at improvising and satisficing, impart them to one another, and condense them in unofficial, self-made documentation (McDonald et al. 2002). Seen from the outside, a defining characteristic of such informal work systems would be routine nonconformity. But from the inside, the same behavior is a mark of expertise, fueled by professional and inter-peer pride. And of course, informal work systems emerge and thrive in the first place because procedures are inadequate to cope with local challenges and surprises, and because procedures' conception of work collides with the scarcity, pressure, and multiple goals of real work.

What are some of the problems that we can observe in the application of procedures in practice?

- Operational work takes place in a context of limited resources and multiple goals and pressures. Procedures assume that there is time to do them in, certainty (of what the situation is), and sufficient information available (e.g., about whether tasks are accomplished according to the procedure). This already keeps rules at a distance from actual tasks, because real work seldom meets those criteria. Work-to-rule strikes show how it can be impossible to follow the rules and get the job done at the same time.

- Some of the safest complex, dynamic work not only occurs despite the procedures—such as aircraft line maintenance—but without procedures altogether. Rochlin et al. (1987, p. 79), commenting on the introduction of ever-heavier and capable aircraft onto naval aircraft carriers, noted that "there were no books on the integration of this new hardware into existing routines and no other place to practice it but at sea... Moreover, little of the process was written down, so that the ship in operation is the only reliable manual." Work is "neither standardized across ships nor, in fact, written down systematically and formally anywhere." Yet naval aircraft carriers, with inherent high-risk operations, have a remarkable safety record, like other so-called high reliability organizations.

- Procedure-following can also be antithetical to safety. In the 1949 US Mann Gulch disaster, wildland firefighters who perished were the ones sticking to the organizational mandate to carry their tools everywhere (Weick 1993). In this case, as in others, people faced the choice between following the procedure or surviving (Dekker 2001).

- There is always a distance between a written rule and an actual task. Documentation cannot present any close relationship to situated action because of the unlimited uncertainty and ambiguity involved in the activity. Especially where normal work mirrors the uncertainty and criticality of emergencies, rules emerge from practice and experience rather than preceding it. Procedures, in other words, end up following work instead of specifying action beforehand.

Human factors has so far been unable to trace and model such coevolution of human and system, of work and rules. Instead, it has typically imposed a mechanistic, static view of one best practice from the top down. This, then, is the tension. Procedures are seen as an investment in safety—but it turns out that they not always are. Procedures are thought to be required to achieve safe practice—yet they are not always necessary, nor likely ever sufficient for creating safety. Procedures spell out how to do the job safely—yet following all the procedures can lead to an inability to get the job done. Though a considerable practical problem, such tensions are underreported and underanalyzed in the human factors literature. Ethnographer Ed Hutchins has pointed out how procedures are not just externalized cognitive tasks (i.e., the task has been transplanted from the head onto the world, for example, onto a checklist). Rather, following a procedure itself requires cognitive tasks that are not specified in the procedure. Transforming the written procedure into activity requires cognitive work (Hutchins 1995). Procedures are inevitably incomplete specifications of action: they contain abstract descriptions of objects and actions that relate only loosely to particular objects and actions that are encountered in the actual situation (Suchman 1987).

LUBRICATING A JACKSCREW

Take as an example the lubrication of the jackscrew on MD-80s—something that was done incompletely and at increasingly greater intervals before the crash of Alaska 261 (see *Drift into Failure*, Dekker 2011b). This is part of the written procedure that describes how the lubrication work should be done (NTSB 2002, pp. 29–30):

A. Open access doors 6307, 6308, 6306, and 6309
B. Lube per the following…
   3. JACKSCREW
      Apply light coat of grease to threads, then operate mechanism through full range of travel to distribute lubricant over length of jackscrew.
C. Close doors 6307, 6308, 6306, and 6309

This leaves a lot to the imagination, or to the mechanic's initiative. How much is a "light" coat? Do you do apply the grease with a brush (if a "light coat" is what you need), or do you pump it onto the parts directly with the grease gun? How often should the mechanism (jackscrew plus nut) be operated through its full range of travel during the lubrication procedure? None of this is specified in the written guidance. It is little wonder that: "Investigators observed that different methods were used by maintenance personnel to accomplish certain steps in the lubrication procedure, including the manner in which grease was applied to the acme nut fitting and the acme screw and the number of times the trim system was cycled to distribute the grease immediately after its application" (NTSB 2002, p. 116).

In addition, actually carrying out the work is difficult enough. The access panels of the horizontal stabilizer were just large enough to allow a hand through, which would then block the view of anything that went on inside. As a mechanic, you can either look at what you have to do or what you have just done, or actually do it. You cannot do both at the same time, since the access doors are too small. This makes judgments about how well the work is being done rather difficult. The investigation discovered as much when they interviewed the mechanic responsible for the last lubrication of the accident airplane: "When asked how he determined whether the lubrication was being accomplished properly and when to stop pumping the grease gun, the mechanic responded, 'I don't'" (NTSB 2002, p. 31).

The time the lubrication procedure took was also unclear, as there was ambiguity about which steps were "included" in the procedure. Where does the procedure begin and where does it end? After access has been created to the area, or before? And is closing the panels part of it as well, as far as time estimates are concerned? Having heard that the entire lubrication process takes "a couple of hours," investigators learned from the mechanic of the accident airplane that

> "...the lubrication task took 'roughly...probably an hour' to accomplish. It was not entirely clear from his testimony whether he was including removal of the access panels in his estimate. When asked whether his 1-hour estimate included gaining access to the area, he replied, 'No, that would probably take a little—well, you've got probably a dozen screws to take out of the one panel, so that's—I wouldn't think any more than an hour.' The questioner then stated, 'including access?', and the mechanic responded, 'Yeah'" (NTSB 2002, p. 32).

As the procedure for lubricating the MD-80 jackscrew above indicates, formal documentation can neither be relied on, nor is normally available in a way that supports a close relationship to action. Sociologist Carol Heimer makes a distinction between universalistic and particularistic rules: universalistic rules are very general proscriptions (e.g., "Apply light coat of grease to threads") but remain at a distance from their actual application. In fact, all universalistic rules or general proscriptions develop into particularistic rules as experience accumulates. With experience, people encounter the conditions under which universalistic rules need to be applied and become increasingly able to specify those conditions. As a result, universalistic rules assume appropriate local expressions through practice.

Remember of course, how our faith in procedures comes out of the scientific management tradition, where their main purpose was minimization of human variability and maximization of predictability, a rationalization of work (Wright and McCarthy 2003). Aviation and many other industries contain a strong heritage: procedures represent and allow a routinization that makes it possible to conduct safety-critical work with perfect strangers. Procedures are a substitute for knowing coworkers. In aviation, for example, the actions of a copilot are predictable not because the copilot is known (in fact, you may never have flown with her or him) but because the procedures make them predictable. Without such standardization, it would be impossible to cooperate safely and smoothly with unknown people.

In the spirit of scientific management, human factors also assumes that order and stability in operational systems are achieved rationally and mechanistically and that control is implemented vertically (e.g., through task analyses that produce prescriptions of work-to-be-carried out). In addition, the strong influence of information processing psychology on human factors has reinforced the idea of procedures as IF–THEN rule following, where procedures are akin to a program in a computer that in turn serves as input signals to the human information processor. The algorithm specified by the procedure becomes the software on which the human processor runs. But it isn't that simple. Following procedures in the sense of applying them in practice requires more intelligence. It requires additional cognitive work. This brings us to the second model of procedures and safety.

PROCEDURE APPLICATION AS SUBSTANTIVE COGNITIVE ACTIVITY

PROCEDURES ARE NOT THE JOB

People at work must interpret procedures with respect to a collection of actions and circumstances that the procedures themselves can never fully specify (e.g., Suchman 1987). In other words, procedures are not the work itself. Work, especially that in complex, dynamic workplaces, often requires subtle, local judgments with regard to timing of subtasks, relevance, importance, prioritization, and so forth. For example, there is no technical reason why a before-landing checklist in a commercial aircraft could not be automated. The kinds of items on such a checklist (e.g., hydraulic pumps OFF, gear down, and flaps selected) are mostly mechanical and could be activated on the basis of predetermined logic without having to rely on, or constantly remind, a human to do so. Yet no before-landing checklist is fully automated today. The reason is that approaches for landing differ—they can differ in terms of timing, workload, priorities, and so forth. Indeed, the reason is that the checklist is not the job itself. The checklist is, to repeat Suchman (1987), a resource for action; it is one way for people to help structure activities across roughly similar yet subtly different situations. Variability in this is inevitable. Circumstances change, or are not as was foreseen by those who designed the procedures. This shows the outlines of a different model:

- Safety is not the result of rote rule following; it is the result of people's insight into the features of situations that demand certain actions and people being skillful at finding and using a variety of resources (including written guidance) to accomplish their goals. This suggests a second model on procedures and safety:
- Procedures are resources for action. Procedures do not specify all circumstances to which they apply. Procedures cannot dictate their own application.
- Applying procedures successfully across situations can be a substantive and skillful cognitive activity.
- Procedures can, in themselves, not guarantee safety. Safety results from people being skillful at judging when and how (and when not) to adapt procedures to local circumstances.
- For progress on safety, organizations must monitor and understand the reasons behind the gap between procedures and practice. Additionally, organizations must develop ways that support people's skill at judging when and how to adapt.

While there is always a distance between the logics dictated in written guidance and real actions to be taken in the world, prespecified guidance is especially inadequate in the face of novelty and uncertainty. Adapting procedures to fit unusual circumstances is a substantive cognitive activity. Take, for instance, the crash of a large passenger aircraft near Halifax, Nova Scotia in 1998. After an uneventful departure, a burning smell was detected and, not much later, smoke was reported inside the cockpit. Newspaper accounts characterized (perhaps unfairly or too stereotypically) the two pilots as respective embodiments of the models of procedures and safety: the copilot preferred a rapid descent and suggested dumping fuel early so that the aircraft would not be too heavy to land. But the captain told the copilot, who was flying the plane, not to descend too fast, and insisted they cover applicable procedures (checklists) for dealing with smoke and fire. The captain delayed a decision on dumping fuel. With the fire developing, the aircraft became uncontrollable and crashed into the sea, taking all 229 lives onboard with it. There were many good reasons for not immediately diverting to Halifax: neither pilot was familiar with the airport; they would have to fly an approach procedure that they were not very proficient at; applicable charts and information on the airport were not easily available, and an extensive meal service had just been started in the cabin (TSB 2003).

FUNDAMENTAL PROCEDURAL DOUBLE BIND

Part of the example illustrates a fundamental double bind for those who encounter surprise and have to apply procedures in practice (Woods and Shattuck 2000):

* If rote rule following persists in the face of cues that suggests procedures should be adapted, this may lead to unsafe outcomes. People can get blamed for their inflexibility; their application of rules without sensitivity to context.
* If adaptations to unanticipated conditions are attempted without complete knowledge of circumstance or certainty of outcome, unsafe results may occur too. In this case, people get blamed for their deviations; their nonadherence.

In other words, people can fail to adapt, or attempt adaptations that may fail. Rule following can become a desynchronized and increasingly irrelevant activity; decoupled from how events and breakdowns are really unfolding and multiplying throughout a system. In the Halifax crash, as is often the case, there was uncertainty about the very need for adaptations (how badly ailing was the aircraft, really?) as well as uncertainty about the effect and safety of adapting: How much time would the crew have to change their plans? Could they skip fuel dumping and still attempt a landing? Potential adaptations, and the ability to project their potential for success, were not necessarily supported by specific training or overall professional indoctrination. Civil aviation, after all, tends to emphasize model 1: stick with procedures and you will most likely be safe (e.g., Lautman and Gallimore 1987). Tightening procedural adherence, through threats of punishment or other supervisory interventions, does not remove the double bind. In fact, it may tighten the double bind—making it more difficult for people to develop judgment at how and when to adapt. Increasing the pressure to comply increases the probability of failures to adapt—compelling people to adopt a more conservative response criterion. People will require more evidence for the need to adapt, which takes time, and time may be scarce in cases that call for adaptation (as in the crash above). Merely stressing the importance of following procedures can increase the number of cases in which people fail to adapt in the face of surprise. Letting people adapt without adequate skill or preparation, on the other hand, can increase the number of failed adaptations. One way out of the double bind is to develop people's skill at adapting. This means giving them the ability to balance the risks between the two possible types of failure: failing to adapt or attempting adaptations that may fail. It requires the development of judgment about local conditions and the opportunities and risks they present, as well as an awareness of larger goals and constraints that operate on the situation. Development of this skill could be construed, to paraphrase Rochlin, as planning for surprise. Indeed, as Rochlin (1999, p. 1549) has observed: the culture of safety in high reliability organizations anticipate and plan for possible failures in "the continuing expectation of future surprise."

Progress on safety also hinges on how an organization responds in the wake of failure (or even the threat of failure). Postmortems can quickly reveal a gap between procedures and local practice, and hindsight inflates the causal role played by unofficial action (McDonald et al. 2002). The response, then, is often to try to forcibly close the gap between procedures and practice, by issuing more procedures or policing practice more closely. The role of informal patterns of behavior and what they represent (e.g., resource constraints, organizational deficiencies or managerial ignorance, countervailing goals, peer pressure, professionalism, perhaps even better ways of working) all go misunderstood. Real practice, as done in the vast informal work systems, is driven and kept underground. Even though failures offer each sociotechnical system an opportunity for critical self-examination, accident stories are developed in which procedural deviations play a major, evil role, and are branded as deviant and causal. The "official" reading of how the system works or is supposed to work is once again reinvented: Rules mean safety, and people should follow them. High reliability organizations, in contrast, distinguish themselves by their constant investment in trying to monitor and understand the gap between procedures and practice. The common reflex is not to try to close the gap, but to understand why it exists. Such understanding provides insight into the grounds for informal patterns of activity and opens ways to improve safety by sensitivity to people's local operational context.

REGULATOR: MODERNIST HIERARCHICAL CONTROL OR GUIDANCE FROM THE SIDE?

CHECKING WHETHER RULES ARE FOLLOWED

That there is always a tension between centralized guidance and local practice creates a clear dilemma for those tasked with regulating safety-critical industries. An important regulatory instrument consists of rules and checking that they are followed. But forcing operational people to stick to rules can lead to ineffective, unproductive, or even unsafe local actions. For various jobs, following the rules and getting the task done are mutually exclusive. On the other hand, letting people adapt their local practice in the face of pragmatic demands can make that they sacrifice global system goals or miss other constraints or vulnerabilities that operate on the system. Helping people solve this fundamental trade-off is not a matter of pushing the criterion one way or the other. Discouraging people's attempts at adaptation can increase the number of failures to adapt in situations where adaptation was necessary. Allowing procedural leeway without encouraging organizations to invest in people's skills at adapting, on the other hand, can increase the number of failed attempts at adaptation.

This means that the gap between rule and task, between written procedure and actual job, needs to be bridged by the regulator as much as by the operator. Inspectors who work for regulators need to "apply" rules as well: find out what exactly the rules mean and what their implications are when imposed upon a field of practice. The development from universalism to particularism applies to regulators too. This raises questions about the role that inspectors should play. Should they function as police—checking to what extent the market is abiding by the laws they are supposed to uphold? In that case, should they apply a black-and-white judgment (which would ground a number of companies immediately)? Or, if there is a gap between procedure and practice that inspectors and operators share and both need to bridge, can inspectors be partners in joint efforts toward progress on safety? The latter role is one that can only develop in good faith, though such good faith may be the very by-product of the development of a new kind of relationship, or partnership, toward progress on safety. Mismatches between rules and practice are no longer seen as the logical conclusion of an inspection, but rather as the starting point; the beginning of joint discoveries about real practice and the context in which it occurs. What are the systemic reasons (organizational, regulatory, resource related) that help create and sustain the mismatch?

OUTSIDE AND INSIDE AT THE SAME TIME

The basic criticism of an inspector's role as partner or guide is easy to anticipate: regulators should not come too close to the ones they regulate, lest their relationship becomes too cozy and objective judgment of safety criteria becomes impossible. And their job is not to be a consultant either. But regulators need to come close to those they regulate in any case. Regulators (or their inspectors) need to be "insiders" in the sense of speaking the language of the organization they inspect, understanding the kind of business they are in, in order to gain the respect and credibility of the informants they need most. At the same time, regulators need to be outsiders—resisting getting integrated in the worldview of the one they regulate. Once on the inside of that system and its worldview, it may be increasingly difficult to discover the potential drift into failure. What is normal to the operator is normal to the inspector.

The tension between having to be an insider and an outsider at the same time is difficult to resolve. The conflictual, adversarial model of safety regulation has in many cases not proven productive. It leads to window dressing and posturing on part of the operator during inspections and to secrecy and obfuscation of safety- and work-related information at all other times. As airline maintenance testifies, real practice is easily driven underground. Even for regulators who apply their power as police rather than as partner, the struggle of having to be an insider and outsider at the same time is not automatically resolved. Issues of access to information (the relevant information about how people do their work, even when the inspector is not there) and inspector credibility demand that there is a kind of relationship between regulator and operator that allows such access and credibility to develop.

Organizations (including regulators) who wish to make progress on safety with procedures would instead need to

- Monitor the gap between procedure and practice and try to understand why it exists (and resist trying to close it by simply telling people to comply)
- Help people develop skills to judge when and how to adapt (and resist only telling people they should follow procedures)

But many organizations or industries do neither. They may not even know, or want to know (or be able to afford to know) about the gap. Take aircraft maintenance again. A variety of workplace factors (communication problems, physical or hierarchical distance, and industrial relations) obscure the gap. For example, continued safe outcomes of existing practice give supervisors no reason to question their assumptions about how work is done (if they are safe, they must be following procedures down there). There is wider industry ignorance, however (McDonald et al. 2002). In the wake of failure, informal work systems typically retreat from view, gliding out of investigators' reach. What goes misunderstood, or unnoticed, is that informal work systems compensate for the organization's inability to provide the basic resources (e.g., time, tools, and documentation with a close relationship to action) needed for task performance. Satisfied that violators got caught and that formal prescriptions of work were once again amplified, the organizational system changes little or nothing. It completes another "cycle of stability," typified by a stagnation of organizational learning and no progress on safety (McDonald et al. 2002).

GOAL CONFLICTS AND PROCEDURAL DEVIANCE

SAFETY IS NOT THE PRIORITY

A major driver behind routine divergence from written guidance is the need to pursue multiple goals simultaneously. Multiple goals mean goal conflicts. As Dietrich Dörner remarked: "Contradictory goals are the rule, not the exception, in complex situations" (Dörner 1989, p. 65). In a study of flight dispatchers, for example, Smith (2001) illustrated the basic dilemma. Would bad weather hit a major hub airport or not? What should the dispatchers do with all the airplanes en route? Safety (by making aircraft divert widely around the weather) would be a pursuit that "tolerates a false alarm but deplores a miss" (Smith 2001, p. 361). In other words, if safety is the major goal, then making all the airplanes divert even if the weather would not end up at the hub (a false alarm) is much better than not making them divert and sending them headlong into bad weather (a miss). Efficiency, on the other hand, severely discourages the false alarm, while it can actually deal with a miss. This is the essence of most operational systems. Though safety is a (stated) priority, these systems do not exist to be safe. They exist to provide a service or product, to achieve economic gain, to maximize capacity utilization. But still they have to be safe. One starting point, then, for understanding a driver behind routine deviations, is to look deeper into these goal interactions, these basic incompatibilities in what people need to strive for in their work. Of particular interest is how people themselves view these conflicts from inside their operational reality, and how this contrasts with management (and regulator) views of the same activities.

FASTER, BETTER, CHEAPER

NASA's "Faster, Better, Cheaper" organizational philosophy in the late nineties epitomized how multiple, contradictory goals are simultaneously present and active in complex systems. The loss of the Mars Climate Orbiter and the Mars Polar Lander in 1999 was ascribed in large part to the irreconcilability of the three goals (faster and better and cheaper) that drove down the cost of launches, made for shorter, aggressive mission schedules, eroded personnel skills and peer interaction, limited time, reduced the workforce, and lowered the level of checks and balances normally found (Report on project management in NASA, by the Mars Climate Orbiter Mishap Investigation Board 2000). People argued that NASA should pick any two from the three goals. Faster and cheaper would not mean better. Better and cheaper would mean slower. Faster and better would be more expensive. Such reduction, however, obscures the actual reality facing operational personnel in safety-critical settings. These people are there to pursue all three goals simultaneously—fine-tuning their operation, as Starbuck and Milliken say, to "render it less redundant, more efficient, more profitable, cheaper, or more versatile" (Starbuck and Milliken 1988, p. 323). Fine-tuning, in other words, to make it faster, better, and cheaper.

The 2003 Space Shuttle Columbia accident focused attention on the maintenance work that was done on the Shuttle's external fuel tank, once again revealing the differential pressures of having to be safe and getting the job done (better, but also faster and cheaper). A mechanic working for the contractor, whose task it was to apply the insulating foam to the external fuel tank, testified that it took just a couple of weeks to learn how to get the job done, thereby pleasing upper management and meeting production schedules. An older worker soon showed him how he could mix the base chemicals of the foam in a cup and brush it over scratches and gouges in the insulation, without reporting the repair. The mechanic soon found himself doing this hundreds of times, each time without filling out the required paperwork. Scratches and gouges that were brushed over with the mixture from the cup basically did not exist as far as the organization was concerned. And those that did not exist could not hold up the production schedule for the external fuel tanks. Inspectors often did not check. A company program that once had paid workers hundreds of dollars for finding defects had been watered down, virtually inverted by incentives for getting the job done now.

Goal interactions are critical in such experiences, which contain all the ingredients of procedural fluidity, maintenance pressure, the meaning of "incidents" worth reporting, and their connections to drift into failure. As in most operational work, the distance between formal, externally dictated logics of action and actual work is bridged with the help of those who have been there before, who have learned how to get the job done (without apparent safety consequences), and who are proud to share their professional experience with younger, newer workers. Actual practice by newcomers settles at a distance from the formal description of the job. Deviance becomes routinized. This is part of the vast informal networks characterizing much maintenance work, including informal hierarchies of teachers and apprentices, informal documentation of how to actually get work done, informal procedures and tasks, and informal teaching practices. Inspectors did not check, did not know, or did not report. Managers were happy that production schedules were met and happy that fewer defects were being discovered. Normal people, doing normal work in a normal organization. Or that is what it seemed to everybody at the time. Once again, the notion of an "incident," of something that was worthy of reporting (a defect), got blurred against a background of routine nonconformity. What was normal versus what was deviant was no longer so clear. Goal conflicts between safer, better, and cheaper were reconciled by doing the work more cheaply, superficially better (brushing over gouges), and apparently without cost to safety. As long as Orbiters kept coming back safely, the contractor must have been doing something right. Understanding the potential side effects was very difficult given the historical mission success rate. Lack of failures was seen as a validation that current strategies to prevent hazards were sufficient. Could anyone foresee, in a vastly complex system, how local actions as trivial as brushing chemicals from a cup could one day align with other factors to push the system over the edge? Paraphrasing Weick, what could not be believed, could not be seen. Past success was taken as guarantee of continued safety.

INTERNALIZATION OF EXTERNAL PRESSURE

Some organizations pass on their goal conflicts to individual practitioners quite openly. Some airlines, for example, pay their crews a bonus for on-time performance. An aviation publication reported on one of those operators (a new airline called Excel, flying from England to holiday destinations):

> As part of its punctuality drive, Excel has introduced a bonus scheme to give employees a bonus should they reach the agreed target for the year. The aim of this is to focus everyone's attention on keeping the aircraft on schedule (Airliner World November 2001, p. 79).

Such plain acknowledgement of goal priorities, however, is not common. Most important goal conflicts are never made so explicit, arising rather from multiple irreconcilable directives from different levels and sources, from subtle and tacit pressures, from management or customer reactions to particular trade-offs. Organizations often resort to "conceptual integration, or plainly put, doublespeak" (Dörner 1989, p. 68). For example, the operating manual of another airline opens by stating that

1. Our flights shall be safe.
2. Our flights shall be punctual.
3. Our customers will find value for money.

Conceptually, this is Dörner's doublespeak; documentary integration of incompatibles. It is impossible, in principle, to do all three simultaneously, as with NASA's faster, better, cheaper. While incompatible goals arise at the level of an organization and its interaction with its environment, the actual managing of goal conflicts under uncertainty gets pushed down into local operating units—control rooms, cockpits, and the like. There, the conflicts are to be negotiated and resolved in the form of thousands of little and larger daily decisions and trade-offs. These are no longer decisions and trade-offs made by the organization, but by individual operators or crews. It is this insidious delegation, this handover, where the internalization of external pressure takes place. Crews of one airline describe their ability to negotiate these multiple goals while under the pressure of limited resources as "the blue feeling" (referring to the dominant color of their fleet). This "feeling" represents the willingness and ability to put in the work to actually deliver on all three goals simultaneously (safety, punctuality, and value for money). This would confirm that practitioners do pursue incompatible goals of faster, better, and cheaper all at the same time and are aware of it too. In fact, practitioners take their ability to reconcile the irreconcilable as source of considerable professional pride. It is seen as a strong sign of their expertise and competence.

The internalization of external pressure, this integration of organizational goal conflicts by individual crews or operators, is not well described or modeled yet. This, again, is a question about the dynamics of the macro–micro connection that we saw in Chapter 2. How is it that a global tension between efficiency and safety seeps into local decisions and trade-offs by individual people or groups? These macrostructural forces that operate on an entire company find their most prominent expression in how local work groups make assessments about opportunities and risks (see also Vaughan 1996). Institutional pressures are reproduced, or perhaps really manifested in what individual people do, not by the organization "as a whole." But how does this connection work? Where do external pressures become internal? When do the problems and interests of an organization under pressure of resource scarcity and competition become the problems and interests of individual actors at several levels within that organization?

"OPERATING MANUAL WORSHIPPERS"

The connection between external pressure and its internalization is relatively easy to demonstrate when an organization explicitly advertises how operators' pursuit of one goal will lead to individual rewards (a bonus scheme to keep everybody focused on the priority of schedule). But such cases are probably rare, and it is doubtful whether they represent actual internalization of a goal conflict. It becomes more difficult when the connection and the conflicts are more deeply buried in how operators transpose global organizational aims onto individual decisions. For example, the "blue feeling" signals aircrews' strong identification with their organization (which flies blue aircraft) and what it and its brand stand for (safety, reliability, value for money). Yet it is a "feeling" that only individuals or crews can have; a "feeling" because it is internalized. Insiders point out how some crews or commanders have the blue feeling while others do not. It is a personal attribute, not an organizational property. Those who do not have the blue feeling are marked by their peers—seldom supervisors—for their insensitivity to, or disinterest in, the multiplicity of goals and for their unwillingness to do substantive cognitive work necessary to reconcile the irreconcilable. These practitioners do not reflect the corps' professional pride since they will always make the easiest goal win over the others (e.g., "don't worry about customer service or capacity utilization, it's not my job"), choosing the path of least resistance and least work in the eyes of their peers. In the same airline, those who try to adhere to minute rules and regulations are called "Operating Manual Worshippers"—a clear signal that their way of dealing with goal contradictions is not only perceived as cognitively cheap (just go back to the book, it will tell you what to do), but as hampering the collective ability to actually get the job done, diluting the blue feeling. The blue feeling, then, is also not just a personal attribute, but an inter-peer commodity that affords comparisons, categorizations, and competition among members of the peer group, independent of other layers or levels in the organization. Similar inter-peer pride and perception operate as subtle engine behind the negotiation among different goals in other professions too—for example, flight dispatchers, air-traffic controllers, or aircraft maintenance workers (McDonald et al. 2002).

The latter group (aircraft maintenance) has incorporated even more internal mechanisms to deal with goal interactions. The demand to meet technical requirements clashes routinely with time or other resource constraints such as inadequate time, personnel, tools, parts, or functional work environment (McDonald et al. 2002). The vast internal, sub-surface networks of routines, illegal documentation, and "shortcuts," which from the outside would be seen as massive infringement of existing procedures, are a result of the pressure to reconcile and compromise. Actual work practices constitute the basis for technicians' strong professional pride and sense of responsibility for delivering safe work that exceeds even technical requirements. Seen from the inside, it is the role of the technician to apply judgment founded on his or her knowledge, experience, and skill—not on formal procedure. Those most adept at this are highly valued for their productive capacity even by higher organizational levels. Yet upon formal scrutiny (e.g., an accident inquiry), informal networks and practices often retreat from view, yielding only a bare-bones version of work in which the nature of goal compromises and informal activities is never explicit, acknowledged, understood, or valued. Similar to the British Army on the Somme, management in some maintenance organizations occasionally decides (or pretends) that there is no local confusion; that there are no contradictions or surprises. In their official understanding, there are rules and people who follow the rules, and there are safe outcomes as a result. They believe that people who do not follow the rules are more prone to causing accidents, as the hindsight bias inevitably points out. To people on the work floor, in contrast, management does not even understand the fluctuating pressures on their work, let alone the strategies necessary to accommodate those (McDonald et al. 2002).

Both cases above (the blue feeling and maintenance work) challenge human factors' traditional reading of violations as deviant behavior. Human factors wants work to mirror prescriptive task analyses or rules, and violations breach vertical control implemented through such managerial or design directive. Seen from the inside of people's own work, however, violations become compliant behavior. Cultural understandings (e.g., expressed in notions of a "blue feeling") affect interpretative work, so that even if people's behavior is objectively deviant, they will see their own conduct as conforming (Vaughan 1999). Their behavior is compliant with the emerging, local, internalized ways to accommodate multiple goals important to the organization (maximizing capacity utilization but doing so safely; meeting technical requirements as well as deadlines). It is compliant, also, with a complex of peer pressures and professional expectations in which unofficial action yields better, quicker ways to do the job, in which unofficial action is a sign of competence and expertise, where unofficial action can override or outsmart hierarchical control and compensate for higher-level organizational deficiencies or ignorance.

ROUTINE NONCONFORMITY

NORMALIZATION OF DEVIANCE

The gap between procedures and practice is not constant. After the creation of new work (e.g., through the introduction of new technology), time can go by before applied practice stabilizes, likely at a distance from the rules as written for the system on the shelf. Social science has characterized this migration from tightly coupled rules to more loosely coupled practice variously as "fine-tuning" (Starbuck and Milliken 1988) or "practical drift" (Snook 2000). Through this shift, applied practice becomes the pragmatic imperative; it settles into a system as normative. Deviance (from the original rules) becomes normalized; nonconformity becomes routine (Vaughan 1996). The literature has identified important ingredients in the normalization of deviance, which can help organizations understand the nature of the gap between procedures and practice:

- Rules that are overdesigned (written for tightly coupled situations, for the worst case) do not match actual work most of the time. In real work, there is slack: time to recover, opportunity to reschedule and get the job done better or more smartly (Starbuck and Milliken 1988). This mismatch creates an inherently unstable situation that generates pressure for change.
- Emphasis on local efficiency or cost-effectiveness pushes operational people to achieve or prioritize one goal or a limited set of goals (e.g., customer service, punctuality, capacity utilization). Such goals are typically easily measurable (e.g., customer satisfaction, on-time performance), whereas it is much more difficult to measure how much is borrowed from safety.
- Past success is taken as guarantee of future safety. Each operational success achieved at incremental distances from the formal, original rules can establish a new norm. From here, a subsequent departure is once again only a small incremental step. From the outside, such fine-tuning constitutes incremental experimentation in uncontrolled settings—on the inside, incremental nonconformity is an adaptive response to scarce resources, multiple goals, and often competition.
- Departures from the routine become routine. Seen from the inside of people's own work, violations become compliant behavior. They are compliant with the emerging, local ways to accommodate multiple goals important to the organization (maximizing capacity utilization but doing so safely; meeting not only technical requirements but also deadlines). They are compliant, also, with a complex of peer pressures and professional expectations

in which unofficial action yields better, quicker ways to do the job, in which unofficial action is a sign of competence and expertise, where unofficial action can override or outsmart hierarchical control and compensate for higher-level organizational deficiencies or ignorance.

Although a gap between procedures and practice always exists, there are different interpretations of what this gap means and what to do about it. Human factors may see the gap between procedures and practice as a sign of complacency—operators' self-satisfaction with how safe their practice or their system is or a lack of discipline. Psychologists may see routine nonconformity as expressing a fundamental tension between multiple goals (production and safety) that pull workers in opposite directions: getting the job done but also staying safe. Others highlight the disconnect that exists between distant supervision or preparation of the work (as laid down in formal rules) on the one hand and local, situated action on the other. Sociologists may see in the gap a political lever applied on management by the work floor, overriding or outsmarting hierarchical control and compensating for higher-level organizational deficiencies or ignorance. To the ethnographer, routine nonconformity would be interesting not only because of what it says about the work or the work context but also because of what it says about what the work means to the operator.

The distance between procedures and practice can create widely divergent images of work. Is routine nonconformity an expression of elitist operators who consider themselves to be above the law, of people who demonstrate a willingness to ignore the rules? Work in that case is about individual choices, supposedly informed choices between doing that work well or badly, between following the rules or not. Or is routine nonconformity a systematic by-product of the social organization of work, where it emerges from the interactions between organizational environment (scarcity and competition), internalized pressures, and the underspecified nature of written guidance? In that case, work is seen as fundamentally contextualized, constrained by environmental uncertainty and organizational characteristics, and influenced only to a small extent by individual choice. People's ability to balance these various pressures and influences on procedure following depends in large part on their history and experience. And, as Wright and McCarthy (2003) pointed out, there are currently very few ways in which this experience can be given a legitimate voice in the design of procedures.

STUDY QUESTIONS

1. What is Cartesian–Newtonian about error counting, and how does it represent the assumption that people are the main problem we need to control? Is there information surrounding the supposed "errors" that are observed and counted?
2. Consult Chapter 6 for this question. How is the observation of errors, their categorization, tabulation, and quantification an ultimate act of ontological alchemy? Whom do we implicitly believe to be right or wrong in this (the observer vs. the observed operator), and what does that say about power in our workplaces?

3. What is the "job perception gap" and how can this become obvious in what others might see as procedural violations?
4. What is the fundamental bind involved in adapting procedures to a particular situation? Can practitioner training be designed to support them in handling the double bind? What would that training look like?
5. Consider your own organization or operating world. If safety is not the priority, then what is? Or is there little stability in the sorts of priorities that govern operational work?
6. Do you have examples from your own organization or operational world where you see the internalization of external pressures and the normalization of (what others would call) deviance?

4 Danger of Losing Situation Awareness

CONTENTS

KEY POINTS

- Situation awareness relies on a Cartesian–Newtonian worldview in which the mind is a mirror of the world. It is always possible to prove that there was less in the mind than in the world. This is then called a "loss of situation awareness."
- Even though it is a folk construct, situation awareness is endowed with causal powers by much human factors research, and "loss of SA" has recently been used in inquiries and court cases to blame (and convict) operators.
- A focus on the "awareness" rather than on the "situation" takes safety thinking back to its prehistoric era, where safety interventions target the human (and their supposed attention deficit), rather than the system or situation surrounding people.
- Empirical research and common experience show that perception does not start with elements that are subsequently mentally processed into something meaningful. Meaning comes effortlessly or prerationally, and human performance is driven by achieving plausible, coherent accounts of what is going on.
- A radical empiricist view of situation awareness proposes that the mapping between the world and some inside representation of it is not the issue, but rather that the experienced world is the only one there is: if there were an "objective" reality, we could not know it.

CONVICTED FOR LOSING SITUATION AWARENESS

Not long ago, a criminal court case was concluded against a professional operator who, in the words of the prosecution (the Crown in this case), had "lost situation awareness" and collided his vehicle with a structure. He had been carrying out the duties of his job as normal that evening, but a whole array of circumstances—weather, equipment distractions, other vehicles, darkness, fatigue—came into play to cause a collision. As the Crown alleged, it was his "loss of situation awareness" that caused an accident. It killed two people. The Crown charged the operator with criminal negligence. The man had already been fired and was now sentenced to four years in jail. Here was a man, accused of something that didn't even exist 20 years before. Accused of something that the human factors community made up. Of something that we believed (and many of us still believe) to be a researchable property of human performance—but no more, no less.

WHY LOSING SA IS DANGEROUS

Peering wearily into the cradle of the concept in the mid-nineties, aviation safety veteran Charlie Billings wondered in front of a large audience of human factors scientists whether the intercession of "situation awareness" was necessary to explain what causes people to see, or miss, or remember seeing something. "It's a construct!" he said in the keynote at a foundational conference on situation awareness. "Constructs cannot cause anything!" (Billings 1996). But apparently they do. Charlie Billings passed away in 2010. By then, loss of situation awareness (i.e., loss of a construct) had become the favored cause for an epidemic of automation and human performance-related mishaps in aviation, shipping, manufacturing, and other settings. Many reports produced by Safety investigators (ATSB 1996) contain references to a "loss of situation awareness." Investigators at the US National Transportation Safety Board, too, have used "loss of situation awareness" in all its putative causal power more than once. For example, it allowed them to "explain" why a regional airline crew took off from the wrong runway at Lexington airport in Kentucky in 2006, resulting in the deaths of 49 people including the captain (NTSB 2007b). Apparently, our constructs contain enough causal power for them to blame the dead. Or the living. The coroner who investigated a friendly fire incident that killed three British soldiers in Afghanistan in 2007 rendered the verdict that the crew of an American fighter jet had lost "situational awareness" and were looking at the wrong village when they dropped the bomb (Bruxelles 2010, p. 1).

With the scientific legitimation that researchers and editors are willing to lend to this sort of constructs, we can hardly blame laypeople from using them as a "convenient explanation that [they] easily grasp and embrace" (Flach 1995, p. 155). As explained earlier in the book, looking at events after the fact, outside-in, we can always show that there was more in the world (that we now know) than there was in somebody else's mind, because in hindsight, anybody can show that. And we can then call that difference the practitioner's "loss of situation awareness." We can even blame his or her complacency or biases for causing that loss. A recently proposed model in one of our high-ranking journals links complacency to attentional bias and a loss of situation awareness (Parasuraman and Manzey 2010). If you are complacent, in other words, you pay attention to some things but not to others, which means you lose situation awareness. Words, words, words. Constructs, constructs, constructs. Your loss of situation awareness is the difference between what *you* knew *then* versus what *I* know *now*. Which is also what you should have known, but you didn't because you were complacent, or negligent. Or are you complacent or negligent because you didn't know? John Flach warned against such indelible circularity (Flach 1995):

- Why did the operators lose situation awareness?
- Because they were complacent.
- How do we know they were complacent?
- Because they lost situation awareness.

How do we help defend the practitioner who is accused (implicitly or explicitly) of losing situation awareness? Given what our literature has published by now, it may be quite difficult. Consider the first sentence of a recent article that introduces a particular model of situation awareness to anesthesia (Schulz et al. 2013, p. 729): "Accurate situation awareness (SA) of medical staff is integral for providing optimal performance during the treatment of patients." Imagine the following exchange that may show up in medical liability, medical indemnity, or even criminal negligence cases not long from its publication:

AN IMAGINED CROSS-EXAMINATION

Q. Wouldn't you agree, doctor, that accurate situation awareness by medical staff like yourself is integral for providing optimal performance during the treatment of patients? This is what the leading journal in your specialty claims. See, here it says so [Counsel points to exhibit. Accused reads exhibit].

A. Uh, I'd have to agree.

Q. Would you say, doctor, that your performance in this case, in which your patient died as a result of the care you provided, was optimal?

A. Uh, we all hoped for a different outcome.

Q. Were you, or were you not aware of the situation that this particular drug X, when used in combination with Y and Z, had produced problems for this patient eighteen years before, a situation that occurred when she was living in another State?

A. I was not aware of that at the time, no.

Q. Yet you agreed that accurate situation awareness is integral for providing optimal performance during the treatment of patients?

A. ... [silence]

Q. No further questions.

The hindsight bias has ways of getting entrenched in human factors thinking. One such way is our vocabulary. "Losing situation awareness" or "deficient situation awareness" has become a legitimate characterization of cases where, we believe, people did not exactly know where they were or what was going on around them. In many applied as well as some scientific settings, it is acceptable to submit "loss of situation awareness" as an explanation for why people ended up where they should not have ended up, or why they did what they, in hindsight, should not have done. Navigational incidents and accidents in transportation are one category of cases where the temptation to rely on situation awareness as elucidatory construct appears irresistible. If people end up where they should not, or where they did not intend to end up, it is easy to see that as a deficient awareness of the cues and indications around them. It is easy to blame a "loss of situation awareness." As Moray and Inagaki warned in 2000:

This point leads to a consideration of the problem of "situation awareness." It... is poorly specified. What is the situation of which the operator is required to be aware? In most of the experimental or field studies, it is not well defined or specific. Rather, and for obvious reasons, it is almost as though the investigators want the operator to be aware of "anything in the environment which might be of importance if it should change unexpectedly." This is, of course, not the phrase used in research. However, clearly, it is at the back of investigators' minds. The pilot of an aircraft needs a keen situation awareness because if anything abnormal occurs he or she must notice it and respond appropriately. However, there are, logically, an infinite number of events which may occur. Perhaps one, therefore, expects the operator to be aware of the status of just those variables which, if ever they change, represent significant events. But, what are those? If one does not define exactly what the set of events is that the operator must monitor, how can he or she devise an optimal or eutactic monitoring strategy? It is logically quite unreasonable to ask merely for "situation awareness" as such, since if the set of events of which operators are to be aware is not defined, it is unreasonable to expect them to monitor the members of an undefined set; whilst if one defines a set, there is always the possibility that the dangerous event is not a member of the set, in which case the operators should not have been monitoring it (Moray and Inagaki 2000, p. 360).

DUALISM IN SITUATION AWARENESS

The kinds of notations that are popular in various parts of the situation awareness literature are one indication that we quickly stop investigating, researching any further, once we have found "human error" under that new guise. Venn diagrams (Figure 4.1), for example, help us point out the mismatch between actual and ideal situation awareness. They illustrate the difference between what people were aware of in a particular situation, and what they could (or should) ideally have been aware of.

Once we have found a mismatch between what we now know about the situation (the large circle) and what people back then apparently knew about it (the small one), that in itself is explanation enough. They did not know, but they could or should have known. This does not apply only in retrospect, by the way. Even design problems can be clarified through this notation, and performance predictions can be made on its

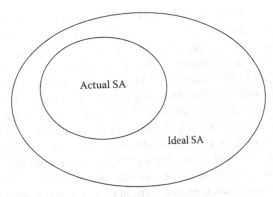

FIGURE 4.1   A common, normative characterization of situation awareness. The large circle depicts ideal awareness (or potential awareness) while the smaller circle represents actual awareness.

basis. When the aim is "designing for situation awareness," the Venn diagram can show what people should pick up in a given setting, versus what they are likely to actually pick up. In both cases, awareness is a relationship between that which is objectively available to people in the outside world on the one hand and what they take in, or understand about it, on the other hand. Terms such as "deficient situation awareness" or "loss of situation awareness" confirm human factors' dependence on a kind of subtractive model of awareness. Here is the Venn diagram notation expressed in an equation:

$$\text{Loss of SA} = f(\text{large circle} - \text{small circle}) \tag{4.1}$$

In this equation, "loss of SA" is equal to "deficient SA" and SA stands for situation awareness. This also reveals the continuing normativist bias in our understanding of human performance. Normativist theories aim to explain mental processes by reference to ideals or normative standards that describe optimal strategies. The large Venn circle is the norm, the standard, the ideal. Situation awareness is explained by reference to that ideal: actual situation awareness is a subtraction from that ideal, a shortfall, a deficit. Indeed, a "loss." This is what Equation 4.1 becomes:

$$\text{Loss of SA} = f(\text{what I know now} - \text{what you knew then}) \tag{4.2}$$

Loss of situation awareness, in other words, is the difference between what I know about a situation now (especially the bits highlighted by hindsight) and what somebody else apparently knew about that situation then. Interestingly, situation awareness is nothing by itself, then. It can only be expressed as a relative, normativist function. For example, the difference between what people apparently knew back then and what they could or should have known (or what we know now). This may explain in part why few researchers venture beyond the "loss of situation awareness." Other than detailing the mismatch between the large and the little circle in a Venn diagram; other than revealing the "elements" that were not seen but could or should have been seen and understood, there is little there. Discourse and research around situation awareness may so far have shed little light on the actual processes of attentional dynamics. Rather, situation awareness has given us a new normativist lexicon that provides large, comprehensive notations for perplexing data (how could they not have noticed? Well, they lost SA). Situation awareness has legitimized the proliferation of the hindsight bias under the pretext of adding to the knowledge base. None of this may help our understanding of "awareness" in complex, dynamic situations. And it certainly does not help us understand the situation from the perspective of those who were inside of it—we already made up our own minds, after all, about what was important and what was not in that situation. And then we blamed the people inside for not seeing what is now so obvious to us.

Constructs such as situation awareness both represent and reify a dualist ontology. Ontology is the branch of philosophy that studies the nature of being, the nature of our world. Dualist, as was explained earlier in the book, means that there is a division of reality into two separate parts—mind and matter in this case. In a dualist ontology, there is a world and a mind, and the mind is an (imperfect) mirror of the world. This has also been called the "correspondence view of knowledge," where knowledge in the mind is understood as a correspondence, a mapping, a simile, to what is in the world out there. So if we can urge people to be less complacent or less biased, to try a little harder, then the mirror in their minds can become a little less imperfect. It can correspond more closely to the world out there, become a better mapping. We can reason our way to this from the other direction too. If in hindsight people turn out to have had an imperfect mental mirror of the world (a loss of situation awareness), we know that because the outcome of their actions was bad. In hindsight, we can easily point to the exact few critical elements that were missing from their mental picture. We, or others, can then blame their deficient motivation for this imperfection.

This makes situation awareness into more than just a causal construct that exists as an agent in the mind of a human operator (Flach 1995). Situation awareness, in these cases, represents a duty of care, the deontological commitment expected of practitioners who are responsible for patients or passengers, whose actions can influence the lives of others. When other people demonstrate the loss of such situation awareness (which is very easy), it represents an absence of a duty of care; a breach of the fiduciary relationship to patients, passengers, colleagues, collateral; a failure to live up to the deontological commitment. It represents a possibly prosecutable crime. This takes our constructs into worlds way beyond the safe, seemingly objective operationalization in laboratories populated with undergraduate student subjects. These are worlds where our words attain representational powers that go way beyond the innocuous operationalism we might have intended for them. Worlds in which operators are put in harm's way by what we come up with. Worlds where our words matter. Worlds where our words have consequences. Our words, our constructs, help conjure up worlds for other people—journalists, investigators, prosecutors, judges, politicians, juries, coroners. People with stories to sell, with legal battles to win; people with prosecutorial ambitions to satisfy, with insurance payouts to reap; people with expensive designs to defend, with manufacturer liability to deny.

"NUMEROUS OPPORTUNITIES" TO NOT LOSE SA

One such accident happened to the *Royal Majesty*, a cruise ship that was sailing from Bermuda to Boston in the summer of 1995. It had more than 1000 people onboard. Instead of in Boston, the *Royal Majesty* ended up on a sandbank close to the Massachusetts shore. Without the crew noticing, it had drifted 17 miles off course during a day and a half of sailing (see Figure 4.2).

Investigators discovered afterward that the ship's autopilot had defaulted to DR (Dead Reckoning) mode (from NAV, or Navigation mode) shortly after departure. DR mode does not compensate for the effects of wind and other drift (waves, currents), which NAV mode does. A northeasterly wind pushed the ship steadily off its course, to the side of its intended track. The US National Transportation Safety Board investigation into the accident judged that "despite repeated indications, the crew failed to recognize numerous opportunities to detect that the vessel had drifted off track" (NTSB 1995, p. 34). But "numerous opportunities" to detect the nature of the real situation become clear only in hindsight. With hindsight, once we know the outcome, it becomes easy to pick out exactly those clues and indications that would have shown people where they were actually headed. If only they had focused on this piece of data, or put less confidence in that indication, or had invested just a little more energy in examining this anomaly, then they would have seen that they were going in the wrong direction. In this sense, situation awareness is a highly functional or adaptive term for us, for those struggling to come to terms with the rubble of a navigational accident. Situation awareness is a notation that assists us in organizing the evidence available to people at the time, and can provide a starting point for understanding why this evidence was looked at differently, or not at all, by those people. Unfortunately, we hardly ever push ourselves to such understanding. "Loss of situation awareness" is accepted as sufficient explanation too quickly too often and in those cases amounts to nothing more than saying "human error" under a fancier label.

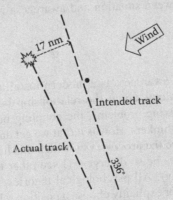

FIGURE 4.2 The difference between where the *Royal Majesty* crew thought they were headed (Boston) and where they actually ended up: a sandbank near Nantucket.

RIVER OF CONSCIOUSNESS

Psychologist William James once illustrated our lack of such understanding with a metaphor. It applies as aptly to situation awareness as it did to the traditional psychology of his time:

> What must be admitted is that the definite images of traditional psychology form but the very smallest part of our minds as they actually live. The traditional psychology talks like one who should say a river consists of nothing but pailsful, spoonsful, quartpotsful, barrelsful and other moulded forms of water. Even were the pails and the pots all actually standing in the stream, still between them the free water would continue to flow. It is just this free water of consciousness that psychologists resolutely overlook. Every definite image in the mind is steeped and dyed in the free water that flows around it. With it goes the sense of its relations, near and remote, the dying echo of whence it came to us, the dawning sense of whither it is to lead. The significance, the value, of the image is all in this halo or penumbra that surrounds and escorts it, or rather that is fused into one with it and has become bone of its bone and flesh of its flesh; leaving it, it is true, an image of the same thing it was before, but making it an image of that thing newly taken and freshly understood (James 1890, p. 255).

The dynamics of experience, of awareness, as captured by William James' metaphor, still represents a significant epistemological problem. Snapshots of short-term memory that are taken in popular situation awareness measurement techniques today are the spoonsful of James' metaphor. These free the researcher to resolutely overlook the context of awareness, of consciousness. James tried to convey how the river of consciousness, of awareness, is complex, active, adaptive, and self-organizing. The separation between observer and observed makes little sense in his metaphor: space and time are intrinsic properties of experience. They are not "out there" in a situation waiting for a mind to become aware of them. The mind is not an observer at all, but rather a central player in, and a creator of, a perceptual cycle (Neisser 1976). The perception of the environment is continually being created by cycles of expectation and action, by the questions asked of it. The dynamic transaction, or conversation, between situation and awareness is the only relevant reality (Flach et al. 2008).

MATTER VERSUS MIND

Discourse about situation awareness is a modern installment of an ancient debate in philosophy and psychology, about the relationship between matter and mind. Surely one of the most vexing problems, the coupling between matter and mind, has occupied centuries of thinkers. How is it that we get data from the outside world inside our minds? What are the processes by which this happens? And how can the products of these processes be so divergent (I see other things than you, or other things than I saw yesterday)? All psychological theories, including those of situation awareness, implicitly or explicitly choose a position relative to the mind–matter problem.

Virtually all theories of situation awareness rely on the idea of correspondence— a match, or correlation, between an external world of stimuli (elements) and an internal world of mental representations (which gives meaning to those stimuli). The relationship between matter and mind, in other words, is one of letting the mind create a mirror, a mental simile, of matter on the outside. This allows a further elaboration of the Venn diagram notation: instead of "ideal" versus "actual" situation awareness, the captions of the circles in the diagram could read "matter" (the large circle) and "mind" (the little circle). Situation awareness is the difference between what is out there in the material world (matter) and what the observer sees or understands (mind). Equation 4.1 can be rewritten once again:

$$\text{Loss of SA} = f(\text{matter} - \text{mind}) \qquad (4.3)$$

Equation 4.3 describes how a loss of situation awareness, or deficient situation awareness, is a function of whatever was in the mind subtracted from what was available matter. The portion of matter that did not make it into mind is "lost"; it is deficient awareness.

GROUND TRUTH AND APERSPECTIVAL OBJECTIVITY

Such thinking is of course profoundly normativist. Normativism is the premise that it is possible to generate a "true" and "objective" characterization of the practitioner's experience. For example, Parasuraman and colleagues have this to say about situation awareness: that "there is a 'ground truth' against which its accuracy can be assessed (e.g., the objective state of the world or the objective unfolding of events that are predicted)" (Parasuraman et al. 2008). This idea, common in much situation awareness work, says that the operator's understanding of the world can be contrasted against (and found more or less deficient relative to) an objectively available state of that world. This rests on the belief that the world is objectively available and apprehensible, that there is such a thing as aperspectival objectivity. It requires that researchers are able to take a "view from nowhere" (Nagel 1992), a value-free, background-free, position-free view that is true (the "ground truth"). With that as the starting point, researchers may learn less about why people saw what they did and why that made sense to them, because what they saw was "wrong" relative to what was decided or predicted to be the "ground truth." This position may also be problematic because it assumes the ability to provide a full accounting of goal-directed actions in the pursuit of task goals. That ability can be claimed only by an omniscient, normative arbiter who knows completely and accurately the values and interdependencies of all contextually dependent variables (Smith and Hancock 1995). Such an arbiter would not only have to have privileged access to the activity of practitioners' minds but also have to have to a complete and infallible view of the unfolding of events (also known as "the big picture"). Such arbiters (e.g., situation awareness researchers) apply norms, implicitly asserting that their view of the world is "correct" (or the ground truth) and their subjects' view is deficient.

Such thinking is also quite purely Cartesian. It separates "mind" from "matter" as if they are distinct entities: a res cogitans and a res extensa. Both exist as separate essentials of our universe, and one serves as the echo or imitation of the other. One problem of such dualism and normativism lies, of course, in the assumptions it makes. An important assumption is what Feyerabend called the "autonomy principle": that facts exist in some objective world, equally accessible to all observers. The autonomy principle is what allows researchers to draw the large circle of the Venn diagram: it consists of matter available to, as well as independent from, any observer, whose awareness of that matter takes the form of an internal simile. This assumption is heavily contested by, for example, radical empiricists. Is matter something "out there," independent of (the minds of) observers, as something that is open to enter the awareness of anyone?

IF YOU LOSE SA, WHAT REPLACES IT?

If you lose situation awareness, what replaces it? No theories of cognition today can easily account for a mental vacuum, for empty-headedness. Rather, people always form an understanding of the situation unfolding around them, even if this understanding can, in hindsight, be shown to have diverged from some "ground truth." This does not mean that this mismatch has any relevance in explaining human performance at the time. For people doing work in a situated context, there seldom is a mismatch, if ever. Performance is driven by the desire to construct plausible, coherent accounts, a "good story" of what is going on. Weick (1995) reminds us that such "sensemaking" is not about accuracy, about achieving an accurate mapping between some objective, outside world and an inner representation of that world. What matters to people is not to produce a precise internalized simile of an outside situation, but to account for their sensory experiences in a way that supports action and goal achievement. This converts the challenge of understanding situation awareness. From studying the mapping accuracy between an external and internal world, it requires the investigation of why people thought they were in the right place, or had the right assessment of the situation around them. What made that so? The adequacy or accuracy of an insider's representation of the situation cannot be called into question: it is what counts for him or her, and it is that which drives further action in that situation. The internal, subjective world is the only one that exists. If there is an objective, external reality, the radical empiricist says, we could not know it.

GETTING LOST AT THE AIRPORT

Now to a simple case, to see how these things play out. Runway incursions (aircraft taxiing onto runways for which they did not have a clearance) are an acute category of such cases in transportation today. Runway incursions are seen as a serious and growing safety problem worldwide, especially at large, controlled airports (where air traffic control organizes and directs traffic movements). Hundreds of incursions occur every year, some leading to fatal accidents. Apart from straying onto a runway without a clearance, the risk of colliding with something else at an airport is considerable. Airports are tight and dynamic concentrations of cars, buses, carts, people, trucks, trucks plus aircraft, and aircraft, all moving at speeds varying from a few knots to hundreds of knots. (And then fog can settle over all of that.) The number of things to hit is much larger on the ground than it is in the air, and the proximity to those things is much closer. And because of the layout of taxiways and ramps, navigating an aircraft across an airport can be considerably more difficult than navigating it in flight.

When runway incursions occur, it can be tempting to blame a "loss of situation awareness." Here is one such case, not a runway incursion, but a taxiway incursion. This case is illustrative not only because it is relatively simple but also because all regulations had been followed in the design and layout of the airport. Safety cases had been conducted for the airport, and it had been certified as compliant with all relevant rules. Incidents in such an otherwise safe system can happen even when everybody follows the rules. This incident happened at Stockholm Arlanda (the international airport) in October 2002. A Boeing 737 had landed on runway 26 (the opposite of runway 08, which can be seen in Figure 4.3) and was directed by air traffic control to taxi to its gate via taxiways ZN and Z (called Zulu November and Zulu, in aviation speak). The reason for taking ZN was that a tow truck with an aircraft was coming from the other direction. It had been cleared to use ZP (Zulu Papa) and then to turn right onto taxiway X (X-ray). But the 737 did not take ZN. To the horror of the tow truck driver, it carried on following ZP instead, almost straight into the tow truck. The pilots saw the truck in time, however, and managed to stop.

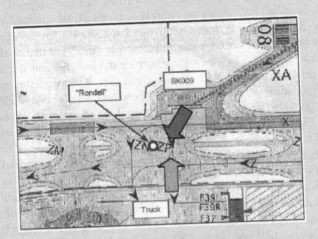

FIGURE 4.3 The site of the near collision between a Boeing 737 and a tow truck. The 737 came off the runway after landing and had been cleared to take ZN. It took ZP instead, almost running into a tow truck (which was pulling another aircraft across ZP). (From SHK, Tillbud mellan flygplanet LN-RPL och en bogsertraktor på Stockholm/Arlanda flygplats, AB län, den 27 oktober 2002 [Rapport RL2003:47] [Incident between aircraft LN-RPL and a tow truck at Stockholm/Arlanda airport, October 27, 2002]. Stockholm, Statens Haverikommission [Swedish Accident Investigation Board], 2003.)

The truck driver had to push his aircraft backward in order to clear up the jam. Did the pilots "lose situation awareness"? Was their situation awareness "deficient"? There were signs pointing out where taxiway ZN ran, and those could be seen from the cockpit. Why did the crew not take these cues into account when coming off the runway?

Such questions consistently pull us toward the position of retrospective outsider, looking down onto the developing situation from a God's-eye point of view. From there, we can see the mismatch grow between where people were and where they thought they were. From there, we can easily draw the circles of the Venn diagram, pointing out a deficiency or a shortcoming in the awareness of the people in question. But none of that explains much. The mystery of the matter–mind problem is not going to go away just because we say that other people did not see what we now know they should have seen. The challenge is to try to understand why the crew of the 737 thought that they were right—that they were doing exactly what air traffic control had told them to do: follow taxiway ZN to Z. The commitment of an anti-dualist position is to try to see the world through the eyes of the protagonists, as there is no other valid perspective. The challenge with navigational incidents, indeed, is not to point out that people were not in the spot they thought they were, but to explain why they thought they were right.

The first clue can be found in the response of the 737 crew after they had been reminded by the tower to follow ZN (they had now stopped, facing the tow truck head-on). "Yeah, it's the chart here that's a little strange," says one of the pilots (SHK 2003, p. 8, translated from Swedish). If there was a mismatch, it was not between the actual world and the crew's model of that world. Rather, there was a mismatch between the chart in the cockpit and the actual airport layout. As can be seen in Figure 4.3, the taxiway layout contained a little island, or roundabout, between taxiways Zulu and X-ray. ZN and ZP were the little bits going between X-ray to Zulu, around the roundabout. However, as is shown in Figure 4.4, the chart available in the cockpit had no little island on it. It showed no roundabout.

Even here, no rules had been broken. The airport layout had recently changed (with the addition of the little roundabout) in connection with the construction of a new terminal pier. It takes time for the various charts to be updated, and this simply had not happened yet at the company of the crew in question. Still, how could the crew of the 737 have ended up on the wrong side of that area (ZP instead of ZN), whether there was an island shown on their charts or not? Figure 4.5 contains more clues. It shows the roundabout from the height of a car (which is lower than a 737 cockpit, but from the same direction as an aircraft coming off of runway 26). The crew in question went left of the roundabout, where it should have gone right. The roundabout is covered in snow, which makes it inseparable from the other (real) islands separating taxiways Zulu and X-ray. These other islands consist of grass,

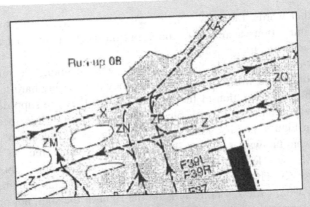

FIGURE 4.4 The chart available in the Boeing 737 cockpit at the time. It shows no little roundabout, or island, between taxiways ZN and ZP. (From SHK, Tillbud mellan flygplanet LN-RPL och en bogsertraktor på Stockholm/Arlanda flygplats, AB län, den 27 oktober 2002 [Rapport RL2003:47] [Incident between aircraft LN-RPL and a tow truck at Stockholm/Arlanda airport, October 27, 2002]. Stockholm, Statens Haverikommission [Swedish Accident Investigation Board], 2003.)

FIGURE 4.5 The roundabout as seen coming off runway 26. The Boeing 737 went left around the roundabout, instead of right. (From SHK, Tillbud mellan flygplanet LN-RPL och en bogsertraktor på Stockholm/Arlanda flygplats, AB län, den 27 oktober 2002 [Rapport RL2003:47] [Incident between aircraft LN-RPL and a tow truck at Stockholm/Arlanda airport, October 27, 2002]. Stockholm, Statens Haverikommission [Swedish Accident Investigation Board], 2003.)

whereas the roundabout, with a diameter of about 20 meters, is the same tarmac as that of the taxiways. Shuffle snow onto all of them, however, and they look indistinguishable. The roundabout is no longer a circle painted on the tarmac: it is an island like all others. Without a roundabout on the cockpit chart, there is only one plausible explanation for what the island in Figure 4.3 ahead of the aircraft is: it must be the grassy island to the right of taxiway

ZN. In other words, the crew "knew" where they were, on the basis of the cues and indications available to them and on the basis of what these cues plausibly added up to.

The signage does not help either. Taxiway signs are among the most confusing directors in the world of aviation, and they are terribly hard to turn into a reasonable representation of the taxiway system they are supposed to help people navigate on. The sign visible from the direction of the Boeing 737 is enlarged in Figure 4.6. The black part of the sign is the "position" part (this indicates what taxiway it is), and the yellow part is the "direction" part: this taxiway (ZN) will lead to taxiway Z, which happens to run at about a right angle across ZN. These signs are placed to the left of the taxiway they belong to. In other words, the ZN taxiway is on the right of the sign, not on the left. But put the sign in the context of Figure 4.5, and things become more ambiguous. The black ZN part is now leaning toward the left side of the roundabout, not the right side. Yet the ZN part belongs to the piece of tarmac on the right of the roundabout (which the crew never saw as such. For them, given their chart, the roundabout was the island to the right of ZN).

Why not swap the black ZN and the yellow Z part? Current rules for taxiway signage will not allow it (rules can indeed stifle innovation and investments in safety). And not all airports comply with this religiously either. There may be exceptions when there is no room or when visibility of the sign would be obstructed if placed on the left side. To make things worse, regulations state that taxiway signs leading to a runway need to be placed on both sides of the taxiway. In those cases, the black parts of the signs are often actually adjacent to the taxiway, and not removed from it, as in Figure 4.5. Against the background of such ambiguity, few pilots actually know or remember that taxiway signs are supposed to be on the left side of the taxiway they belong to. In fact, very little time in pilot training is used to get pilots to learn to navigate around airports, if any. It is a peripheral activity, a small portion of mundane, pedestrian work, that merely leads up to, and concludes the real work: flying from A to B. When rolling off the runway, and going to taxiway Zulu and then

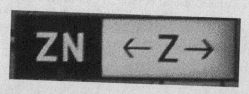

FIGURE 4.6 The sign on the roundabout that is visible for aircraft coming off runway 26. (From SHK, Tillbud mellan flygplanet LN-RPL och en bogsertraktor på Stockholm/Arlanda flygplats, AB län, den 27 oktober 2002 [Rapport RL2003:47] [Incident between aircraft LN-RPL and a tow truck at Stockholm/Arlanda airport, October 27, 2002]. Stockholm, Statens Haverikommission [Swedish Accident Investigation Board], 2003.)

to their gate, this crew "knew" where they were. Their indications (cockpit chart, snow-covered island, taxiway sign) compiled into a plausible story: ZN, their assigned route, was the one to the left of the island, and that was the one they were going to take. Until they ran into a tow truck. But nobody in this case "lost situation awareness." The pilots "lost" nothing. On the basis of the combination of cues and indications observable by them at the time, they had a plausible story of where they were. Even if a mismatch can be shown between how the pilots saw their situation and how retrospective, outside observers now see that situation, this has no bearing on understanding how the pilots made sense of their world at the time.

Seeing situation awareness as a measure of the accuracy of correspondence between some outer world and an inner representation carries with it a number of irresolvable problems that have always been connected to such a dualist position. Taking the mind–matter problem apart by separating the two means that the theory needs to connect the two again. Theories of situation awareness typically rely on a combination of two schools in psychological thought to reconstitute this tie, to make this bridge. One is empiricism, a traditional school in psychology that makes claims on how knowledge is chiefly, if not uniquely, based on experience. The second is the information processing school in cognitive psychology, still popular in large areas of human factors. None of these systems of thought, however, are particularly successful in solving the really hard questions about situation awareness, and may in fact be misleading in certain respects. We will look at both of them in turn here. Once that is done, the chapter will briefly develop the counter-position on the mind–matter question: an anti-dualist one (as related to situation awareness). This position will be worked out further in the rest of this chapter, using the *Royal Majesty* case as example.

EMPIRICISM AND THE PERCEPTION OF ELEMENTS

Most theories of situation awareness actually leave the processes by which matter makes it into mind to the imagination. A common denominator, however, appears to be the perception of "elements" in the environment. Elements are the starting point of perceptual and meaning-making processes. It is on the basis of these elements that we gradually build up an understanding of the situation, by processing such elementary stimulus information through multiple stages of consciousness or awareness ("levels of SA"). Theories of situation awareness borrow from empiricism (particularly British empiricism), which assumes that the organized character and the meaningfulness of our perceptual world are achieved by matching incoming stimuli with prior experience through a process called association. In other words, the world as we experience it is disjointed (consisting of "elements") except when mediated by previously stored knowledge. Correspondence between mind and matter is made by linking incoming impressions through earlier associations.

Empiricism in its pure form is nothing more than saying that the major source of knowledge is experience; that we do not know about the world except through making contact with that world with our sensory organs. Among Greek philosophers around five centuries BC, empiricism was accepted as a guide to epistemology, as a way of understanding the origin of knowledge. Questions already arose, however, on whether all psychic life could be reduced to sensations. Did the mind have a role to play at all in turning perceptual impressions into meaningful percepts? The studies of perception by Johannes Kepler (1571–1630) would come to suggest that the mind had a major role, even though he himself left the implications of his findings up to other theoreticians. Studying the eyeball, Kepler found that it actually projects an inverted image on the retina at the back. Descartes, himself dissecting the eye of a bull to see what image it would produce, saw the same thing. If the eye inverts the world, how can we see it the right way up? There was no choice but to appeal to mental processing. Not only is the image inverted, it is also two-dimensional, and it is cast onto the backs of two eyes, not one. How does all that get reconciled in a single coherent, upright percept? The experiments boosted the notion of impoverished, meaning-deprived stimuli entering our sensory organs, in need of some serious mental processing work from there on. Further credence to the perception of "elements" was given by the 19th century discovery of photoreceptors in the human eye. This mosaic of retinal receptors appeared to chunk up any visual percept coming into the eyeball. The resulting fragmented neural signals had to be sent up the perceptual pathway in the brain for further processing and scene restoration.

British empiricists such as John Locke (1632–1704) and George Berkeley (1685–1753), though not privy to 19th century findings, were confronted with the same epistemological problems that their Greek predecessors had struggled with. Rather than knowledge being innate, or the chief result of reasoning (as claimed by rationalists of that time), what role did experience have in creating knowledge? Berkeley, for example, wrestled with the problem of depth perception (not a negligible problem when it comes to situation awareness). How do we know where we are in space, in relation to objects around us? Distance perception, to Berkeley, though created through experience, was itself not an immediate experience. Rather, distance and depth are additional aspects of visual data that we learn about through combinations of visual, auditory, and tactile experiences. We understand distance and depth in current scenes by associating incoming visual data with these earlier experiences. Berkeley reduced the problem of space perception to more primitive psychological experiences; decomposing the perception of distance and magnitude into constituent perceptual elements and processes (e.g., lenticular accommodation, blurring of focus). Such deconstruction of complex, intertwined psychological processes into elementary stimuli turned out to be a useful tactic. It encouraged many after him, among them Wilhelm Wundt and latter-day situation awareness theorists, to analyze other experiences in terms of elements as well.

Interestingly, neither all prehistoric empiricists nor all British empiricists could be called dualists in the same way that situation awareness theorists can be. Protagoras, a contemporary of Plato around 430 BC already said that "man is the measure of all things." An individual's perception is true to him, and cannot be proven untrue (or inferior or superior) by some other individual. Today's theories of situation awareness, with their emphasis on the accuracy of the mapping between matter and mind, are very much into inferiority and superiority ("deficient SA" vs. "good SA"), as something that can be objectively judged. This would not have worked for some of the British empiricists either. To Berkeley, who disagreed with earlier characterizations of an inner versus an outer world, people can actually never know anything but our experiences. The world is a plausible but unproved hypothesis. In fact, it is a fundamentally untestable hypothesis, since we can only know our own experience. Like Protagoras before him, Berkeley would not have put much stock in claims of the possibility of superior or "ideal" situation awareness, as such a thing is logically impossible. There are no superlatives when it comes to knowledge through experience. For Berkeley too, this meant that even if there is an objective world out there (the large circle in the Venn diagram), we could never know it. It also meant that any characterization of such an objective world with the aim of understanding somebody's perception, somebody's situation awareness, would have been nonsense.

EXPERIMENTS, EMPIRICISM, AND SITUATION AWARENESS

Wilhelm Wundt is credited with founding the first psychological laboratory in the world at the University of Leipzig in the late 1870s. The aim of his laboratory was to study mental functioning by deconstructing it into separate elements. These could then be combined to understand perceptions, ideas, and other associations. Wundt's argument was simple and compelling, and versions of it are still used in psychological method debates today. While the empirical method had been developing all around psychology, it was still occupied with grand questions of consciousness, soul, destiny, and it tried to gain access to these issues through introspection and rationalism. Wundt argued that these were questions that should perhaps be asked at the logical end point of psychology, but not at the beginning. Psychology should learn to crawl before it tries to walk. This justified the appeal of the elementarist approach: chopping the mind and its stimuli up into minute components, and studying them one by one. But how to study them?

Centuries before, Descartes had argued that mind and matter were not only entirely separate, but should also be studied using different methods as well. Matter should be investigated using methods from natural science (i.e., the experiment), while mind should be examined through processes of meditation, or introspection. Wundt did both. In fact, he combined the natural science tradition with the introspective one, molding them into a novel brand of psychological experimentation that still governs much of human factors research to this day. Relying on complicated sets of stimuli, Wundt investigated sensation and perception, attention, feeling, and association. Using intricate measurements of reaction times, the Leipzig laboratory hoped they would one day be able to achieve a chronometry of mind (which was not long thereafter dismissed as infeasible).

Rather than just counting on quantitative experimental outcomes, Wundt asked his subjects to engage in introspection, to reflect on what had happened inside their minds during the trials. Wundt's introspection was significantly more evolved and demanding than the experimental "report" psychologists ask their subjects for today. Introspection was a skill that required serious preparation and expertise, because the criteria for gaining successful access to the elementary makeup of mind were set very high. As a result, Wundt mostly used his assistants. Realizing that the contents of awareness are in constant flux, Wundt produced rigid rules for the proper application of introspection: (1) the observer, if at all possible, must be in a position to determine when the process is to be introduced; (2) he must be in a state of "strained attention"; (3) the observation must be capable of being repeated several times; (4) the conditions of the experiment must be such that they are capable of variation through introduction or elimination of certain stimuli and through variation of the strength and quality of the stimuli.

Wundt thus imposed experimental rigor and control on introspection. Similar introspective rigor, though different in some details and prescriptions, is applied in various methods for studying situation awareness today. Some techniques involve "blanking" or freezing of displays, with researchers then going in to elicit what participants remember about the scene. This requires active introspection. Wundt would have been fascinated, and he probably would have had a thing or two to say about the experimental protocol. If subjects are not allowed to say when the blanking or freezing is to be introduced, for example (Wundt's first rule), how does that compromise their ability to introspect? In fact, the blanking of displays and handing out a situation awareness questionnaire is more akin to the Würzburg school of experimental psychology that started to compete with Wundt in the late 19th century. The Würzburgers pursued "systematic experimental introspection" by having subjects pursue complex tasks that involved thinking, judging, and remembering. They would then have their subjects render a retrospective report of their experiences during the original operation. The whole experience had to be described time period by time period, thus chunking it up. In contrast to Wundt, and like situation awareness research participants today, Würzburg subjects did not know in advance what they were going to have to introspect.

Others today disagree with Descartes' original exhortation and remain fearful of the subjectivist nature of introspection. They favor the use of clever scenarios in which the outcome, or behavioral performance of people, will reveal what they understood the situation to be. This is claimed to be more of a natural science approach that stays away from the need to "introspect." It relies on objective performance indicators instead. Such an approach to studying situation awareness could be construed as neo-behaviorist, as it equates the study of behavior with the study of consciousness. Mental states are not themselves the object of investigation: performance is. If desired, such performance can faintly hint at the contents of mind (situation awareness). But that itself is not the aim; it cannot be, because through such pursuits psychology (and human factors) would descend into subjectivism and ridicule. Watson, the great proponent of behaviorism, would himself have argued along these lines. Additional arguments in favor of performance-oriented approaches include the assertion that introspection cannot possibly test the contents of awareness, as it necessarily appeals to a situation or stimulus from the past. The situation on which people are asked to reflect has already disappeared. Introspection thus simply probes people's memory. Indeed, if you want to study situation awareness, how can you take away the "situation" by blanking or freezing their world, and still hope you have relevant "awareness" left to investigate by introspection? Wundt, as well as many of today's situation awareness researchers, may in part have been studying memory, rather than the contents of consciousness.

Wundt was, and still remains, one of the chief representatives of the elementarist orientation, pioneered by Berkeley centuries before and perpetuated in modern theories of situation awareness. But if we perceive "elements," if the eyeball deals in two-dimensional, fragmented, inverted, meaning-deprived stimuli, then how does order in our perceptual experience come about? What theory can account for our ability to see coherent scenes, objects? The empiricist answer of association is one way of achieving such order, of creating such inter-elementary connections and meaning. Order is an end product, it is the output of mental or cognitive work. This is also the essence of information processing, the school of thought in cognitive psychology that accompanied and all but colonized human factors since its inception during the closing days of the Second World War. Meaning and perceptual order are the end result of an internal trade in representations; representations that get increasingly filled out and meaningful as a result of processing in the mind.

INFORMATION PROCESSING

Information processing did not follow neatly on empiricism, nor did it accompany the initial surge in psychological experimentation. Wundt's introspection did not immediately fuel the development of theoretical substance to fill the gap between elementary matter and the mind's perception of it. Rather, it triggered an anti-subjectivist response that would ban the study of mind and mental processes for decades to come, especially in North America. John Watson, a young psychologist, introduced psychology to the idea of behaviorism in 1913 and aimed to conquer psychology as a purely objective branch of natural science. Introspection was to be disqualified, and any references to, or investigations of, consciousness were proscribed. The introspective method was seen as unreliable and unscientific, and psychologists had to turn their focus exclusively to phenomena that could be registered and described objectively by independent observers. This meant that introspection had to be replaced by tightly controlled experiments that varied subtle combinations of rewards and punishments in order to bait organisms (anything from mice to pigeons to humans) into particular behaviors over others. The outcome of such experiments was there for all to see, with no need for introspection.

Behaviorism became an early 20th century embodiment of the Baconian ideal of universal control, this time reflected in a late Industrial Revolution obsession with manipulative technology and domination. It appealed enormously to an optimistic, pragmatic, rapidly developing, and result-oriented North America. Laws extracted from simple experimental settings were thought to carry over to more complex settings and to more experiential phenomena as well, including imagery, thinking, and emotions. Behaviorism was thus fundamentally nomothetic: deriving general laws thought to be applicable across people and settings. All human expressions, including art and religion, were reduced to no more than conditioned responses. Behaviorism turned psychology into something wonderfully Newtonian: a schedule of stimuli and responses, of mechanistic, predictable, and changeable couplings between inputs and outputs. The only legitimate characterization of psychology and mental life was one that conformed to the Newtonian framework of classical physics and abided by its laws of action and reaction.

Then, as related in the first chapter, came the Second World War, and the behaviorist bubble was deflated. No matter how clever a system of rewards and punishments psychologists set up, radar operators monitoring for German aircraft intruding into Britain across the Channel would still lose their vigilance over time. They would still have difficulty distinguishing "signals" from "noise," independent of the possible penalties. Pilots would get controls mixed up, and radio operators were evidently limited in their ability to hold information in their heads while getting ready for the next transmission. Where was behaviorism? It could not answer to the new pragmatic appeals. Thus came the first cognitive revolution. The cognitive revolution reintroduced "mind" as a legitimate object of study. Rather than manipulating the effect of stimuli on overt responses, it concerned itself with "meaning" as the central concept of psychology. Its aims were, as Bruner (1990) recalls, to discover and describe meanings that people created out of their encounters with the world and then to propose hypotheses for what meaning-making processes were involved. The very metaphors, however, that legitimized the reintroduction of the study of mind also began to immediately corrupt it. The first cognitive revolution fragmented and became unduly technicalized.

The radio and the computer, two technologies accelerated by developments during the Second World War, quickly captured the imagination of those once again studying mental processes. These were formidable similes of mind, able to mechanistically fill the black box (which behaviorism had kept shut) between stimulus and response. The innards of a radio showed filters, channels, and limited capacities through which information flowed. Not much later, all those words appeared in cognitive psychology. Now the mind had filters, channels, and limited capacities too. The computer was even better, containing a working memory, a long-term memory, various forms of storage, input and output, and decision modules. It did not take long for these terms, too, to appear in the psychological lexicon. What seemed to matter most was the ability to quantify and compute mental functioning. Information theory, for example, could explain how elementary stimuli (bits) would flow through processing channels to produce responses. A processed stimulus was deemed informative if it reduced alternative choices, no matter whether the stimulus had to do with Faust or a digit from a statistical table.

As Bruner recollects, computability became the necessary and sufficient criterion for cognitive theories. Mind was equated to program. Through these metaphors, the "construction of meaning" quickly became the "processing of information." Newton and Descartes simply would not let go. Once again, psychology was reduced to mechanical components and linkages, and exchanges of energy between and among them. Testing the various components (sensory store, memory, decision making) in endless series of fractionalized laboratory experiments, psychologists hoped, and many are still hoping, that more of the same will eventually add up to something different, that profound insight into the workings of the whole will magically emerge from the study of constituent components.

MECHANIZATION OF MIND

Information processing has been a deeply Newtonian–Cartesian answer to the mind–matter problem. It is the ultimate mechanization of mind. The basic idea is the human (mind) is an information processor that takes in stimuli from the outside world and gradually makes sense of those stimuli by combining them with things already stored in the mind. For example, I see the features of a face, but through coupling them to what I have in long-term memory, I recognize the face as that of my youngest son. Information processing is loyal to the biological psychological model that sees the matter–mind connection as a physiologically identifiable flow of neuronal energy from periphery to center (from eyeball to cortex), along various nerve pathways. The information processing pathway of typical models mimics this flow, by taking a stimulus and pushing it through various stages of processing, adding more meaning the entire time. An appropriate response can be generated (through a backflow from center to periphery; brain to limbs) once the processing system has understood what the stimulus means (or stimuli mean), which in turn creates more (new) stimuli to process.

The Newtonian as well as dualist intimations of the information processing model are a heartening sight for those with Cartesian anxiety. Thanks to the biological model underneath it, the mind–matter problem is one of a Newtonian transfer (conversion as well as conservation) of energy at all kinds of levels (from photonic energy to nerve impulses, from chemical releases to electrical stimulation, from stimulus to response at the overall organismic level). Both Descartes and Newton can be recognized in the componential explanation of mental functioning (memory, for example, is typically parsed up in iconic memory, short-term memory, and long-term memory): end products of mental processing can be exhaustively explained on the basis of interactions between these and other components. And finally of course, information processing is mentalist: it neatly separates res cogitans from res extensa by studying what happens in the mind entirely separately from what happens in the world. The world is a mere adjunct, truly a res extensa, employed solely to lob the next stimulus at the mind (which is where the really interesting processes take place).

The information processing model might work marvelously for the simple laboratory experiments that brought it to life. Laboratory studies of perception and decision making and reaction time reduce stimuli to single snapshots, fired off at the human processing mechanism as one-stop triggers. The Wundtian idea of awareness as a continually flowing phenomenon is artificially reduced, chunked up, and frozen by the very stimuli that subjects are to become aware of. Such dehumanization of the settings in which perception takes place, as well as of the models by which such perception comes about, has given rise to considerable critique. If people are seen to be adding meaning to impoverished, elementary stimuli, then this is because they are given impoverished, elementary, meaningless stimuli in their laboratory tasks! None of that says anything about natural perception or the processes by which people perceive or construct meaning in actual settings. The information processing model may be true (though even that is judged as unlikely by most), but only for the constrained, spartan laboratory settings that keep cognition in captivity. If people are seen to struggle in their interpretation of elements, then this may have something to do with the elementary stimuli given to them.

GESTALT

Even Wundt was not without detractors in this respect. The Gestalt movement was launched in part as a response or protest to Wundtian elementarism. Gestaltists claimed that we actually perceive meaningful wholes; that we immediately experience those wholes. We cannot help but to see these patterns, these wholes. Max Wertheimer (1880–1934), one of the founding fathers of Gestaltism, illustrates this as such:

> I am standing at the window and see a house, trees, sky. And now, for theoretical purposes, I could try to count and say: there are... 327 nuances of brightness [and hue]. Do I see "327"? No; I see sky, house, trees (Wertheimer 1923/1950; translated from the original by Sarter [Woods et al. 2010]).

The gestalts that Wertheimer sees (house, trees, sky) are primary to their parts (their elements), and they are more than the sum of their parts. There is an immediate orderliness in experiencing the world. Wertheimer inverts the empiricist claim and information processing assumption: rather than meaning being the result of mental operations on elementary stimuli, it actually takes painstaking mental effort (counting 327 nuances of brightness and hue) to reduce primary sensations to their primitive elements. We do not perceive elements: we perceive meaning. Meaning comes effortlessly, prerationally. In contrast, it takes cognitive work to see elements. In the words of William James' senior Harvard colleague Chauncey Wright, there is no antecedent chaos that requires some intrapsychic glue to prevent percepts from falling apart.

MEANING-MAKING

Empiricism does not recognize the immediate orderliness of experience because it does not see relations as real aspects of immediate experience (Heft 2001). Relations, according to empiricists, are a product of mental (information) processing. This is true for theories of situation awareness. For them, relations between elements are mental artifacts. They get imposed through stages of processing. Subsequent "levels of SA" add relationships to elements by linking those elements to current "meanings" and future projections. The problem of the relationship between matter and mind is not at all solved through empiricist responses. But perhaps engineers and designers, as well as many experimental psychologists, are happy to hear about "elements" (or 327 nuances of brightness and hue). For those can be manipulated in a design prototype and experimentally tested on subjects. Wundt would have done the same thing. Not unlike Wundt 100 years before him, Ulrich Neisser warned in 1976 that psychology was not quite ready for grand questions about consciousness. Neisser feared that models of cognition would treat consciousness as if it were just a particular stage of processing in a mechanical flow of information. His fears were justified in the mid-1970s, as many psychological models did exactly that. Now they have done it again. Awareness, or consciousness, is equated to a stage of processing along an intrapsychic pathway ("levels of SA"). As Neisser points out, this is an old idea in psychology. The three levels of SA in vogue today were anticipated by Freud, who even supplied flow charts and boxes in his *Interpretation of Dreams* to map the movements from unconscious (level 1) to preconscious (level 2) to conscious (level 3). The popularity of finding a home, a place, a structure for consciousness in the head is irresistible, says Neisser, as it allows psychology to nail down its most elusive target (consciousness) to a box in a flow chart. There is a huge cost, though. Along with the deconstruction and mechanization of mental phenomena comes their dehumanization.

SECOND COGNITIVE REVOLUTION

Information processing theories have lost much of their appeal and credibility. Researchers in safety and human factors realize how they might have corrupted the spirit of the post-behaviorist cognitive revolution by losing sight of humanity and meaning-making. Empiricism (or British empiricism) has slipped into history as another school of thought at the beginning of psychological theorizing. But both form apparently legitimate offshoots in current understandings of SA. Notions similar to those of empiricism and information processing are reinvented under new guises, which reintroduces the same type of foundational problems, while leaving some of the really hard problems unaddressed. The problem of the nature of stimuli is one of those, and associated with it is the problem of meaning-making. How does the mind "make sense" of those stimuli? Is meaning the end product of a processing pathway that flows from periphery to center? These are enormous problems in the history of psychology, all of them problems of the relationship between mind and matter, and all essentially still unresolved. Perhaps they are fundamentally unsolvable within the dualist tradition that psychology has inherited from Descartes and Newton.

Some movements in human factors are increasingly pulling away from the experimental psychological domination. The idea of distributed cognition has renewed the status of the environment as active, constituent participant in cognitive processes, closing the gap between res cogitans and res extensa. Other people, artifacts and even body parts are in fact all part of the res cogitans. How is it otherwise that a child learns to count on his hand? Or a soccer player "thinks with her feet"? Concomitant interest in cognitive work analysis and cognitive systems engineering see such joint cognitive systems as units of analysis, not the constituent human or machine components. Qualitative methods such as ethnography are increasingly legitimate and critical for understanding distributed cognitive systems. These movements have triggered and embodied what has now become known as the second cognitive revolution, recapturing and rehabilitating the impulses that brought to life the first. How do people make meaning? In order to begin to answer such aboriginal questions, it is now increasingly held as justifiable and necessary to throw the human factors net wider than experimental psychology. Other forms of social inquiry can shed more light on how we are goal-driven creatures in actual, dynamic environments, not passive recipients of snapshot stimuli in a sterile laboratory.

The concerns of these thinkers overlap with functionalist approaches, which formed yet another psychology of protest against Wundt's elementarism. The same protest works equally well against the mechanization of mind by the information processing school. A century ago, functionalists (William James was one of their foremost exponents) pointed out how people are integrated, living organisms engaged in goal-directed activities, not passive element-processors locked into laboratory headrests, buffeted about by one-shot stimuli from an experimental apparatus. The environment in which real activities play out helps shape the organism's responses. Psychological functioning is adaptive: it helps the organism survive and adjust, by incrementally modifying and tweaking its composition or its behavior to generate greater gains on whatever dimension is relevant. Such ecological thinking is now even beginning to seep into approaches to safety, which has so far also been dominated by mechanistic, structuralist models. James was not just a functionalist, however. In fact, he was one of the most all-round psychologists ever. His views on radical empiricism are one great way to access novel thinking about situation awareness and sensemaking, and only appropriate against a background of increasing interest in the role of ecological psychology in human factors.

RADICAL EMPIRICISM

REJECTING DUALISM

Radical empiricism is one way of circumventing the insurmountable problems associated with psychologies based on dualistic traditions, and William James introduced it as such at the beginning of the 20th century. Radical empiricism rejects the notion of separate mental and material worlds; it rejects dualism. James adheres to an empiricist philosophy, which holds that our knowledge comes (largely) from our discoveries, our experience. But, as Heft (2001) points out, James' philosophy is radically empiricist. What is experienced, according to James, is not elements, but relations—meaningful relations. Experienceable relations are what perception is made up of. Such a position can account for the orderliness of experience, as it does not rely on subsequent, or a posteriori mental processing. Orderliness is an aspect of the ecology, of our world as we experience it and act in it. The world as an ordered, structured universe is experienced, not constructed through mental work. James deals with the matter–mind problem by letting the knower and the known coincide during the moment of perception (which itself is a constant, uninterrupted flow, rather than a moment). Ontologies (our being in the world) are characterized by continual transactions between knower and known. Order is not imposed on experience but is itself experienced.

Variations of this approach have always represented a popular countermove in the history of psychology of consciousness. Rather than containing consciousness in a box in the head, it is seen as an aspect of activity. Weick (1995) uses the term *enactment* to indicate how people produce the environment they face and are aware of. By acting in the world, people continually create environments that in turn constrain their interpretations and consequently constrain their next possible actions. This cyclical, ongoing nature of cognition and sensemaking has been recognized by many (see Neisser 1976) and challenges common interpretations rooted in information processing psychology where stimuli precede meaning-making and (only then) action, and where frozen snapshots of environmental status can be taken as legitimate "input" to the human processing system. Instead, activities of individuals are only partially triggered by stimuli, because the stimulus itself is produced by activity of the individual.

This moved Weick (1995) to comment that sensemaking never starts; that people always are in the middle of things. While we may look back on our own experience as consisting of discrete "events," the only way to get this impression is to step out of that stream of experience and look down on it from a position of outsider, or retrospective outsider. It is only possible, really, to pay direct attention to what already exists (that which has already passed). "Whatever is now, at the present moment, under way will determine the meaning of whatever has just occurred" (Weick 1995, p. 27). Situation awareness is in part about constructing a plausible story of the process by which an outcome came about, and the reconstruction of immediate history probably plays a dominant role in this. Few theories of situation awareness acknowledge this role, actually, instead of directing their analytics to the creation of meaning from elements and the future projection of that meaning.

Radical empiricism does not take the stimulus as its starting point, as does information processing, and neither does it need a posteriori processes (mental, representational) to impose orderliness on sensory impressions. We already experience orderliness and relationships through ongoing, goal-oriented transactions of acting and perceiving. Indeed, what we experience during perception is not some cognitive end product "in the head." Neisser reminded us of this long-standing issue in 1976 too: can it be true that we see our own retinal images? The theoretical distance that then needs to be bridged is too large. For if we see that retinal image, who does the looking? Homunculus explanations were unavoidable (and often still are in information processing). Homonculi do not solve the problem of awareness, they simply relocate it. Rather than a little man in our heads looking at what we are looking at, we ourselves are aware of the world, and its structure, in the world. As Edwin Holt put it, "Consciousness, whenever it is localized at all in space, is not in the skull, but is 'out there' precisely where it appears to be" (Heft 2001, p. 59). James, and the entire ecological school after him, anticipated this. What is perceived, according to James, is not a replica, not a simile of something out there. What is perceived is already out there. There are no intermediaries between perceiver and percept; perceiving is direct. This position forms the groundwork of ecological approaches in psychology and human factors.

If there is no separation between matter and mind, then there is no gap that needs bridging; there is no need for reconstructive processes in the mind that make sense of elementary stimuli. The Venn diagram with a little and a larger circle that depict "actual" and "ideal SA" is superfluous too. Radical empiricism allows human factors to stick closer to the anthropologist's ideal of describing and capturing insider accounts. If there is no separation between mind and matter, between "actual" and "ideal SA," then there is no risk of getting trapped in judging performance by use of extrogenous criteria; criteria imported from outside the setting (informed by hindsight or some other source of omniscience about the situation which opens up that delta, or gap, between what the observer inside the situation knew and what the researcher knows). What the observer inside the situation knows must be seen as canonical—it must be understood not in relation to some normative ideal. For the radical empiricist, there would not be two circles in the Venn diagram, but rather different rationalities, different understandings of the situation—none of them right or wrong or necessarily better or worse, but all of them coupled directly to the interests, expectations, knowledge, and goals of the respective observer.

DRIFTING OFF TRACK: REVISITING A CASE OF "LOST SITUATION AWARENESS"

Let us go back to the *Royal Majesty*. Traditionalist ideas about a lack of correspondence between a material and a mental world get a boost from this sort of case. A crew ended up 17 miles off track, after a day and a half of sailing. How could this happen? As said before, hindsight makes it easy to see where people were, versus where they thought they were. In hindsight, it is easy to point to the cues and indications that these people should have picked up in order to update or correct or even form their understanding of the unfolding situation around them. Hindsight has a way of exposing those "elements" that people missed and a way of amplifying or exaggerating their importance. The key question is not why people did not see what we now know was important. The key question is how they made sense of the situation the way they did. What must the crew in question at the time have seen? How could they, on the basis of their experiences, construct a story that was coherent and plausible? What were the processes by which they became sure that they were right about their position? Let us not question the accuracy of the insider view. Research into situation awareness already does enough of that. Instead, let us try to understand why that insider view was plausible for people at the time; why it was, in fact, the only possible view.

DEPARTURE FROM BERMUDA

The *Royal Majesty* departed Bermuda, bound for Boston at 12:00 noon on June 9, 1995. The visibility was good, the winds light, and the sea calm. Before departure, the navigator checked the navigation and communication equipment. He found it in "perfect operating condition." About half an hour after departure, the harbor pilot disembarked and the course was set toward Boston. Just before 13:00, there was a cutoff in the signal from the GPS (Global Positioning System) antenna, routed on the fly bridge (the roof of the bridge), to the receiver—leaving the receiver without satellite signals. Post-accident examination showed that the antenna cable had separated from the antenna connection. When it lost satellite reception, the GPS promptly defaulted to dead reckoning (DR) mode. It sounded a brief aural alarm and displayed two codes on its tiny display: DR and SOL. These alarms and codes were not noticed. (DR means that the position is estimated, or deduced, hence "ded," or now "dead," reckoning. SOL means that satellite positions cannot be calculated.) The ship's autopilot would stay in DR mode for the remainder of the journey.

Why was there a DR mode in the GPS in the first place, and why was a default to that mode neither remarkable, nor displayed in a more prominent way on the bridge? When this particular GPS receiver was manufactured (during the 1980s), the GPS satellite system was not as reliable as it is today. The receiver could, when satellite data were unreliable, temporarily use a DR mode in which it estimated positions using an initial position, the gyrocompass for course input and a log for speed input. The GPS thus had two modes, "normal" and DR. It switched autonomously between the two depending on the accessibility of satellite signals.

By 1995, however, GPS satellite coverage was pretty much complete and had been working well for years. The crew did not expect anything out of the ordinary. The GPS antenna was moved in February, since parts of the superstructure occasionally would block the incoming signals, which caused temporary and short (a few minutes, according to the captain) periods of DR navigation. This was to a great extent remedied by the antenna move, as the Cruise Line's electronics technician testified. People on the bridge had come to rely on GPS position data and considered other systems to be backup systems. The only times the GPS positions could not be counted on for accuracy were during these brief, normal episodes of signal blockage. Thus, the whole bridge crew was "aware" of the DR mode option and how it worked, but none of them ever imagined or were prepared for a sustained loss of satellite data caused by a cable break—no previous loss of satellite data had ever been so swift and so absolute and so long lasting.

When the GPS switched from normal to DR on this journey in June 1995, an aural alarm sounded and a tiny visual mode annunciation appeared on the display. The aural alarm sounded like that of a digital wristwatch and was less than a second long. The time of the mode change was a busy time (shortly after departure), with multiple tasks and distractors competing for the crew's attention. A departure involves complex maneuvering; there are several crew members on the bridge and there is a great deal of communication. When a pilot disembarks, the operation is highly time-constrained and risky. In such situations, the aural signal could easily have been drowned out. No one was expecting a reversion to DR mode, and thus the visual indications were not seen either. From an insider perspective, there was no alarm, as there was not going to be a mode default. There was neither a history, nor an expectation of its occurrence.

Yet even if the initial alarm was missed, the mode indication was continuously available on the little GPS display. None of the bridge crew saw it, according to their testimonies. If they had seen it, they knew what it meant, literally translated—dead reckoning means no satellite fixes. But as we have seen before, there is a crucial difference between data that in hindsight can be shown to have been available and data that were observable at the time. The indications on the little display (DR and SOL) were placed between two rows of numbers (representing the ship's latitude and longitude) and were about one sixth the size of those numbers. There was no difference in the size and character of the position indications after the switch to DR. The size of the display screen was about 7.5 by 9 centimeters, and the receiver was placed at the aft part of the bridge on a chart table, behind a curtain. The location is reasonable, since it places the GPS, which supplies raw position data, next to the chart, which is normally placed on the chart table. Only in combination with a chart does the GPS data make sense, and furthermore the data were forwarded to the integrated bridge system and displayed there (quite a bit more prominent) as well.

For the crew of the *Royal Majesty*, this meant that they would have to leave the forward console, actively look at the display, and expect to see more than large digits representing the latitude/longitude. Even then, if they had seen the two-letter code and translated it into the expected behavior of the ship, it is not a certainty that the immediate conclusion would have been "this ship is not heading toward Boston anymore," since temporary DR reversions in the past had never led to such dramatic departures from the planned route. When the officers did leave the forward console to plot a position on the chart, they looked at the display and saw a position, and nothing but a position, because that is what they were expecting to see. It is not a question of them not attending to the indications. They were attending to the indications, the position indications, since plotting the position is the professional thing to do. For them, the mode change did not exist.

But if the mode change was so nonobservable on the GPS display, why was it not shown more clearly somewhere else? How could one small failure have such an effect—were there no backup systems? The *Royal Majesty* had a modern integrated bridge system, of which the main component was the navigation and command system (NACOS). The NACOS consisted of two parts, an autopilot part to keep the ship on course and a map construction part, where simple maps could be created and displayed on a radar screen. When the *Royal Majesty* was being built, the NACOS and the GPS receiver were delivered by different manufacturers, and they, in turn, used different versions of electronic communication standards.

Because of these differing standards and versions, valid position data and invalid DR data sent from the GPS to the NACOS were both "labeled" with the same code (GP). The installers of the bridge equipment were not told, nor did they expect, that position data (GP-labeled) sent to the NACOS would be anything but valid position data. The designers of the NACOS expected that if invalid data were received, it would have another format. As a result, the GPS used the same "data label" for valid and invalid data, and thus the autopilot could not distinguish between them. Since the NACOS could not detect that the GPS data were invalid, the ship sailed on an autopilot that was using estimated positions until a few minutes before the grounding.

A principal function of an integrated bridge system is to collect data such as depth, speed, and position from different sensors, which are then shown on a centrally placed display to provide the officer of the watch with an overview of most of the relevant information. The NACOS on the *Royal Majesty* was placed at the forward part of the bridge, next to the radar screen. Current technological systems commonly have multiple levels of automation with multiple mode indications on many displays. An adaptation of work strategy is to collect these in the same place and another solution is to integrate data from many components into the same display surface. This presents an integration problem for shipping in particular, where quite often components are delivered by different manufacturers.

The centrality of the forward console in an integrated bridge system also sends the implicit message to the officer of the watch that navigation may have taken place at the chart table in times past, but the work is now performed at the console. The chart should still be used, to be sure, but only as a backup option and at regular intervals (customarily every half-hour or every hour). The forward console is perceived to be a clearing house for all the information needed to safely navigate the ship.

As mentioned, the NACOS consisted of two main parts. The GPS sent position data (via the radar) to the NACOS in order to keep the ship on track (autopilot part) and to position the maps on the radar screen (map part). The autopilot part had a number of modes that could be manually selected; NAV and COURSE. NAV mode kept the ship within a certain distance of a track, and corrected for drift caused by wind, sea, and current. COURSE mode was similar but the drift was calculated in an alternative way. The NACOS also had a DR mode, in which the position was continuously estimated. This backup calculation was performed in order to compare the NACOS DR with the position received from the GPS. To calculate the NACOS DR position, data from the gyro compass and Doppler log were used, but the initial position was regularly updated with GPS data. When the *Royal Majesty* left Bermuda, the navigation officer chose the NAV mode and the input came from the GPS, normally selected by the crew during the three years the vessel had been in service.

If the ship had deviated from her course more than the preset limit, or if the GPS position differed from the DR position calculated by the autopilot, the NACOS would have sounded an aural, and clearly shown a visual, alarm at the forward console (position-fix alarm). There were no alarms since the two DR positions calculated by the NACOS and the GPS were identical. The NACOS DR, which was the perceived backup, was using GPS data, believed to be valid, to refresh its DR position at regular intervals. This is because the GPS was sending DR data, estimated from log and gyro data, but labeled as valid data. Thus, the radar chart and the autopilot were using the same inaccurate position information and there was no display or warning of the fact that DR positions (from the GPS) were used. Nowhere on the integrated display could the officer on watch confirm what mode the GPS was in, and what effect the mode of the GPS was having on the rest of the automated system, not to mention the ship.

In addition to this, there were no immediate and perceivable effects on the ship since the GPS calculated positions using the log and the gyrocompass. It cannot be expected that a crew should become suspicious of the fact that the ship actually is keeping her speed and course. The combination of a busy departure, an unprecedented event (cable break) together with a non-event (course keeping), and the change of the locus of navigation (including the intrasystem communication difficulties) shows that it made sense, in the situation and at the time, that the crew did not know that a mode change had occurred.

THE OCEAN VOYAGE

Even if the crew did not know about a mode change immediately after departure, there was still a long voyage at sea ahead. Why did none of the officers check the GPS position against another source, such as the Loran-C receiver that was placed close to the GPS? (Loran-C is a radio navigation system that relies on land-based transmitters). Until the very last minutes before the grounding, the ship did not act strangely and gave no reason for suspecting that anything was amiss. It was a routine trip, the weather was good, and the watches and watch changes uneventful.

Several of the officers actually did check the displays of both Loran and GPS receivers, but only used the GPS data (since those had been more reliable in their experience) to plot positions on the paper chart. It was virtually impossible to actually observe the implications of a difference between Loran and GPS numbers alone. Also, there were other kinds of cross-checking. Every hour, the position on the radar map was checked against the position on the paper chart, and cues in the world (e.g., sighting of the first buoy) were matched with GPS data. Another subtle reassurance to officers must have been that the master on a number of occasions spent several minutes checking the position and progress of the ship, and did not make any corrections.

Before the GPS antenna was moved, the short spells of signal degradation that led to DR mode also caused the radar map to "jump around" on the radar screen (the crew called it chopping) since the position would change erratically. The reason chopping was not observed on this particular occasion was that the position did not change erratically, but in a manner consistent with dead reckoning. It is entirely possible that the satellite signal was lost before the autopilot was switched on, thus causing no shift in position. The crew had developed a strategy to deal with this occurrence in the past. When the position-fix alarm sounded, they first changed modes (from NAV to COURSE) on the autopilot and then they acknowledged the alarm. This had the effect of stabilizing the map on the radar screen so that it could be used until the GPS signal returned. It was an unreliable strategy, since the map was being used without knowing the extent of error in its positioning on the screen. And it also led to the belief that, as mentioned earlier, the only time the GPS data were unreliable was during chopping. Chopping was more or less alleviated by moving the antenna, which means that by eliminating one problem, a new pathway for accidents was created. The strategy of using the position-fix alarm as a safeguard no longer covered all or most of the instances of GPS unreliability.

This locally efficient procedure would almost certainly not be found in any manuals, but gained legitimacy through successful repetition becoming common practice over time. It may have sponsored the belief that a stable map is a good map, with the crew concentrating on the visible signs instead of being wary of the errors hidden below the surface. The chopping problem had been resolved for about four months, and trust in the automation grew.

FIRST BUOY TO GROUNDING

Looking at the unfolding sequence of events from the position of a retrospective outsider, it is once again easy to point to indications missed by the crew. Especially toward the end of the journey, there appears to be a larger number of cues that could potentially have revealed the true nature of the situation. There was an inability of the first officer to positively identify the first buoy that marked the entrance of the Boston sea lanes (such lanes form a separation scheme delineated on the chart to keep meeting and crossing traffic at safe distance and to keep ships away from dangerous areas). A position error was still not suspected, even with the vessel close to the shore. The lookouts reported red lights and later blue and white water, but the second officer did not take any action. Smaller ships in the area broadcasted warnings on the radio, but nobody on the bridge of the *Royal Majesty* interpreted those to concern their vessel. The second officer failed to see the second buoy along the sea lanes on the radar, but told the master that it had been sighted. In hindsight, there were numerous opportunities to avoid the grounding, which the crew consistently failed to recognize (NTSB 1997).

Such conclusions are based on a dualist interpretation of situation awareness. What matters there is the accuracy of the mapping between an external world that can be pieced together in hindsight (and which contains shopping bags full of epiphanies never opened by those who most needed them) and people's internal representation of that world. This internal representation (or situation awareness) can be shown to be clearly deficient, as falling far short of all the cues that were available. But making claims about the awareness of other people at another time and place requires us to put ourselves in their shoes and limiting ourselves to what they knew. We have to find out why people thought they were in the right place or had the right assessment of the situation around them. What made that so? Remember, the adequacy or accuracy of an insider's representation of the situation cannot be called into question: it is what counts for them, and it is that which drives further action in that situation. Why was it plausible for the crew to conclude that they were in the right place? What did their world look like to them (not: how does it look not to retrospective observers)?

The first buoy ("BA") in the Boston traffic lanes was passed at 19:20 on June 10, or so the chief officer thought (The buoy identified by the first officer as the "BA" later turned out to be the "AR" buoy placed about 15 miles to the west–southwest of the "BA."). To the chief officer, there was a buoy on the radar. And it was where he expected it to be; it was where it should be. It made sense to the first officer to identify it as the correct buoy since the echo on the radar screen coincided with the mark on the radar map that signified the "BA." Radar map and radar world matched. We now know that the overlap between radar map and radar return was a mere stochastic fit. The map showed the BA buoy, and the radar showed a buoy return. A fascinating coincidence was the sun glare on the ocean surface that made it impossible to visually identify the "BA." But independent cross-checking had already occurred: the first officer probably verified his position by two independent means; the radar map and the buoy.

The officer, however, was not alone in managing the situation, or in making sense of it. An interesting aspect of automated navigation systems in real workplaces is that several people typically use it, in partial overlap and consecutively, like the watchkeeping officers on a ship. At 20:00, the second officer took over the watch from the chief officer. The chief officer must have provided the vessel's assumed position, as is good watchkeeping practice. The second officer had no reason to doubt that this was a correct position. The chief officer had been at sea for 21 years, spending 30 of the last 36 months on board the *Royal Majesty*. Shortly after the takeover, the second officer reduced the radar scale from 12 to 6 nautical miles. This is normal practice when vessels come closer to shore or other restricted waters. By reducing the scale, there is less clutter from the shore and an increased likelihood to see anomalies and dangers.

When the lookouts later reported lights, the second officer had no expectation that there was anything wrong. To him, the vessel was safely in the traffic lane. Also, lookouts are liable to report everything indiscriminately; it is always up to the officer of the watch to decide whether to take action or not. There is also a cultural and hierarchical gradient between the officer and the lookouts; they come from different nationalities and backgrounds. At this time, the master also visited the bridge, and just after he left, there was a radio call. This escalation of work may well have distracted the second officer from considering the lookouts' report, even if he had wanted to.

After the accident investigation was concluded, it was discovered that two Portuguese fishing vessels had been trying to call the *Royal Majesty* on the radio to warn her of the imminent danger. The calls were made not long before the grounding, at which time the *Royal Majesty* was already 16.5 nautical miles from where the crew knew her to be. At 20:42, one of the fishing vessels called "fishing vessel, fishing vessel call cruise boat" on channel 16 (an international distress channel for emergencies only). Immediately following this first call in English, the two fishing vessels started talking to each other in Portuguese. One of the fishing vessels tried to call again a little later, giving the position of the ship he was calling. Calling on the radio without positively identifying the intended receiver can lead to mix-ups. Or in this case, if the second officer heard the first English call and the ensuing conversation, he most likely disregarded it since it seemed to be two other vessels talking to each other. Such an interpretation makes sense: if one ship calls without identifying the intended receiver, and another ship responds and consequently engages the first caller in conversation, the communication loop is closed. Also, as the officer was using the 6-mile scale, he could not see the fishing vessels on his radar. If he heard the second call and checked the position, he might well have decided that the call was not for him, as it appeared that he was far from that position. Whomever the fishing ships were calling, it could not have been him, because he was not there.

At about this time, the second buoy should have been seen, and around 21:20, it should have been passed, but was not. The second officer assumed that the radar map was correct when it showed that they were on course. To him the buoy signified a position, a distance traveled in the traffic lane, and reporting that it had been passed may have amounted to the same thing as reporting that they had passed the position it was (supposed to have been) in. The second officer did not, at this time, experience an accumulation of anomalies, warning him that something was going wrong. In his view, this buoy, which was perhaps missing or not picked up by the radar, was the first anomaly and not perceived as a large one. Paraphrasing the "Bridge Procedures Guide," it is said that the master should be called when (a) something unexpected happens, (b) when something expected does not happen (e.g., a buoy), and (c) at any other time of uncertainty. It is all very well to define an unexpected event, but when it happens, people tend to quickly rationalize it. This is even more so in the case of not seeing what was expected: "well, I guess the X isn't doing Y...," and clearly an act of local rationality. The NTSB report, on the other hand, lists at least five actions that the officer should have taken. He did not take any of these actions, because he was not missing opportunities to avoid the grounding. He was navigating the vessel safely to Boston.

The master visited the bridge just before the radio call, telephoned the bridge about one hour after it, and made a second visit around 22:00. The times at which he chose to visit the bridge were calm and uneventful, and did not prompt the second officer to voice any concerns, nor did they trigger the master's interest in more closely examining the apparently safe handling of the ship. Five minutes before the grounding, a lookout reported blue and white water. For the second officer, these indications alone were no reason for taking action. They were no warnings of anything about to go amiss, because nothing was going to go amiss. The crew knew where they were. Nothing in their situation suggested to them that they were not doing enough or that they should question the "accuracy" of their awareness of the situation.

At 22:20, the ship started to veer, which brought the captain to the bridge. The second officer, still certain that they were in the traffic lane, believed that there was something wrong with the steering. This interpretation would be consistent with his experiences of cues and indications during the trip so far.

The master, however, came to the bridge and saw the situation differently, but was too late to correct the situation. The *Royal Majesty* ran aground east of Nantucket at 22:25, at which time she was 17 nautical miles from her planned and presumed course. None of the over 1000 passengers were injured, but repairs and lost revenues cost the company $7 million.

With a discrepancy of 17 miles at the premature end to the journey of the *Royal Majesty*, and a day and a half to discover the growing gap between actual and intended track, the case of "loss of SA" or "deficient SA" looks like it is made. But the "elements" that make up all the cues and indications that the crew should have seen, and should have understood, are mostly products of hindsight; products of our ability to look at the unfolding sequence of events from the position of retrospective outsiders. In hindsight, we wonder how these repeated "opportunities to avoid the grounding," these repeated invitations to undergo some kind of epiphany about the real nature of the situation were never experienced by the people who most needed it. But the revelatory nature of the cues and the structure or coherence that they apparently have in retrospect are not products of the situation itself or the actors in it. They are retrospective imports.

When looked at from the position of retrospective outsider, a "loss of SA" can look so very real, so compelling. They failed to notice, they did not know, they should have done this or that. But from the point of view of people inside the situation, as well as other potential observers, these deficiencies do not exist in and of themselves; they are artifacts of hindsight, "elements" removed retrospectively from a stream of action and experience. To people on the inside, it is often nothing more than normal work. If we want to begin to understand why it made sense for people to do what they did, we have to put ourselves in their shoes. What did they know? What was their understanding of the situation? Rather than construing the case as a "loss of SA" (which simply judges other people for not seeing what we, in our retrospective omniscience, would have seen), there is more explanatory leverage in seeing the crew's actions as normal processes of sensemaking—of transactions between goals, observations and actions. As Weick (1995) points out, sensemaking is

> … something that preserves plausibility and coherence, something that is reasonable and memorable, something that embodies past experience and expectations, something that resonates with other people, something that can be constructed retrospectively but also can be used prospectively, something that captures both feeling and thought… In short, what is necessary in sensemaking is a good story. A good story holds disparate elements together long enough to energize and guide action, plausibly enough to allow people to make retrospective sense of whatever happens, and engagingly enough that others will contribute their own inputs in the interest of sensemaking (Weick 1995, p. 61).

Even if one does make the concessions to the existence of "elements," as Weick does in the above quote, it is only for the role they play in constructing a plausible story of what is going on, not for building an accurate mental simile of an external world somewhere "out there."

STUDY QUESTIONS

1. Explain how situation awareness is a dualist concept, and why it is impossible to defend oneself effectively against charges of "loss of SA."
2. Not only is SA dualist, it is essentially Cartesian–Newtonian in almost all other respects. Please explain how linearity and reductionism (as well as knowledge-as-correspondence) all help form the basis for the concept.
3. Please describe the first and the second cognitive revolution and what they did with our thinking about the relationship between matter and mind.
4. How was the Gestalt movement a response to Wilhelm Wundt's elementarism? Yet how have elementarism and mentalism been reintroduced to human factors by the 1970s information processing school and today's situation awareness?
5. Find a case study that has been attributed to the operator's "loss of SA" and build a radically empiricist narrative of the events instead.
6. If you lose SA, what replaces it?

5 Accidents

CONTENTS

KEY POINTS

- Most people agree that their organizations are complex. But typical models and methods for the control of risk in such organizations do not do justice to that complexity.
- There is a marked Newtonian legacy in understanding accidents: risk is located in components and in the linear progression of failures, producing trajectories toward failure in which there is a proportionality or symmetry between causes and effects.
- Organizational failures can be seen as the result of a drift—long, steady declines into greater risk and an erosion of margins.
- Organizations drift into failure by doing what it takes to produce success in a world with limited resources and multiple goal conflicts in a dynamic environment. It requires ideas about systems thinking, complexity, and emergence.
- Failure breeds opportunistically, nonrandomly, among the very structures designed to protect an organization from failure. Failure, like success, can be seen as emergent—not present in the components that make up the system, but visible as higher-order property. Organizations that have to manage risk should be regarded more as ecological systems, not as machines with breakable components or linkages, machines that are insulated and isolated from their environments.
- Failure is not inevitable. But it does require a new vocabulary: one with leverage points for recognizing and stopping drift in ways that are not componential or structural.

DISAPPEARANCE OF THE ACCIDENTAL

Accidents actually do not happen very often. Many industries have reported a drop in their injury and fatality rates over the past decades. There are even systems in the developed world that are, as measured by fatality risk, even ultra-safe. Their likelihood of a fatal accident is less than 10^{-7}, which means a 1-out-of-10,000,000 chance of death, serious loss of property, or environmental or economic devastation per activity or operation (Amalberti 2001). At the same time, this appears to be a magical frontier. All systems that get there go asymptotic after this. No system has figured out a way of becoming even safer. Progress on safety beyond 10^{-7} is elusive. As René Amalberti has pointed out, linear extensions of current safety efforts (incident reporting, safety and quality management, proficiency checking, standardization and proceduralization, more rules and regulations) seem of little use in breaking the asymptote, even if they are necessary to sustain the 10^{-7} safety level.

More intriguingly still, the accidents that happen at this frontier appear to be of a type that is difficult to predict using the logic that governs safety thinking up to 10^{-7}. It is here that the limitations of the vocabulary from Descartes and Newton become most apparent. Accident models that rely largely on failures, holes, violations, deficiencies, and flaws can have a difficult time accommodating accidents that seem to emerge from (what looks to everybody like) normal people doing normal work in normal organizations. Yet the mystery is that in the hours, days, or even years leading up to an accident beyond 10^{-7}, there may be few report-worthy failures or noteworthy organizational deficiencies. Regulators as well as insiders typically do not see people violating rules, nor do they discover other flaws that would give cause to shut down or seriously reconsider operations. If only it were that easy. And up to 10^{-7} it probably is. But when failures, serious failures, are no longer preceded by serious failures, predicting accidents becomes a lot more difficult. And modeling them with the help of mechanistic, structuralist notions may be of little help.

The impossibility of breaking through the asymptote in progress on safety is vexing at a much deeper level too. Recall from the first chapter how the Enlightenment, and the modernism that followed, promised the West that progress was always possible. Tomorrow would always be better than yesterday; the human race was destined for ever better things. Negative events that befell us could be brought under control (wars, famines, other disasters). By thinking rationally, by careful planning and organization, with ever better science and technology, this was a distinct possibility, generating ever more evidence for itself. The very notion of accident investigation, now faithfully incorporating human factors and other sciences and technologies, makes a profound modernist assumption too: we can stop this from happening. It is an oxymoron, of course—accident investigation. After all, if we truly believed that accidents were accidental, that they were unintentional, unexpected, happening by chance, then what would be there to investigate, and to what end? Chance is hard to predict, and if it is hard to predict, it is even harder to prevent. The thing is, we do not believe in "accidents" anymore, not really. We believe in failures of risk management. That which we see as "accidents" are caused by such failures of risk management, failures that occur somewhere in the chain, somewhere in the organizational hierarchy looming over the time and place where things went wrong. If we can put our finger on that failure, condense our explanation of what went wrong into that broken *Eureka!* part, then we can change it, make it go away, replace it, and once again help the world become a better place; a place where such things do not happen.

Paradoxically, this has landed us back in pre-modern, anti-systems thinking. Behind the mismanagement of risk, after all, there is a person, or multiple people. These are people we should control with more rules, more supervision, oversight, regulation, surveillance, and with more accountability. Indeed, accountability of entire corporations or company directors has become more popular in the last few decades, with legislation in both the common law and Napoleonic law countries. Holding people or entities "accountable" in such ways is of course the same as blaming single components for the failure of the system—a profoundly Newtonian notion. To counter this, a more noncommittal view of accidents have begun to grow in popularity alongside such characterizations in the second half of the 20th century (though not in the field of "accident" investigation). More industries and insurance companies, for example, have begun to refer to "injuries" instead of "accidents." This leaves the question of cause open. After all, "accidents" refer to the cause (which was "accidental"), whereas "injury" refers to the result, the outcome. The outcome is empirically less controversial: it is easier to establish or argue than the supposed cause and also opens up the possibility for different types of intervention that are denied by, or nonsensical because, that which is truly accidental (Burnham 2009).

DRIFTING INTO FAILURE

One residual risk in otherwise safe sociotechnical systems is a drift into failure (Dekker 2011b). Drift into failure is about a slow, incremental movement of systems operations toward the edge of their safety envelope. Take the 2010 Macondo (or Deepwater Horizon) accident as an example.

TRANSOCEAN ON THE MACONDO DISASTER

Transocean was contracted by BP Exploration & Production Inc. (BP) to provide the Deepwater Horizon rig and personnel to drill the Macondo well in the Gulf of Mexico. Drilling started on Feb. 11, 2010, and was completed on April 9, 2010. On April 20, 2010, the blowout of the Macondo well resulted in explosions and an uncontrollable fire onboard the Deepwater Horizon. Eleven people lost their lives, 17 were seriously injured, and 115 of the 126 onboard evacuated. The Deepwater Horizon sank 36 hours later, and the Macondo well discharged hydrocarbons into the Gulf of Mexico for nearly three months before it was contained. The Macondo incident was the result of a succession of interrelated well design, construction, and temporary abandonment decisions that compromised the integrity of the well and compounded the risk of its failure.

BP had been concerned that downhole pressure—whether exerted by heavy drilling mud used to maintain well control or by pumping cement to seal the well—would exceed the fracture gradient and result in losses to the formation.

On April 20, 2010, the actions of the drill crew reflected its understanding that the well had been properly cemented and successfully tested.

The failure of the downhole cement to isolate the reservoir, however, allowed hydrocarbons to enter the wellbore. BP's original well plan called for use of a long-string production casing. While drilling the Macondo well, BP experienced both lost circulation events and kicks and stopped short of its planned total depth. In these delicate conditions, cementing a long-string casing further increased the risk of exceeding the fracture gradient. BP had adopted a technically complex nitrogen foam cement program that left little margin for error, and which was not tested adequately before or after the cementing operation. The integrity of the cement may also have been compromised by contamination, instability, and an inadequate number of devices used to center the casing in the wellbore.

The results of a critical negative pressure test were misinterpreted because of displacement calculation errors, a lack of adequate fluid volume monitoring, and a lack of management of change when the well monitoring arrangement was switched during the test. It is now apparent that the negative pressure test results should not have been approved, but no one involved recognized the errors at the time. The well became underbalanced during the final displacement, and hydrocarbons began entering the wellbore through the faulty cement barrier. None of the individuals monitoring the well, including the Transocean drill crew, initially detected the influx.

With the benefit of hindsight and a thorough analysis of the data available to the investigation team, several indications of an influx during final displacement operations can be identified. Given the death of the members of the drill crew, and the loss of the rig and its monitoring systems, it is not known which information the drill crew was monitoring or why the drill crew did not detect a pressure anomaly until approximately 9:30 p.m. on April 20, 2010. At 9:30 p.m., the drill crew acted to evaluate an anomaly. Upon detecting an influx of hydrocarbons by use of the trip tank, the drill crew undertook well-control activities that were consistent with their training. By the time actions were taken, hydrocarbons had come into the riser, resulting in a massive release of gas and other fluids that overwhelmed the mud–gas separator system and released high levels of gas around the aft deck of the rig. The resulting ignition of this gas cloud was inevitable. The Deepwater Horizon was overcome by conditions created by the extreme dynamic flow, the force of which pushed the drill pipe upward and washed or eroded the drill pipe and other rubber and metal elements (Transocean 2011).

As is typical for cases of drift into failure, the company, Transocean, had been considered to have a "strong overall safety record" over the seven years leading up to the accident. Paradoxically, pride in, and celebration of, low incident and injury rates may have hidden a gradual drift in what was considered acceptable engineering risk. It was, or at least looked to everyone, to be a safe organization. This was perhaps not as true for BP, the client company, which indeed had had significant problems before. BP had procedures in place forbidding employees from carrying coffee in cups without a lid, but it did not have a procedure for conducting a negative pressure test, a critical last step in avoiding a well blowout. This is from a strategy document dated December 2008:

> "It's become apparent," the BP document stated, "that process-safety major hazards and risks are not fully understood by engineering or line operating personnel. Insufficient awareness is leading to missed signals that precede incidents and response after incidents, both of which increases the potential for and severity of process-safety related incidents." The document called for stronger "major hazard awareness" (Elkind and Whitford 2011, p. 4).

Pressures of scarcity and competition typically fuel a drift into failure. Uncertain technology and incomplete knowledge about where the boundaries are can mean that organizations, regulators, and other stakeholders do not stop the drift or do not even see it. Such accidents do not happen just because somebody suddenly errs or something suddenly breaks. There is supposed to be too much built-in protection against the effects of single failures. But what if these protective structures themselves contribute to drift, in ways inadvertent, unforeseen, and hard to detect? What if the organized social complexity surrounding the technological operation, all the engineering calculations, tests, approval guidelines, maintenance committees, working groups, regulatory interventions, approvals, and manufacturer inputs, that all intended to protect the system from breakdown, actually helped to set its course to the edge of the envelope? Barry Turner's 1978 *Man-Made Disasters* showed how accidents in complex, well-protected systems are incubated. The potential for an accident accumulates over time, but this accumulation, this steady slide into disaster, generally goes unrecognized by those on the inside and even those on the outside. Macondo was like this: uncertain technology, gradual adaptations, a drift into failure. It was about the inseparable, mutual influences of mechanical and social worlds, and it puts the inadequacy of our current models in human factors and safety on full display.

> For years BP had touted its safety record, pointing to a steep decline in the number of slips, falls, and vehicle accidents that generate days away from work, a statistic that is closely followed by both the industry and its regulators. BP had established a dizzying array of rules that burnished this record, including prohibitions on driving while speaking on a cellphone, walking down a staircase without holding a handrail, and carrying a cup of coffee around without a lid. Bonuses for BP executives included a component tied to these personal-injury metrics. BP cut its injury rate dramatically after the Amoco merger (the previous owner of the Texas City refinery). But BP's personal-safety achievements masked failures in assuring process safety. In the energy business, process safety generally comes down to a single issue: keeping hydrocarbons contained inside a steel pipe or tank. Disasters don't happen because someone drops a pipe on his foot or bumps his head. They result from flawed ways of doing business that permit risks to accumulate (Elkind and Whitford 2011, p. 7).

A trajectory of drift into failure may be easy to draw in hindsight. It is fascinating to look at, too. The realities they represent, however, were not similarly compelling to those on the inside of the system at the time. The greatest conundrum since Turner

(1978) has been to elucidate why the slide into disaster, so easy to see and depict in retrospect, is missed by those who inflict it on themselves. Judging, after the fact, that there was a failure of foresight is easy: all you need to do is plot the numbers and spot the slide into disaster. Standing amid the rubble, it is easy to marvel at how misguided or misinformed people must have been. Even before Macondo, BP realized how critical procedures should be formalized and carried out with rigor and that it was essential to maintain multiple safeguards against an accident and that it shouldn't change operational plans on the fly. It tried to inculcate the belief that small incidents were warning signs that conditions are ripe for a disaster and that periods without serious accident could encourage a belief in invulnerability. Despite its focus on hard-hat or personal safety, the federal Occupational Safety and Health Administration proposed a record fine against BP in 2009 for its failure to deal with previously cited hazards at Texas City, and cited it for hundreds of new willful safety violations (Elkind and Whitford 2011).

The pressing question is why the conditions conducive to an accident were not acknowledged or acted on as such by those on the inside of the system—those whose job it was to not have such accidents happen. Foresight is not hindsight. There is a profound revision of insight that turns on the present. It converts a once vague, unlikely future into an immediate, certain past. The future, said David Woods (Woods et al. 2010), seems implausible before an accident ("No, that won't happen to us"). But after an accident, the past seems incredible ("How could we not have seen that this was going to happen to us!"). What is now seen as extraordinary was once ordinary. The decisions, trade-offs, preferences, and priorities that seem so out of the ordinary and immoral after an accident were once normal and commonsensical to those who contributed to its incubation.

BANALITY, CONFLICT, AND INCREMENTALISM

Sociological research as well as human factors work and research on safety has begun to sketch the contours of answers to the why of drift (Dekker 2011b; Leveson 2012; Rasmussen and Svedung 2000; Vaughan 1996). Though different in background, pedigree, and much substantive detail, these works converge on important commonalities about the drift into failure. The first is that accidents, and the drift that precedes them, are associated with normal people doing normal work in normal organizations—not with miscreants engaging in immoral deviance. We can call this the banality-of-accidents thesis. Second, most works have at their heart a conflictual model: organizations that involve safety-critical work are essentially trying to reconcile irreconcilable goals (staying safe and staying in business). Third, drifting into failure is incremental. Accidents do not happen suddenly, nor are they preceded by monumentally bad decisions or bizarrely huge steps away from the ruling norm.

The banality-of-accidents thesis says that the potential for having an accident grows as a normal by-product of doing normal business under normal pressures of resource scarcity and competition. No system is immune to the pressures of scarcity and competition, well, almost none. The only transportation system that ever approximated working in a resource-unlimited universe was NASA during the early Apollo years (a man had to be put on the moon, whatever the cost). There was plenty of money, and plenty of highly motivated talent. But even here technology was uncertain, faults and failures not uncommon, and budgetary constraints got imposed quickly and increasingly tightly. Human resources and talent started to drain away. Indeed, even such noncommercial enterprises know resource scarcity: government agencies such as NASA or safety regulators may lack adequate financing, personnel, or capacity to do what they need to do. Referring to drilling for oil in deepwater, the investigation commission reporting to the US President in 2011 "documented the weaknesses and the inadequacies of the federal regulation and oversight, and made important recommendations for changes in legal authority, regulations, investments in expertise, and management." It further wrote that regulatory oversight

... of leasing, energy exploration, and production require reforms even beyond those significant reforms already initiated since the Deepwater Horizon disaster. Fundamental reform will be needed in both the structure of those in charge of regulatory oversight and their internal decisionmaking process to ensure their political autonomy, technical expertise, and their full consideration of environmental protection concerns. Because regulatory oversight alone will not be sufficient to ensure adequate safety, the oil and gas industry will need to take its own, unilateral steps to increase dramatically safety throughout the industry, including self-policing mechanisms that supplement governmental enforcement (Graham et al. 2011, p. vii).

This echoes what we see elsewhere in this book about the role of the regulator (including its limits), and that doing and regulating business under pressures of technological uncertainty and resource scarcity is normal. Scarcity is part and parcel of this (or any kind of) work. Few government regulators anywhere will easily claim that they have adequate time and personnel resources or expertise to carry out their mandates. Yet the fact that resource scarcity is normal does not mean that it has no consequences. Of course, scarcity finds its ways in, seeping into the capillaries of an organization, affecting what will be seen as normal, as acceptable, on a dark and lonely night in a corner of the system that is already beyond its project deadlines and over its budgetary projections. Supervisors write memos. Battles over resources and timelines are fought. Trade-offs are made. Scarcity expresses itself in the common organizational, political wrangles over resources and primacy, in managerial preferences for certain activities and investments over others, and in almost all engineering and operational trade-offs between strength and cost, between efficiency and diligence. In fact, working successfully under pressures, scarcity, and resource constraints is a source of professional pride. Building something that is strong and light, for example, marks the expert in the aeronautical engineer. Procuring and nursing into existence a system that has both low development costs and low operational costs (these are typically each other's inverse) is the dream of most investors and many a manager. Being able to create a program that putatively allows better inspections with fewer inspectors may win a civil servant compliments and chances at promotion, while the negative side effects of the program are felt primarily in some faraway field office.

Yet the major engine of drift hides somewhere in this conflict, in this tension between operating safely and operating at all, between building safely and building at all. This tension provides the energy behind the slow, steady disengagement of practice from earlier established norms or design constraints. This disengagement can eventually become drift into failure. As a system is taken into use, it learns, and as it learns, it adapts:

> Experience generates information that enables people to fine-tune their work: fine-tuning compensates for discovered problems and dangers, removes redundancy, eliminates unnecessary expense, and expands capacities. Experience often enables people to operate a sociotechnical system for much lower cost or to obtain much greater output than the initial design assumed (Starbuck and Milliken 1988, p. 333).

This fine-tuning drift toward operational safety margins is one testimony to the limits of the structuralist systems safety vocabulary in vogue today. We think of safety cultures as learning cultures: cultures that are oriented toward learning from events and incidents. But learning cultures are neither unique (because every open system in a dynamic environment necessarily learns and adapts) nor necessarily positive: Starbuck and Milliken highlighted how an organization can learn to "safely" borrow from safety while achieving gains in other areas. Drift into failure could not happen without learning.

A critical ingredient of this learning is the apparent insensitivity to mounting evidence that, from the position of a retrospective outsider, could have shown how bad the judgments and decisions actually are. This is how it looks from the position of a retrospective outsider: the retrospective outsider sees a failure of foresight. From the inside, however, the abnormal is pretty normal, and making trade-offs in the direction of greater efficiency is nothing unusual. In making these trade-offs, however, there is a feedback imbalance. Information on whether a decision is cost effective or efficient can be relatively easy to get. An early arrival time is measurable and has immediate, tangible benefits. How much is or was borrowed from safety in order to achieve that goal, however, is much more difficult to quantify and compare. If it was followed by a safe landing, apparently it must have been a safe decision. Each consecutive empirical success seems to confirm that fine-tuning is working well: the system can operate equally safely, yet more efficiently. As Weick (1993) pointed out, however, safety in those cases may not at all be the result of the decisions that were or were not made, but rather an underlying stochastic variation that hinges on a host of other factors, many not easily within the control of those who engage in the fine-tuning process. Empirical success, in other words, is not proof of safety. Past success does not guarantee future safety. Borrowing more and more from safety may go well for a while, but you never know when you are going to hit. This moved Langewiesche (1998) to say that Murphy's law is wrong: everything that can go wrong usually goes right, and then we draw the wrong conclusion.

The nature of this dynamic, this fine-tuning, this adaptation, is incremental. The organizational decisions that are seen as "bad decisions" after the accident (even though they seemed like perfectly good ideas at the time) are seldom big, risky, order-of-magnitude steps. Rather, there is a succession of decisions, a long and steady progression of small, incremental steps that unwittingly take an organization toward disaster. Each step away from the original norm that meets with empirical success (and no obvious sacrifice of safety) is used as the next basis from which to depart just that little bit more again. It is this incrementalism that makes distinguishing the abnormal from the normal so difficult. If the difference between what "should be done" (or what was done successfully yesterday) and what is done successfully today is minute, then this slight departure from an earlier established norm is not worth remarking or reporting on. Incrementalism is about continued normalization: it allows normalization and rationalizes it.

DRIFT INTO FAILURE AND INCIDENT REPORTING

Can incident reporting not reveal a drift into failure? This would seem to be a natural role of incident reporting, but it is not so easy. The normalization that accompanies drift into failure severely challenges the ability of insiders to define incidents. What is an incident? Many of the things that would, after the fact, be constructed as "incidents" and worthy of reporting are the normal, everyday workarounds, frustrations, and improvisations needed to get the job done. They are not report-worthy. They do not qualify as incidents. Even if the organization has a reporting culture, even if it has a learning culture, even if it has a just culture so that people would feel secure in sending in their reports without fear of retribution, such "incidents" would not turn up in the system. This is the banality-of-accidents thesis. These are not incidents. In 10^{-7} systems, incidents do not precede accidents. Normal work does. In these systems:

> accidents are different in nature from those occurring in safe systems: in this case accidents usually occur in the absence of any serious breakdown or even of any serious error. They result from a combination of factors, none of which can alone cause an accident, or even a serious incident; therefore these combinations remain difficult to detect and to recover using traditional safety analysis logic. For the same reason, reporting becomes less relevant in predicting major disasters (Amalberti 2001, p. 112).

Even if we were to direct greater analytic force onto our incident-reporting databases, this may still not yield any predictive value for accidents beyond 10^{-7}, simply because the data are not there. The databases do not contain, in any visible format, the ingredients of accidents that happen beyond 10^{-7}. Learning from incidents to prevent accidents beyond 10^{-7} may well be impossible. Incidents are about independent failures and errors, noticed and noticeable by people on the inside. But these independent errors and failures no longer make an appearance in the accidents that happen beyond 10^{-7}. The failure to adequately see the part to be lubricated (that nonredundant, single-point, safety-critical part), the failure to adequately and reliably perform an end-play check—none of this appears in incident reports. But it is deemed "causal" or "contributory" in the accident report. The etiology of accidents in 10^{-7} systems, then, may well be fundamentally different from that of incidents, hidden instead in the residual risks of doing normal business under normal pressures of scarcity and competition. This means that the so-called common-cause hypothesis (which holds that accidents and incidents have common causes and that incidents are qualitatively identical to accidents except for being just one step short) is probably wrong at 10^{-7} and beyond:

... reports from accidents such as Bhopal, Flixborough, Zeebrugge and Chernobyl demonstrate that they have not been caused by a coincidence of independent failures and human errors. They were the effect of a systematic migration of organizational behavior toward accident under the influence of pressure toward cost-effectiveness in an aggressive, competitive environment (Rasmussen and Svedung 2000, p. 14).

Despite this insight, independent errors and failures are still the major return of any accident investigation today. The BP report on the Macondo blowout, following Newtonian–Cartesian logic, spoke of deficiencies in maintenance programs, of shortcomings in regulatory oversight, of responsibilities not fulfilled, of flaws and failures and breakdowns. Hewing to Swiss Cheese orthodoxy, BP modeled the disaster as a linear series of breaches of layers of defense, finding the following ones (BP 2010):

1. The cement barrier did not isolate the hydrocarbons.
2. The shoe track barriers did not isolate the hydrocarbons.
3. The negative-pressure test was accepted although well integrity had not been established.
4. Influx was not recognized until hydrocarbons were in the riser.
5. Well control response actions failed to regain control of the well.
6. Diversion to the mud gas separator resulted in gas venting onto the rig.
7. The fire and gas system did not prevent hydrocarbon ignition.
8. The blowout preventer emergency mode did not seal the well.

If this reads like a line of dominoes falling, it is because that is how the analysts conceived the accident. Of course, in hindsight, a plausible sequence of events (of linear, successive failures) may well look like that. It is neat and Newtonian. And finding faults and failures is fine because it gives people something to fix. Yet the sequence as drawn up by BP above covers only the last moments before the accident, tracing the domino effect from the seafloor upward to the rig and into the air around it. It leaves the organizational story entirely unmentioned, however. Why did nobody at the time see these now so apparent faults and failures for what they (in hindsight) were? This is where the structuralist vocabulary of traditional human factors and safety is most limited, and limiting. The holes found in the layers of defense (the cement, the engineers, the regulator, the manufacturer, the operators, the emergency systems) are easy to discover once the rubble is strewn before one's feet. Indeed, one common critique of structuralist models is that they are good at identifying deficiencies, or latent failures, postmortem.

Yet these deficiencies and failures are not seen as such, nor easy to see as such, or act on, by those on the inside (or even those relatively on the outside, like the regulator) before the accident happens. Indeed, structuralist models can capture the deficiencies that result from drift very well: they accurately identify latent failures or resident pathogens in organizations and can locate holes in layers of defense. But the buildup of latent failures, if that is what you want to call them, is not modeled. The process of erosion, of attrition of safety norms, of drift toward margins, cannot be captured well by structuralist approaches, for those are inherently metaphors for resulting forms, not models oriented at processes of formation. Structuralist models are static. Their ability to begin accounting in interesting ways for the many influences and interactions is virtually zero. The Deepwater commission reporting to the US President quoted the Columbia Space Shuttle Accident investigation when it warned that "complex systems almost always fail in complex ways" (Graham et al. 2011, p. viii). Drawing up a simple, linear story of such a complex event does it no justice. Even BP itself, in its 2010 report, acknowledged that what accounted for the accident was

> ... a complex and interlinked series of mechanical failures, human judgments, engineering design, operational implementation and team interfaces [that] came together to allow the initiation and escalation of the accident. Multiple companies, work teams and circumstances were involved over time (BP 2010, p. 11).

Recall from Chapters 1 and 2 that although the structuralist models of the 1990s are often called "system models" or "systemic models," they are a far cry from what is considered systems thinking. The systems part of structuralist models has so far largely been limited to identifying, and providing a vocabulary for, the upstream structures (blunt ends) behind the production of errors at the sharp end. The systems part of these models is a reminder that there is context, that we cannot understand errors without going into the organizational background from which they hail. All of this is necessary, of course, as errors are still all too often taken as the legitimate conclusion of an investigation (just look at the spoiler case with "breakdown in CRM" as cause). But reminding people of context is no substitute for beginning to explain the dynamics, the subtle, incremental processes that lead to, and normalize, the behavior eventually observed. This requires a different perspective for looking at the messy interior of organizations and a different language to cast the observations in. It requires human factors and system safety to look for ways that move toward real systems thinking, where accidents are seen as an emergent feature of organic, ecological, transactional processes, rather than just the end point of a trajectory through holes in layers of defense. Structuralist approaches, and fixing the things they point us to, may not help much in making further progress on safety:

> ... we should be extremely sensitive to the limitations of known remedies. While good management and organizational design may reduce accidents in certain systems, they can never prevent them ... The causal mechanisms in this case suggest that technical system failures may be more difficult to avoid than even the most pessimistic among us would have believed. The effect of unacknowledged and invisible social forces on information, interpretation, knowledge, and—ultimately—action, are very difficult to identify and to control (Vaughan 1996, p. 416).

Yet the retrospective explanatory power of structuralist models makes them the instruments of choice for those in charge of managing safety. Indeed, the idea of a banality of accidents has not always easily found traction outside academic circles. For one thing, it is scary. It makes the potential for failure commonplace, or relentlessly inevitable (Vaughan 1996). This can make accident models practically useless and managerially demoralizing. If the potential for failure is everywhere, in everything we do, then why try to avoid it? If an accident has no causes in the traditional sense, then why try to fix anything? Such questions are indeed nihilist, fatalist. It is not surprising, then, that resistance against the possible world lurking behind their answers takes many forms. Pragmatic concerns are directed toward control, toward hunting down the broken parts, the bad guys, the violators, the incompetent mechanics. Why did this one technician not perform the last lubrication of the accident airplane jackscrew as he should have? Pragmatic concerns are about finding the flaws, identifying the weak areas and trouble spots, and fixing them before they cause real problems. But those pragmatic concerns find neither a sympathetic ear nor a constructive lexicon in the misers about drift into failure, for drift into failure is hard to spot, certainly from the inside.

SYSTEMS THINKING

If we want to understand failures beyond 10^{-7}, we have to stop looking for failures. It is no longer failures that go into creating these failures—it is normal work. Thus, the banality of accidents makes their study philosophically philistine. It shifts the object of examination away from the darker sides of humanity and unethical corporate governance, and toward pedestrian, everyday decisions of normal, everyday people under the influence of normal, everyday pressures. The study of accidents is rendered dramatic or fascinating only because of the potential outcome, not because of the processes that incubate it (which in itself can be fascinating, of course).

Having studied the Challenger Space Shuttle disaster extensively, Diane Vaughan (1996) was forced to conclude that this type of accident is not caused by a series of component failures, even if component failures are the result. Instead, together with other sociologists, she pointed to an indigenousness of mistake, to mistakes and breakdown as systematic, normal by-products of an organization's work processes:

> Mistake, mishap, and disaster are socially organized and systematically produced by social structures. No extraordinary actions by individuals explain what happened: no intentional managerial wrongdoing, no rule violations, no conspiracy. These are mistakes embedded in the banality of organizational life and facilitated by environments of scarcity and competition, uncertain technology, incrementalism, patterns of information, routinization and organizational structures (Vaughan 1996, p. xiv).

If we want to understand, and become able to prevent, failure beyond 10^{-7}, this is where we need to look. Forget wrongdoing. Forget rule violations. Forget errors. Safety, and the lack of it, is an emergent property. What we need to study instead is patterns of information, the uncertainties in operating complex technology and the ever-evolving and imperfect sociotechnical systems surrounding it to make that operation happen, the influence of scarcity and competition on those systems, and how they set in motion an incrementalism (itself an expression of organizational learning or adaptation under those pressures). To understand safety, an organization needs to capture the dynamics in the banality of its organizational life and begin to see how the emergent collective moves toward the boundaries of safe performance.

SYSTEMS AS DYNAMIC RELATIONSHIPS

Capturing and describing the processes by which organizations drift into failure require systems thinking. Systems thinking is about relationships and integration.

It sees a sociotechnical system not as a structure consisting of constituent departments, blunt ends and sharp ends, deficiencies and flaws, but as a complex web of dynamic, evolving relationships and transactions. Instead of building blocks, the systems approach emphasizes principles of organization. Understanding the whole is quite different from understanding an assembly of separate components. Instead of mechanical linkages between components (with a cause and an effect), it sees transactions—simultaneous and mutually interdependent interactions. Such emergent properties are destroyed when the system is dissected and studied as a bunch of isolated components (a manager, department, regulator, manufacturer, operator). Emergent properties do not exist at lower levels; they cannot even be described meaningfully with languages appropriate for those lower levels. Consider the discordant mix of inputs and expectations in the dynamic environment in which BP was working when it decided to engage in deepwater drilling:

> BP needed "elephants"—monster fields that could make a big difference in its bottom line. To find them, the geologists concluded, it should drill in new places, where BP's diplomatic skill, ingenuity, and daring gave it an edge... For the first time, technology permitted drilling below more than a mile of water, a hostile environment of total darkness, crushing pressures, and brutal temperatures. Sea-floor operations were carried out with remotely operated vehicles, known as ROVs: unmanned subs with stubby arms that could connect sections of drill pipe or wield tools to cut through steel. Advances in seismic imaging made it possible to locate hidden oil deposits. Drilling in deepwater was risky and expensive; a single well could cost more than $100 million. But the payoffs were huge. The gulf had particular appeal: The U.S. offered a stable democracy, low taxes, and minimal regulation, as well as nearby refineries and an insatiable market. BP rushed in, acquiring offshore leases, becoming the biggest player there. Over time, the deepwater gulf emerged as the engine of new U.S. oil production. All this was encouraged by American politicians, who cut taxes for offshore drilling and opened up new swaths of ocean for exploration. Meanwhile the government's Minerals Management Service—the drillers' chief regulator—operated like a promotional arm of the industry. Just about everybody, it seemed, liked offshore drilling (Elkind and Whitford 2011, p. 6).

Confronting such an environment with the right technologies and expertise, and knowing that they will work, is virtually impossible. In the wake of the Macondo disaster, an executive of a large contractor commented that equipment and rig designs exist to meet the challenges for which they are designed. Government approval follows if these things can be shown to meet the guidelines, but they don't really know what the guidelines are or whether they are applicable to deepwater scenarios. And most important of all, the pieces of kit that get certified and approved are not necessarily integrated. So what are the rules for certifying a new piece of kit so that it will not drift into failure? What should the standards be? The introduction of a new technology in a field of practice filled with other technologies is followed by experimentation, by negotiation, by surprise and discovery, by the creation of new relationships and rationalities. "Technical systems turn into models for themselves," said Weingart (1991), "the observation of their functioning, and especially their malfunctioning, on a real scale is required as a basis for further technical development" (p. 8). Rules, guidelines and standards do not exist as unequivocal, aboriginal markers against a tide of incoming operational data (and if they do, they are quickly proven useless or out of date). Rather, rules and standards are the constantly updated products of the processes of conciliation, of give and take, of the detection and rationalization of new data. As Wynne (1988) said:

> Beneath a public image of rule-following behavior and the associated belief that accidents are due to deviation from those clear rules, experts are operating with far greater levels of ambiguity, needing to make expert judgments in less than clearly structured situations. The key point is that their judgments are not normally of a kind—how do we design, operate and maintain the system according to "the" rules? Practices do not follow rules, rather, rules follow evolving practices (Wynne 1988, p. 153).

Setting up the various teams, engineering groups, and committees in systems like this is a way of bridging the gap between building and maintaining a system, between producing it and operating it. Bridging the gap is about adaptation—adaptation to newly emerging data (e.g., surprising wear rates) about an uncertain technology. But adaptation can mean drift. And drift can mean failure, breakdown.

MODELING LIVE SOCIOTECHNICAL SYSTEMS

What kind of safety model could capture such adaptation and predict its eventual collapse? Structuralist models are limited. Were there holes in layers of defense? Absolutely. If that is the way you want to analyze it. But such metaphors do not help you look for where the "hole" occurred, or why, let alone that they remind you that the hole is really your own retrospective construction. In the case of Macondo, there was something complexly organic about the vast web of clients, contractors, suppliers, something ecological. Something is lost when we model each of them as a layer of defense with a hole in it; when we see them as a mere deficiency or a latent failure. When we see systems instead as internally plastic, as flexible, as organic, their functioning is controlled by dynamic relations and ecological adaptation, rather than by rigid mechanical structures. They also exhibit self-organization. From year to year—or even more frequently—and from project to project, the organizational makeup of a drill rig is different in response to market, personnel, and environmental changes. The complex of organizations around also exhibit self-transcendence: the ability to reach out beyond currently known boundaries and learn, develop, and perhaps improve. What is needed is not yet another structural account of the end result of organizational deficiency. What is needed instead is a more functional account of living processes that coevolve with respect to a set of environmental conditions and that maintain a dynamic and reciprocal relation with those conditions. Such accounts need to capture what happens within an organization, with the gathering of knowledge and creation of rationality within workgroups, once a technology gets fielded. A functional account could cover the organic organization of maintenance steering groups and committees, whose makeup, focus, problem definition, and understanding coevolved with emerging anomalies and growing knowledge about an uncertain technology.

A model that is sensitive to the creation of deficiencies, not just to their eventual presence, makes a sociotechnical system come alive. It must be a model of processes, not just one of structure. Extending a lineage of cybernetic and systems-engineering research, Leveson (2012) proposed that control models can fulfill part of this task. Control models use the ideas of hierarchies and constraints to represent the emergent interactions of a complex system. In their conceptualization, a sociotechnical system consists of different levels, where each superordinate level imposes constraints on (or controls what is going on in) subordinate levels. Control models are one way to begin to map the dynamic relationships between different levels within a system—a critical ingredient of moving toward true systems thinking (where dynamic relationships and transactions are dominant, not structure and components). Emergent behavior is associated with the limits or constraints on the degrees of freedom of a particular level.

The division into hierarchical levels is an analytic artifact necessary to see how system behavior can emerge from those interactions and relationships. The resulting levels in a control model are of course a product of the analyst who maps the model onto the sociotechnical system. Rather than reflections of some reality out there, the patterns are constructions of a human mind looking for answers to particular questions. In fact, a one-dimensional hierarchical representation (with only up and down along one direction) probably oversimplifies the dynamic web of relationships surrounding (and determining the functioning of) any such multiparty, evolving group like a drill rig. But all models are simplifications, and the levels analogy can be helpful for an analyst who has particular questions in mind (e.g., why did these people at this level or in this group make the decisions they did, and why did they see that as the only rational way to go?).

Control among levels in a sociotechnical system is hardly ever completely effective. In order to control effectively, any controller needs a good model of what it is supposed to control, and it requires feedback about the effectiveness of its control. But such internal models of the controllers easily become inconsistent with, and no longer match, the system to be controlled (Leveson 2012). Buggy control models are true especially with uncertain, emerging technology (like deepwater drilling) and the engineering and regulatory requirements surrounding them. Feedback about the effectiveness of control is incomplete and can be unreliable too. A lack of incidents may provide the illusion that risk control is effective, whereas the paucity of risk actually depends on factors quite different from those in the organization's safety management systems. In this sense, the imposition of constraints on the degrees of freedom is mutual between levels and not just top-down. If subordinate levels generate imperfect feedback about their functioning, then higher-order levels do not have adequate resources (degrees of freedom) to act as would be necessary. Thus, the subordinate level imposes constraints on the superordinate level by not telling (or not being able to tell) what is really going on. Such a dynamic has been noted in various cases of drift into failure, including the Challenger Space Shuttle disaster (see Feynman 1988)

DRIFT INTO FAILURE AS EROSION OF CONSTRAINTS AND EVENTUAL LOSS OF CONTROL

Nested control loops can make a model of a sociotechnical system come alive more easily than a line of layers of defense. In order to model drift, it has to come alive. Control theory sees drift into failure as a gradual erosion of the quality or the enforcement of safety constraints on the behavior of subordinate levels. Drift results from either missing or inadequate constraints on what goes on at other levels. Modeling an accident as a sequence of events, in contrast, is really only modeling the end product of such erosion and loss of control. If safety is seen as a control problem, then events (just like the holes in layers of defense) are the results of control problems, not the causes that drive a system into disaster. A sequence of events, in other words, is at best the starting point of modeling an accident, not the analytic conclusion. The processes that generate these weaknesses are in need of a model.

One type of erosion of control occurs because original engineering constraints are loosened in response to the accumulation of operational experience. A variety of Starbuck and Milliken's (1988) "fine-tuning," in other words. This does not mean that the kind of ecological adaptation in system control is fully rational, or that it makes sense even from a global perspective on the overall evolution and eventual survival of the system. It does not. Adaptations occur, adjustments get made, and constraints get loosened in response to local concerns with limited time-horizons. They are all based on uncertain, incomplete knowledge. Often it is not even clear to insiders that constraints have become less tight as a result of their decisions in the first place, or that it matters if it is. And even when it is clear, the consequences may be hard to foresee, and judged to be a small potential loss in relation to the immediate gains. As Leveson (2012) put it, experts do their best to meet local conditions, and in the busy daily flow and complexity of activities, they may be unaware of any potentially dangerous side effects of those decisions. It is only with the benefit of hindsight or omniscient oversight (which is utopian) that these side effects can be linked to actual risk. Jensen (1996) describes it as such:

> We should not expect the experts to intervene, nor should we believe that they always know what they are doing. Often they have no idea, having been blinded to the situation in which they are involved. These days, it is not unusual for engineers and scientists working within systems to be so specialized that they have long given up trying to understand the system as a whole, with all its technical, political, financial, and social aspects (Jensen 1996, p. 368).

Being a member of a system, then, can make systems thinking all but impossible. Perrow (1984) made this argument very persuasively, and not just for the system's insiders. An increase in system complexity diminishes the system's transparency: diverse elements interact in a greater variety of ways that are difficult to foresee, detect, or even comprehend. Influences from outside the technical knowledge base (those "political, financial, and social aspects" of Jensen 1996, p. 368) exert a subtle but powerful pressure on the decisions and trade-offs that people make, and constrain what is seen as a rational decision or course of action at the time (Vaughan 1996). Thus, even though experts may be well educated and motivated, a "warning of an incomprehensible and unimaginable event cannot be seen, because it cannot be believed" (Perrow 1984, p. 23). How can experts and other decision makers inside organizational systems make sense of the available indicators of system safety performance? Making sure that experts and other decision makers are well informed is in itself an empty pursuit. What well informed really means in a complex organizational setting is infinitely negotiable, and clear criteria for what constitutes enough information are impossible to obtain. As a result, the effect of beliefs and premises on decision making and the creation of rationality can be considerable. Weick (1995, p. 87) pointed out that "seeing what one believes and not seeing that for which one has no beliefs are central to sensemaking. Warnings of the unbelievable go unheeded." That which cannot be believed will not be seen. This confirms the earlier pessimism about the value of incident reporting beyond 10^{-7}. Even if relevant events and warnings end up in the reporting system (which is doubtful because they are not seen as warnings even by those who would do the reporting), it is even more generous to presume that further expert analysis of such incident databases could succeed in coaxing the warnings into view. The difference, then, between expert insight at the time and hindsight (after an accident) is tremendous. With hindsight, the internal workings of the system may become lucid: the interactions and side effects are rendered visible. And with hindsight, people know what to look for, where to dig around for the rot, the missing connections.

Yet, as said, seeing holes and deficiencies in hindsight is not an explanation of the generation or continued existence of those deficiencies. It does not help predict or prevent failure. Instead, the processes by which such decisions come about, and by which decision makers create their local rationality, are one key to understanding how safety can erode on the inside of a complex, sociotechnical system. Why did these things make sense to organizational decision makers at the time? Why was it all normal, why was it not report-worthy, not even for the regulator tasked with overseeing these processes? The questions hang in the air. Little evidence is available from, for example, the BP investigation on such interorganizational processes or how they produced a particular conceptualization of risk. The report, like others, is testimony to the structuralist, mechanistic tradition in accident probes to date, applied even to investigative forays into social–organizational territory.

CREATION OF LOCAL RATIONALITY

The question is, how do insiders make those numerous little and larger trade-offs that together contribute to erosion, to drift? How is it that these seemingly harmless decisions can incrementally move a system to the edge of disaster? As indicated earlier, a critical aspect of this dynamic is that people in decision-making roles on the inside of a sociotechnical system miss or underestimate the global side effects of their locally rational decisions. These decisions are sound when set against local judgment criteria, given the time and budget pressures and short-term incentives that shape behavior. Given the knowledge, goals, and attentional focus of the decision makers and the nature of the data available to them at the time, it made sense. It is in these normal, day-to-day processes where we can find the seeds of organizational failure and success. And it is these processes we must turn to in order to find leverage for making further progress on safety. As Rasmussen and Svedung (2000) put it:

To plan for a proactive risk management strategy, we have to understand the mechanisms generating the actual behavior of decision-makers at all levels... an approach to proactive risk management involves the following analyses:

- A study of normal activities of the actors who are preparing the landscape of accidents during their normal work, together with an analysis of the work features that shape their decision making behavior.
- A study of the present information environment of these actors and the information flow structure, analyzed from a control theoretic point of view (Rasmussen and Svedung 2000, p. 14).

Reconstructing or studying the "information environment" in which actual decisions are shaped, in which local rationality is constructed, can help us penetrate processes of organizational sensemaking. These processes lie at the root of organizational learning and adaptation, and thereby at the source of drift into failure. The two space shuttle accidents (Challenger in 1986 and Columbia in 2003) are highly instructive here, if anything because the Columbia Accident Investigation Board (CAIB), as well as later analyses of the Challenger disaster (e.g., Vaughan 1996) represent significant (and, to date, rather unique) departures from the typical structuralist probes into such accidents. These analyses take normal organizational processes toward drift seriously, applying and even extending a language that helps us capture something essential about the continuous creation of local rationality by organizational decision makers.

One critical feature of the information environment in which NASA engineers made decisions about safety and risk was "bullets." Richard Feynman, who participated in the original Rogers Presidential Commission investigating the Challenger disaster, already fulminated against them and the way they collapsed engineering judgments into crack statements: "Then we learned about 'bullets'—little black circles in front of phrases that were supposed to summarize things. There was one after another of these little goddamn bullets in our briefing books and on the slides" (Feynman 1988, p. 127).

Eerily, "bullets" appear again as an outcropping in the 2003 Columbia accident investigation. With the proliferation of commercial software for making "bulletized" presentations since Challenger, bullets proliferated as well. This too may have been the result of locally rational (though largely unreflective) trade-offs to increase efficiency: bulletized presentations collapse data and conclusions and are dealt with more quickly than technical papers. But bullets filled up the information environment of NASA engineers and managers at the cost of other data and representations. They dominated technical discourse and, to an extent, dictated decision making, determining what would be considered as sufficient information for the issue at hand. Bulletized presentations were central in creating local rationality and central in nudging that rationality ever further away from the actual risk brewing just below.

Edward Tufte analyzed one Columbia slide in particular, from a presentation given to NASA by a contractor in February 2003 (CAIB 2003). The aim of the slide was to help NASA consider the potential damage to heat tiles created by ice debris that had fallen from the main fuel tank. (Damaged heat tiles triggered the destruction of Columbia on the way back into the earth's atmosphere, see Figure 5.1.) The slide was used by the Debris Assessment Team in their presentation to the Mission Evaluation Room. It was entitled "Review of Test Data Indicates Conservatism for Tile Penetration," suggesting, in other words, that the damage done to the wing was not so bad (CAIB 2003, p. 191). But actually, the title did not refer to predicted tile damage at

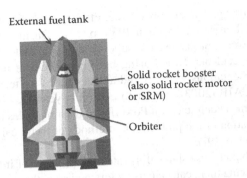

External fuel tank

Solid rocket booster
(also solid rocket motor
or SRM)

Orbiter

FIGURE 5.1 The design of the space shuttle and its external tank and solid rocket boosters.

all. Rather, it pointed to the choice of test models used to predict the damage. A more appropriate title, according to Tufte, would have been "Review of test data indicates irrelevance of two models." The reason was that the piece of ice debris that struck the Columbia was estimated to be 640 times larger than the data used to calibrate the model on which engineers based their damage assessments (later analysis showed that the debris object was actually 400 times larger). So the calibration models were not of much use: they hugely underestimated the actual impact of the debris.

The slide went on to say that "significant energy" would be required to have debris from the main tank penetrate the (supposedly harder) tile coating of the shuttle wing, yet that test results showed that this was possible at sufficient mass and velocity, and that, once the tiles were penetrated, significant damage would be caused. As Tufte observed, the vaguely quantitative word *significant* or *significantly* was used five times on the one slide, but its meaning ranged all the way from the ability to see it using those irrelevant calibration tests, through a difference of 640-fold, to damage so great that everybody onboard would die. The same word, the same token on a slide, repeated five times, carried five profoundly (yes, significantly) different meanings, yet none of those were really made explicit because of the condensed format of the slide. Similarly, damage to the protective heat tiles was obscured behind one little word, *it*, in a sentence that read "Test results show that it is possible at sufficient mass and velocity" (CAIB 2003, p. 191). The slide weakened important material, and the life-threatening nature of the data on it was lost behind bullets and abbreviated statements.

A decade and a half before, Feynman (1988) had discovered a similarly ambiguous slide about Challenger. In his case, the bullets had declared that the eroding seal in the field joints was "most critical" for flight safety, yet that "analysis of existing data indicates that it is safe to continue flying the existing design" (p. 137). The accident proved that it was not. Solid rocket boosters (or SRBs or SRMs) that help the space shuttle out of the earth's atmosphere are segmented, which makes ground transportation easier and has some other advantages. A problem that was discovered early in the shuttle's operation, however, was that the solid rockets did not always properly seal at these segments and that hot gases could leak through the rubber O-rings in the seal, called blow-by. This eventually led to the explosion of Challenger in 1986. The pre-accident slide picked out by Feynman had declared that while the lack of a secondary seal in a joint (of the solid rocket motor) was "most critical," it was still "safe to continue flying." At the same time, efforts needed to be "accelerated" to eliminate SRM seal erosion (Feynman 1988, p. 137). During Columbia as well as Challenger, slides were not just used to support technical and operational decisions that led up to the accidents. Even during both post-accident investigations, slides with bulletized presentations were offered as substitutes for technical analysis and data, causing the CAIB (2003), similar to Feynman years before, to grumble that: "The Board views the endemic use of PowerPoint briefing slides instead of technical papers as an illustration of the problematic methods of technical communication at NASA" (CAIB 2003, p. 191).

The overuse of bullets and slides illustrates the problem of information environments and how studying them can help us understand something about the creation of local rationality in organizational decision making. NASA's bulletization shows how organizational decision makers are configured in an "epistemic niche" (Hoven 2001). That which decision makers can know is generated by other people, and gets distorted during transmission through a reductionist, abbreviationist medium. The narrowness and incompleteness of the niche in which decision makers find themselves can come across as disquieting to retrospective observers, including people inside and outside the organization. It was after the Columbia accident that the Mission Management Team "admitted that the analysis used to continue flying was, in a word, 'lousy.' This admission—that the rationale to fly was rubber-stamped—is, to say the least, unsettling" (CAIB 2003, p. 190). "Unsettling" it may be, and probably is—in hindsight. But from the inside, people in organizations do not spend a professional life making "unsettling" decisions. Rather, they do mostly normal work. Again, how can a manager see a "lousy" process to evaluate flight safety as normal, as not something that is worthy reporting or repairing? How could this process be normal? The CAIB (2003) itself found clues to answers in pressures of scarcity and competition:

The Flight Readiness process is supposed to be shielded from outside influence, and is viewed as both rigorous and systematic. Yet the Shuttle Program is inevitably influenced by external factors, including, in the case of STS-107, schedule demands. Collectively, such factors shape how the Program establishes mission schedules and sets budget priorities, which affects safety oversight, workforce levels, facility maintenance, and contractor workloads. Ultimately, external expectations and pressures impact even data collection, trend analysis, information development, and the reporting and disposition of anomalies. These realities contradict NASA's optimistic belief that pre-flight reviews provide true safeguards against unacceptable hazards (CAIB 2003, p. 191).

Perhaps there is no such thing as "rigorous and systematic" decision making based on technical expertise alone. Expectations and pressures, budget priorities and mission schedules, contractor workloads, and workforce levels all influence technical decision making. All these factors determine and constrain what will be seen as possible and rational courses of action at the time. This dresses up the epistemic niche in which decision makers find themselves in hues and patterns quite a bit more varied than dry technical data alone. But suppose that some decision makers would see through all these dressings on the inside of their epistemic niche, and alert others to it. Tales of such whistleblowers exist. Even if the imperfection of an epistemic niche (the information environment) would be seen and acknowledged from the inside at the time, that still does not mean that it warrants change or improvement. The niche, and the way in which people are configured in it, answers to other concerns and pressures that are active in the organization—efficiency and speed of briefings and decision-making processes, for example. The impact of this imperfect information, even if acknowledged, is underestimated because seeing the side effects, or the connections to real risk, quickly glides outside the computational capability of organizational decision makers and mechanisms at the time.

Studying information environments, how they are created, sustained, and rationalized, and in turn how they help support and rationalize complex and risky decisions, is one route to understanding organizational sensemaking. More will be said on these processes of sensemaking elsewhere in this book. It is a way of making what sociologists call the macro–micro connection. How is it that global pressures of production and scarcity find their way into local decision niches, and how is it that they there exercise their often invisible but powerful influence on what people think and prefer; what people then and there see as rational or unremarkable? Although the intention was that NASA's flight safety evaluations be shielded from those external pressures, these pressures nonetheless seeped into even the collection of data, analysis of trends and reporting of anomalies. The information environments thus created for decision makers were continuously and insidiously tainted by pressures of production and scarcity (and in which organization are they not?), prerationally influencing the way people saw the world. Yet even this "lousy" process was considered "normal"—normal or inevitable enough, in any case, to not warrant the expense of energy and political capital on trying to change it. Drift into failure can be the result.

STUDY QUESTIONS

1. Why are structuralist models based on Newtonian ideas of linearity, cause–effect, and closed systems not very well suited for understanding accidents in complex organizations?
2. Why, in order to understand drift, is it more powerful to conceive of systems as dynamic relationships, rather than as structural layers or components?
3. How is understanding local rationality linked to studying the information environment of an organization's decision makers? What does this have to do with a radical–empiricist view of situation awareness from the previous chapter?
4. How does it help your answer to question 3 to see an organization as a complex system of relationships rather than a collection of components and layers?
5. Explain why, in already safe organizations, incident reporting alone cannot prevent a drift into failure, and might even offer the false idea that risk is under control.
6. If drift into failure is preceded by an erosion of safety constraints, and an eventual loss of control over the system, can you think of an example from your own industry or organization that would fit that description?
7. What are some of the steps an organization can take to prevent drifting into failure?

6 Methods and Models

CONTENTS

KEY POINTS

- Human factors and safety research has strong Cartesian–Newtonian influences, particularly because it lives at the intersection of psychology and engineering as well as pragmatic interests.
- These influences are visible in what human factors and safety research typically accepts as method and evidence, and in how the mental and social world is seen to work (as a Newtonian system).
- Social or psychological constructs are assumed to have causal power (e.g., complacency), but they are folk models—substitutive descriptions rather than explanations, immune against falsification and easily overgeneralized.

- Human factors and safety research with humans involves a double herme-neutic: (1) self-interpretations among the people studied, and (2) interpreta-tions by human researchers constituted in a particular context that offers a particular set of constructs, methods, and techniques.
- A key concern for safety interventions and design changes is anticipating how they will affect future human performance. Research methods aimed at this have to deal with evolution, change, and the emergence of new capa-bilities and complexities.

NEWTONIAN SOCIAL PHYSICS IN HUMAN FACTORS AND SAFETY RESEARCH

The challenge for a new era is to go beyond seeing the world as a Cartesian–Newtonian machine. Rather, it is a less stable, less predictable, and more complex place—active, adaptive, self-organizing, resilient. Its behavior is organic; emergent from the interac-tion between its many components and processes, not resultant from single compo-nents or processes. What we do in that world influences almost everything, yet controls almost nothing. Over the past few decades, science in general has begun to shed high modernist idealism. It has come to emphasize nonlinearity over linearity, complexity over simplification, holism over reductionism, and interpretation over accuracy. Such relinquishment of rigorous determinism presents new challenges and new opportu-nities for what we do in human factors and safety. One challenge, among many, is this. There has always been a kind of utilitarian, pragmatic instrumentalism that has characterized human factors research work and that has helped determine what passes for results in our domain. From its earliest beginnings, the field has been organized more around problems (what is it that we need to solve, attack?) than around solutions (these are the ways you need to go about solving problems if you want to call yourself a human factors researcher or safety scientist). In order to understand both the challenges and opportunities better, let's first briefly revisit the Cartesian–Newtonian precepts around which the field has been organized. We will unpack these to get a better idea about what might count as models and methods in our field.

CAUSALITY AND LINEAR THINKING

As shown throughout the book, Newtonian–Cartesian precepts, especially the idea of causality, have long been dominant in human factors and safety thinking and in many places still are. It has led us to explain and think about human action in terms of linear sequence(s) of causes and effects. It has offered our research and our investigations enormous analytical and explanatory leverage, seemingly matching the way the world works. But it has the potential of becoming a relatively simplistic form of social physics like that captured by the First Axiom of Industrial Safety (which stems from the 1930s):

The occurrence of an injury invariably results from a completed sequence of factors—the last one of these being the accident itself. The accident in turn is invariably caused or permitted by the unsafe act of a person and/or a mechanical or physical hazard (Heinrich et al. 1980, p. 47).

The Cartesian–Newtonian view of physics is evident in this axiom, as it believes action in the world can be described as a set of causal laws, with time reversibility, symmetry between cause and effect, and a preservation of the total amount of energy in the system. If enough is known about the initial conditions and parameters of a system, when Newtonian laws are derived for this system, it will be possible to predict all its future states. It does not matter whether a system is human or natural: the only limiting factor is complexity of the system. Yet greater complexity will merely increase the time and intellectual effort required to discover the laws. As has been argued for both behaviorism and empiricism earlier in this book, this is precisely what has given our models and methods their legitimacy: they reproduced and instantiated Newtonian physics.

Accident models and theory based on classical physics, like the Swiss Cheese Model, have been popular ever since. These models are attractive (and seductive) in part because they allow us to think safety and behavior in terms of causal series rather than complex relationships. What makes this particularly attractive is the reduction and translation of risk into energy. Energy is a critical property of Newtonian physics and much attention is given in this model to the accumulation, the often dangerous buildup, of energy, as well as its unintended, uncontrolled transfer, or release. Risk can be handled through a variety of forms of control. For example, in classic models of safety, an object can be protected by separating it from the source of hazard. This is typically achieved by a system of barriers: multiple layers whose function it is to diffuse propagations of dangerous and unintended energy transfers. Better organization of the organizational blunt end, better administrative ordering, helps as well: they can look after the condition of the barriers through safety management systems, loss prevention systems, auditing, counting incidents, violations, and other precursor problems, and implement bureaucratic interventions. Note how moral tags (unsafe acts, violations, complacency, deficiency) have insidiously become part of this model. This is interesting, because Newtonian physics was never concerned with moral valence. But to handle risk, both energy and its release have to be controlled. We expect responsible people to reduce vehicle speeds or the available dosage of a particular drug in its package. Risk has to be regulated as a moral, societal imperative. As discussed in Chapter 8, this imperative has become ensconced in zero visions in industries across the Western world.

Today, "best practice" social analysis challenges the usefulness and validity of retrospective analysis, such as that used in accident investigations. This is because the future cannot be entirely determined by the past—the science of linear equilibria does not apply, nor does the Newtonian symmetry between past and future. This has consequences for how we think about accountability for failure. Accidents where no physical breakage can be found of course heighten suspicions about human error. Given that no components in the engineered system malfunctioned or broke, the fault must lie with the people operating the system (the human factor). Failures of risk management can then be attributed to deficient supervision, ineffective leadership, or lack of appropriate rules and procedures (which points to components that were broken somewhere in the organization). This sanctions reductive thinking, of a kind that still assumes that broken components represent meaningful targets for intervention (e.g., eliminating "errors" in the design or operation or control of a safety-critical process).

REDUCTIONISM AND RELIABILITY

In the Newtonian model, macro properties of a system (e.g., safety) are treated as a straightforward function of the lower-order components or subsystems. This translates into the assumption common to many human factors and safety research projects and their models: safety can be increased by shielding the system from the unreliability of individual system components (human as well as machine). It has become not just part of academic models of safety, it informs folk or lay models as well. When asked how she could help improve patient safety in her hospital, a nurse manager replied that "if only I could get rid of the nurses who make mistakes, things would be a lot safer around here" (Dekker 2011c, p. 1). What informs statements like these is the West's central legitimating myth about safety. Remember from Chapter 1 how the first half of the 20th century was dominated by the idea that we are caretakers of inherently stable systems that require protection from random events like error. Again, this rests on the Newtonian idea of energy and the high-modernist emphasis on control, and it is still popular today:

> ...the elimination of human error is of particular importance in high-risk industries that demand reliability. For example, errors in commercial aviation, the military, and healthcare can be particularly costly in terms of loss of life (Krokos and Baker 2007, p. 175).

The frailties of individuals are seen as an inherent source of risk: they undermine the engineered and managed features of already safe systems. Human error has become seen as a managerial problem, a part of the organizational or regulatory brief. From this, the custodian role of safety management has arisen, forming a central element in today's master narrative about safety and the human factor. Over the past decades, social science has critiqued and attempted to distance itself from the kind of social physics that leads human actors and action to be equated with objects and events in the natural world. Yet disciplines and domains related to safety and human factors are lagging in this—instead introducing concepts such as "loss of SA" and reinvigorating others like "complacency" (e.g., "automation complacency") and imbuing them with Newtonian causal power. This is a constitution of human agency as informed by the instrumental rationality and engineering orientation the field has always had. Indeed, the Newtonian model makes no distinction between "natural" and "human" factors. The result is that the model is not well suited to explain the organizational and sociotechnical factors that lead to system breakdown. Nor, as shown in Chapter 5, does the model provide either an analysis or a language that can meaningfully handle processes of adaptation, risk management, and decision making. In fact, when this model is employed, many of the normal, daily, messy organizational details will be neglected or ignored, in particular the ones that eventually can have huge consequences (recall the banality of accidents thesis, for example).

LABORATORY STUDIES AND THE LEGACY OF LEIPZIG

For human factors, the traditional way of finding out about the world at the interface between humans and technology is to conduct experiments in the laboratory. Say that researchers want to test whether operators can safely use voice-input systems, or whether their interpretation of some target is better on three-dimensional displays. The typical strategy is to build microversions of the future system and expose a limited number of participants to various conditions, some or all of which may contain partial representations of a target system. Through its controlled settings, laboratory research already makes some sort of verifiable step into the future. Empirical contact with a world to be designed is ensured because some version of that future world has been prefabricated in the laboratory. This also leads to problems. Experimental steps into the future are necessarily narrow, which affects the generalizability of research findings. The mapping between test and target situations may miss several important factors.

In part as a result of a restricted integration of context, laboratory studies can yield divergent and eventually inconclusive results. Laboratory research on decision making, for example, has found several biases in how decision makers deal with information presented to them. Can new technology circumvent the detrimental aspects of such biases, which, according to some views, would lead to human error and safety problems? One bias is that humans are generally conservative and do not extract as much information from sources as they optimally should. Another bias, derived from the same experimental research, is that people have a tendency to seek far more information than they can absorb adequately. Such biases would seem to be in direct opposition to each other. It means that reliable predictions of human performance in a future system may be difficult to make on the basis of such research. Indeed, laboratory findings often come with qualifying labels that limit their applicability. Sanders and McCormick (1993, p. 572), for example, advised:

> When interpreting the … findings and conclusions, keep in mind that much of the literature is comprised of laboratory studies using young, healthy males doing relatively unmotivating tasks. The extent to which we can generalize to the general working population is open to question.

Still, experimental human factors research in the laboratory holds a special appeal because it makes mind measurable, and it even allows mathematics to be applied to the results. Quantitativism is good: it helps equate psychology with natural science, shielding it from the unreliable wanderings through mental life using dubious methods like introspection. The University laboratories that became a mainstay of many human factors departments were a 19th-century European invention, pioneered by scientists such as the chemist Justus Liebig. Wundt of course started the trend in psychology with his Leipzig laboratory. Leipzig did psychology a great service: psychophysics and its methods of inquiry introduced psychology as a serious science, as something realist, with numbers, calculations, and equations. The systematization, mechanization, and quantification of psychological research in Leipzig, however, must be seen as an antimovement against earlier introspection and rationalism.

Recall Watson's main concern when launching behaviorism. It was to rescue psychology from vague subjectivist introspection (by which he even meant Wundt's systematic, experimental laboratory research) and plant it firmly within the natural science tradition. Ever since Newton and others established what a scientific investigation was to be, psychology and human factors have struggled to find an acceptance and an acceptability within that conceptualization.

WHICH METHOD CAN HELP SEE INTO THE FUTURE?

IS FUTURE GIVEN?

In 2000, Nobel Laureate Ilya Prigogine gave a lecture on why changes in our descriptions of nature are necessary (Prigogine 2003). He asked the audience at the National Technical University of Athens: is future given? He noted the difficulties with our current descriptions of nature. The first, he said, is that nature leads to unexpected complexity. The second is that the classical view does not correspond to historical time-oriented evolution. Newton's theory, the classical view, is a deterministic one. According to it, time is reversible. Therefore, nothing new can appear. Everything was there and will forever be there. Even Laplace and Einstein believed that we are machines within the cosmic machine. Spinoza, too, argued that we are all machines even if we don't know it. To Prigogine, this seemed unsatisfactory: life, evolution, and change are everywhere around us. The universe is evolving and what we see are mere narrative stages. Collectively, these seem to move toward entropy, a lack of order and predictability. Future, then, is not given. It is not determined. It is full of emergence, events, bifurcations, diversity, and differentiation.

Prigogine was speaking about his field, physics. But the same, he proposed, is true of sociotechnical systems. Let us look at how this plays out in a concrete case—putting to the test the extent to which Newtonian natural-science methods can help see into the future. This, after all, is often a significant concern in safety work: will interventions help or hurt the safety and resilience of the system? Once such question has long been whether air-traffic controllers can "work" without paper flight strips. The paper flight strip is a small slip that contains flight plan data about each controlled aircraft's route, speed, altitude, times over waypoints, and other characteristics. It is used by air-traffic controllers in conjunction with a radar representation of air traffic. While a number of new control centers around the world are doing away with these paper strips, the ergonomics literature has come to no consensus on whether air-traffic control can actually do without them, and if it does, how it succeeds in keeping air traffic under control. Other ATC systems are being developed without flight strips, which generates concern on the part of some regulators because it creates largely unknown future work. Some literature suggests that flight strips are expendable without consequences for safety (e.g., Albright et al. 1996), while others argue that air-traffic control is basically impossible without them (e.g., Hughes et al. 1993). Design guidance that could be extracted from this research base (in its most basic form: do away with the strips, or keep the strips) can go either way.

EXPERIMENTAL SETTINGS

One way to find out if controllers can control air traffic without the aid of flight strips is to test it in an experimental setting. You take a limited number of controllers, and put them through a short range of tasks to see how they do. In their experiments, Albright et al. (1996) deployed a wide array of measurements to find out if controllers perform just as well in a condition with no strips as in a condition with strips. The work they performed was part of an effort by the US Federal Aviation Administration, a regulator (and ultimately the certifier of any future air-traffic control system in the United States). In their study, the existing air-traffic control system was retained, but to compare stripped versus stripless control, the researchers removed the flight strips in one condition:

> The first set of measurements consisted of the following: total time watching the PVD [plan view display, or radar screen], number of FPR [flight plan requests], number of route displays, number of J-rings used, number of conflict alerts activated, mean time to grant pilot requests, number of unable requests, number of requests ignored, number of controller-to-pilot requests, number of controller-to-center requests, and total actions remaining to complete at the end of the scenario (Albright et al. 1996, p. 6).

The assumption that drives most experimental research is that reality (in this case about the use and usefulness of flight strips) is objective and that it can be discovered by the researcher wielding the right measuring instruments. This is consistent with the structuralism and realism of human factors. The more measurements, the better, the more numbers, the more you know. This is assumed to be valid even when an underlying model that would couple the various measurements together into a coherent account of expert performance is often lacking. In experimental work, the number and diversity of measurements can become the proxy indicator of the accuracy of the findings, and of the strength of the epistemological claim (Q: So how do you know what you know? A: Well, we measured this, and this, and this, and that, and....). The assumption is that, with enough quantifiable data, knowledge can eventually be offered that produces an accurate and definitive account of a particular system. More of the same will eventually lead to something different. The strong influence that engineering has had on human factors makes this appear as just common sense. In engineering, technical debates are closed by amassing results from tests and experience; the essence of the craft is to convert uncertainty into certainty. Degrees of freedom are closed through numbers; ambiguity is worked out through numbers; uncertainty is reduced through numbers.

Independent of the number of measurements, each empirical encounter is of necessity limited, in both place and time. In the case of Albright et al. (1996), 20 air-traffic controllers participated in two simulated airspace conditions (one with strips and one without strips) for 25 minutes each. One of the results was that controllers took longer to grant pilot requests when they did not have access to flight strips, presumably because they had to assemble the basis for a decision on the request from other information sources. The finding is anomalous compared to other results, which showed no significant difference between workload and ability to keep control over the traffic situation across the strip–no strip conditions, leading to the conclusion that "the presence or absence of strips had no effect on either performance or perceived workload. Apparently, the compensatory behaviors were sufficient to maintain effective control at what controllers perceived to be a comparable workload" (Albright et al.

1996, p. 11). Albright et al. explain the anomaly as follows: "Since the scenarios were only 25 minutes in length, controllers may not have had the opportunity to formulate strategies about how to work without flight strips, possibly contributing to the delay" (Albright et al. 1996, p. 11).

At a different level, this explanation of an anomalous datum implies that the correspondence between the experimental setting and a future system and setting may be weak. Lacking a real chance to learn how to formulate strategies for controlling traffic without flight strips, it would be interesting to pursue the question of how controllers in fact remained in control over the traffic situation and kept their workload down. It is not clear how this lack of a developed strategy can affect the number of requests granted but not the perceived workload or control performance. Certifiers may, or perhaps should, wonder what 25 minutes of undocumented struggle tells them about a future system that will replace decades of accumulated practice. The emergence of new work and the establishment of new strategies are a fundamental accompaniment to the introduction of new technology, representing a transformation of tasks, roles, and responsibilities. These shifts are not something that could easily be noticed within the confines of an experimental study, even if controllers were studied for much longer than 25 minutes. Albright et al. (1996) resolve this by placing the findings of control performance and workload earlier in their text: "Neither performance nor perceived workload (as we measured them in this study) was affected when the strips were removed" (Albright et al. 1996, p. 8). The qualification that pulled the authority of the results back into the limited time and place of the experimental encounter (how we measured them in this study) was presented parenthetically and thus accorded less central importance. The resulting qualification suggests that comparable performance and workload may be mere artifacts of the way the study was conducted, of how these things were measured at that time and place, with those tools, by those researchers. The qualification, however, was in the middle of the paper, in the middle of a paragraph, and surrounded by other paragraphs adorned with statistical allusions. Nothing of the qualification remained at the end of the paper, where the conclusions presented these localized findings as universally applicable truths.

Rhetoric, in other words, is enlisted to deal with problematic areas of epistemological substance. The transition from localized findings (in this study, the researchers found no difference in workload or performance the way they measured them with these 20 controllers) to generalizable principles (we can do away with flight strips) essentially represents a leap of faith. As such, central points of the argument were left unsaid or were difficult for the reader to track, follow, or verify. By bracketing doubt this way, Albright et al. (1996) communicated that there was nothing, really, to doubt. Authority (i.e., true or accurate knowledge) derives from the replicable, quantifiable experimental approach. As Xiao and Vicente (2000) argued, it is very common for quantitative human factors research not to spend much time on the epistemological foundation of its work. Most often, it moves unreflectively from a particular context (e.g., an experiment) to concepts (not having strips is safe), from data to conclusions, or from the modeled to the model.

The ultimate resolution of the fundamental constraint on empirical work (i.e., each empirical encounter is limited to a time and place) is that more research is always necessary. This is regarded as a highly reasonable conclusion of most quantitative human factors or indeed any experimental work. For example, in the Albright et al. study, one constraint was the 25-minute time limit on the scenarios played. Does flight-strip removal actually change controller strategies in ways that were not captured by the present study? This would seem to be a key question. But again, the reservation was bracketed. Whether or not the study answered this question does not in the end weaken the study's main conclusion: "(Additional research is necessary to determine if there are more substantial long term effects to strip removal)" (Albright et al. 1996, p. 12).

In addition, the empirical encounter of the Albright et al. (1996) study was limited because it only explored one group of controllers (upper airspace). The argument for more research was drafted into service for legitimizing (not calling into question) results of the study: "Additional studies should be conducted with field controllers responsible for other types of sectors (e.g., low altitude arrival, or nonradar) to determine when, or if, controllers can compensate as successfully as they were able to in the current investigation" (Albright et al. 1996, p. 12). The idea is that more of the same, eventually, will lead to something different, that a series of similar studies over time will produce a knowledge increment useful to the literature and useful to the consumers of the research (certifiers in this case). This, once again, is largely taken for granted in the human factors community. Findings will invariably get better next time, and such successive, incremental enhancement is a legitimate route to the logical human factors end point: the discovery of an objective truth about a particular human–machine system and, through this, the revelation of whether it will be safe to use or not.

Experimental work typically relies on the production of quantifiable data. Some of this quantification (with statistical ornaments such as F-values and standard deviations) was achieved in Albright et al. (1996) by converting tick marks on lines of a questionnaire (called the "PEQ," or post-experimental questionnaire) into an ordinal series of digits:

> The form listed all factors with a 9.6 centimeter horizontal line next to each. The line was marked low on the left end and high on the right end. In addition, a vertical mark in the center of the line signified the halfway mark. The controllers were instructed to place an X on the line adjacent to the factor to indicate a response.... The PEQ scales were scored by measuring distance from the right anchor to the mark placed by the controller on a horizontal line (in centimeters)... Individual repeated measures ANOVAs [were then conducted] (Albright et al. 1996, pp. 5–8).

But ANOVAs cannot be used for the kind of data gathered through PEQ scales. The PEQ is made up of so-called ordinal scales. In ordinal scales, data categories are mutually exclusive (a tick mark cannot be at two distances at the same time), they have some logical order, and they are scored according to the amount of a particular characteristic they possess (in this case, distance in centimeters from the left anchor). Ordinal scales, however, do not represent equal differences (a distance of 2 cm does not represent twice as much of the category measured as a distance of 1 cm), as interval and ratio scales do. Besides, reducing complex categories such as "usefulness" or "likeability" to distances along a few lines probably misses out on an interesting ideographic reality beneath all of the tick marks. Put in experimental terms, the operationalization of usefulness as the distance from a tick mark along a line is not particularly high on internal validity. How can the researcher be sure that usefulness means the same thing to all responding controllers? If different respondents have different ideas of what usefulness meant during their particular experimental scenario, and if different respondents have different ideas of how much usefulness a tick mark, say, in the middle of the line represents, then the whole affair is deeply confounded. Researchers do not know what they are asking and do not know what they are getting in reply. Further numeric analysis deals with apples and oranges. This is one of the greater risks of folk modeling in human factors. It assumes that everybody understands what usefulness means, and that everybody has the same definition. But these are generous and untested assumptions. It was only with qualitative inquiry that researchers could ensure that there was some consensus on understandings of usefulness with respect to the controlling task with or without strips. Or they could discover that there was no consensus and then control for it. This would be one way to deal with the confound.

It may not matter, and it may not have been noticed. Numbers are good. Also, the linear, predictable format of research writing and the use of abbreviated statistical curios throughout the results section represent a rhetoric that endows the experimental approach with its authority—authority in the sense of privileged access to a particular layer or slice of empirical reality that others outside the laboratory setting do or do not have admittance to. Other rhetoric invented particularly for the study (e.g., PEQ scales for questions presented to participants after their trials in the Albright et al. [1996] study) certifies the researchers' unique knowledge of this slice of reality. It validates the researcher's competence to tell readers what is really going on there. It may dissuade second-guessing. Empirical results are deemed accurate by virtue of a controlled encounter, a standard reporting format that shows logical progress to objective truths and statements (introduction, method, results, discussion, and summary), and an authoritative dialect intelligible only to certified insiders.

CLOSING THE GAP FROM LABORATORY EXPERIMENT TO FUTURE WORLD

Because of some limited correspondence between the experiment and the system to be designed, quantitative research seemingly automatically closes the gap to the future. The stripless condition in the research (even if contrived by simply leaving out one artifact [the flight strip] from the present) is a model of the future. It is an impoverished model to be sure, and one that offers only a partial window onto what future practice and performance may be like (despite the epistemological reservations about the authenticity of that future discussed earlier). The message from Albright et al.'s (1996) encounter with the future is that controllers can compensate for the lack of flight strips. Take flight strips away, and controllers compensate for the lack of information by seeking information elsewhere (the radar screen, flight-plan readouts, controller-to-pilot requests). Someone might point out that Albright et al. prejudged the use and usefulness of flight strips in the first few sentences of their introduction, that they did not see their data as an opportunity to seek alternative interpretations:

Currently, en route control of high altitude flights between airports depends on two primary tools: the computer-augmented radar information available on the Plan View Display (PVD) and the flight information available on the Flight Progress Strip (Albright et al. 1996, p. 1).

This is not really an enabling of knowledge; it seems to be more the imposition of it. Here, flight strips are not seen as a problematic core category of controller work, whose use and usefulness would be open to negotiation, disagreement, or multiple interpretations. Instead, flight strips function as information-retrieval devices. Framed as such, the data and the argument can really only go one way: by removing one source of information, controllers will redirect their information-retrieving strategies onto other devices and sources. This displacement is possible, it may even be desirable, and it is probably safe: "Complete removal of the strip information and its accompanying strip marking responsibilities resulted in controllers compensating by retrieving information from the computer" (Albright et al. 1996, p. 11). For a certifier, this closes a gap to the future: removing one source of information will result in people finding the information elsewhere (while showing no decrement in performance or increment in workload). The road to automation is open and people will adapt successfully (or will have to), for that has been scientifically proven. Therefore, doing away with the flight strips is (probably) safe, and certifiable as such.

If flight strips are removed, then what other sources of information should remain available? Albright et al. (1996) inquired about what kind of information controllers would minimally like to preserve: route of flight scored high, as did altitude information and aircraft call sign. Naming these categories gives developers the opportunity to envision an automated version of the flight strip that presents the same data in digital format, one that substitutes a computer-based format for the paper-based one, without any consequences for controller performance. Such a substitution, however, may overlook critical factors associated with flight strips that contribute to safe practice, and that would not be incorporated or possible in a computerized version (Mackay 2000).

Any signs of potential ambiguity or ambivalence about what else flight strips may mean to those working with them were not given further consideration beyond a brief mention in the experimental research write-up—not because these signs were actively, consciously stifled, but because they were inevitably deleted as Albright et al. (1996) carried out and wrote up their encounter with empirical reality. Albright et al. explicitly solicited qualitative, richer data from their participants by asking if controllers themselves felt that the lack of strips impaired their performance. Various controllers indicated how strips help them preplan and that, without strips, they cannot preplan. The researchers, however, never unpacked the notion of preplanning or investigated the role of flight strips in it. Again, such notions (e.g., preplanning) are assumed to speak for themselves, taken to be self-evident. They require no deconstruction, no further interpretive work. Paying more attention to these qualitative responses could create noise that confounds experimental accuracy. Comments that preplanning without strips was impossible hinted at flight strips as a deeper, problematic category of controller work. But if strips mean different things to different controllers, or worse, if preplanning with strips means different things to different controllers, then the experimental bedrock of comparing comparable people across comparable conditions would disappear. This challenges the nomothetic averaging out of individual differences. Where individual differences are the nemesis of experimental research, interpretive ambiguity can call into question the legitimacy of the objective scientific enterprise.

INVESTIGATING HOW PEOPLE MAKE SENSE OF THEIR WORK

Rather than looking at people's work from the outside in (as do quantitative experiments), qualitative research tries to understand people's work from the inside out. When taking the perspective of the one doing the work, how does the world look through his or her eyes? What role do tools play for people themselves in the accomplishments of their tasks; how do tools affect their expression of expertise? An interpretive perspective is based on the assumption that people give meaning to their work and that they can express those meanings through language and action. Qualitative research interprets the ways in which people make sense of their work experiences by examining the meanings that people use and construct in light of their situation (Golden-Biddle and Locke 1993).

The criteria and end points for good qualitative research are different from those for quantitative research. As a research goal, accuracy is practically and theoretically unobtainable. Qualitative research is relentlessly empirical, but it rarely achieves finality in its findings. Not that quantitative research ever achieves finality (remember that virtually every experimental report finishes with the exhortation that more research is necessary). But qualitative researchers admit that there is never one accurate description or analysis of a system in question, no definitive account—only versions. What flight strips exactly do for controllers is forever subject to interpretation; it will never be answered objectively or finitely, never be closed to further inquiry. What makes a version good, though, or credible, or worth paying attention to by a certifier, is its authenticity. The researcher has to not only convince the certifier of a genuine field experience in writing up the research account but also make intelligible what went on there. Validation from outside the field emerges from an engagement with the literature (What have others said about similar contexts?) and from interpretation (How well are theory and evidence used to make sense of this particular context?). Field research, though critical to the ethnographic community as a stamp of authenticity, is not necessarily the only legitimate way to generate qualitative data. Surveys of user populations can also be tools that support qualitative inquiry.

FIND OUT WHAT THE USERS THINK

The reason that qualitative research may appeal to certifiers is that it lets the informants, the users, speak—not through the lens of an experiment, but on the users' terms and initiative. Yet this is also where a central problem lies. Simply letting users speak can be of little use. Qualitative research is not (or should not be) plain conversational mappings—a direct transfer from field setting to research account. If human factors would (or continues to) practice and think about ethnography in these terms, doubts about both the method and the data it yields will continue to surface. What certifiers, as consumers of human factors research, care about is not what users say in raw, unpacked form, but about what their remarks mean for work, and especially for future work. As Hughes et al. (1993) put it: "It is not that users cannot talk about what it is they know, how things are done, but it needs bringing out and directing toward the concerns of the design itself" (p. 138). Within the human factors community, qualitative research seldom takes this extra step. What human factors requires is a strong ethnography, one that actually makes the hard analytical move from user statements to a design language targeted at the future.

A qualitative undertaking related to flight strips was the Lancaster University project (Hughes et al. 1993). Many man-months were spent (an index of the authenticity of the research) observing and documenting air-traffic control with flight strips. During this time, the researchers developed an understanding of flight strips as an artifact whose functions derive from the controlling work itself. Both information and annotations on the strip and the active organization of strips among and between controllers were essential: "The strip is a public document for the members of the (controlling) team; a working representation of an aircraft's control history and a work site of controlling. Moving the strips is to organize the information in terms of work activities and, through this, accomplishing the work of organizing the traffic" (Hughes et al. 1993, pp. 132–133). Terms such as *working representation* and *organizing traffic* are concepts, or categories, that were abstracted well away from the masses of deeply context-specific field notes and observations gathered in the months of research. Few controllers would themselves use the term working representation to explain what flight strips mean to them. This is good. Conceptual abstraction allows a researcher to reach a level of greater generality and increased generalizability (Woods 1993; Xiao and Vicente 2000). Indeed, working representation may be a category that can lead to the future, where a designer would be looking to computerize a working representation of flight information, and a certifier would be evaluating whether such a computerized tool is safe to use. But such higher-order interpretive work is seldom found in human factors research. It would separate ethnography and ethnographic argument from research that simply makes claims on the basis of authenticity. Even Hughes et al. (1993) relied on authenticity alone when they told of the various annotations made on flight strips and did little more than parrot their informants:

Amendments may be done by the controller, by the chief, or less often, by one of the "wings." "Attention-getting" information may also be written on the strips, such as arrows indicating unusual routes, symbols designating "crossers, joiners, and leavers" (that is, aircraft crossing, leaving or joining the major traffic streams), circles around unusual destinations, and so on (Hughes et al. 1993, p. 132).

Though serving as evidence of socialization, of familiarity and intimacy, speaking insider language is not enough. By itself, it is not helpful to certifiers who may be struggling with evaluating a version of air-traffic control without paper flight strips. Appeals to authenticity ("Look, I was there, and I understand what the users say") and appeals to future relevance ("Look, this is what you should pay attention to in the future system") can thus pull in opposite directions: the former toward the more context-specific that is hardly generalizable, the latter toward abstracted categories of work that can be mapped onto yet-to-be-fielded future systems and conceptions of work. The burden to resolve the tension should not be on the certifier or the designer of the system, it should be on the researcher. Hughes et al. (1993) agreed that this bridge-building role should be the researcher's:

> Ethnography can serve as another bridge between the users and the designers. In our case, controllers have advised on the design of the display tool with the ethnographer, as someone knowledgeable about but distanced from the work, and, on the one hand able to appreciate the significance of the controllers' remarks for their design implications and, on the other hand, familiar enough with the design problems to relate them to the controllers' experiences and comments (Hughes et al. 1993, p. 138).

Hostage to the Present, Mute about the Future

Hughes et al.'s (1993) research account actually missed the "significance of controller remarks for their design implications" (p. 138). No safety implications were extracted. Instead, the researchers used insider language to forward insider opinions, leaving user statements unpacked and largely underanalyzed. Ethnography essentially gets confused with what informants say and consumers of the research are left to pick and choose among the statements. This is a slightly naive form of ethnography, where what informants can tell researchers is equated or confused with what strong, analytical ethnography (and ethnographic argument) could reveal. Hughes et al. relied on informant statements to the extent they did because of a common belief that the work that their informants did, and the foundational categories that informed it, are for the most part self-evident; close to what we would regard as common sense. As such, they require little, if any, analytic effort to discover. It is an ethnography reduced to a kind of mediated user show-and-tell for certifiers—not as thorough analysis of the foundational categories of work. For example, Hughes et al. concluded that "(flight strips) are an essential feature of 'getting the picture,' 'organising the traffic,' which is the means of achieving the orderliness of the traffic" (Hughes et al. 1993, p. 133).

So flight strips help controllers get the picture. This kind of statement is obvious to controllers and merely repeats what everyone already knows. If ethnographic analysis cannot go beyond common sense, it merely privileges the status quo. As such, it offers certifiers no way out: a system without flight strips would not be safe, so forget it. There is no way for a certifier to circumvent the logical conclusion of Hughes et al. (1993) "The importance of the strip to the controlling process is difficult to overestimate" (p. 133). So is it safe? Going back to Hughes et al.: "For us, such questions were not easily answerable by reference to work which is as subtle and complex as our ethnographic analysis had shown controlling to be" (p. 135).

This might read a little bit like a surrender to the complexity and intricacy of a particular phenomenon—what Dawkins (1986, p. 38) called the "argument from personal incredulity." When faced with highly complicated machinery or phenomena, it is easy to take cover behind our own sense of extreme wonder, and resist efforts at explanation. In the case of Hughes et al., it recalls an earlier reservation: "The rich, highly detailed, highly textured, but nevertheless partial and selective descriptions associated with ethnography would seem to contribute little to resolving the designers problem where the objective is to determine what should be designed and how" (p. 127). Such justification maneuvers an ethnographic enterprise out of the certifier's or designer's view as something not particularly helpful. Synthesizing the complexity and subtlety of a setting should not be the burden of the certifier. Instead, this is the role of the researcher; it is the essence of strong ethnography. That a phenomenon is remarkable does not mean it is inexplicable; so if we are unable to explain it, "we should hesitate to draw any grandiose conclusions from the fact of our own inability" (Dawkins 1986, p. 39).

Informant remarks such as "Flight strips help me get the mental picture" should serve as a starting point for qualitative research, not as its conclusion. But how can researchers move from native category to analytic sense? Qualitative work should be hermeneutic and circular in nature: not aiming for a definitive description of the target system, but rather a continuous reinterpretation and reproblematization of the successive layers of data mined from the field. Data demand analysis. Analysis in turn guides the search for more data, which in turn demand further analysis: categories are continually revised to capture the researcher's (and, hand in hand, the practitioner's) evolving understanding of work. There is a constant interplay between data, concepts, and theory.

The analysis and revision of categories are a hallmark of strong ethnography, and Ross's (1995) study of flight-progress strips in Australia serves as an interesting example. Qualitative in nature, Ross's research relied on surveys of controllers using flight strips in their current work. Surveys are often derided by qualitative researchers for imposing the researcher's understanding of the work onto the data, instead of the other way around (Hughes et al. 1993). Demonstrating that it is not just the empirical encounter or rhetorical appeals to authenticity that matter (through large numbers of experimental probes or months of close observation), the survey results Ross gathered were analyzed, coded, categorized, recoded, and recategorized until the inchoate masses of context-specific controller remarks began to form sensible, generalizable wholes that could meaningfully speak to certifiers.

Following previous categorizations of flight-strip work, Ross (1995) moves down from these conceptual descriptions of controller work and up again from the context-specific details, leaving several layers of intermediate steps. In line with characterizations of epistemological analysis through abstraction hierarchies (see Xiao and Vicente 2000), each step from the bottom up is more abstract than the previous one; each is cast less in domain-bound terms and more in concept-dependent terms than the one before. This is *induction*: reasoning from the particular to the general. One example from Ross (1995, p. 27) concerns domain-specific controller activities such as "entering a pilot report; composing a flight plan amendment." These lower-level, context-specific data are of course not without semantic load themselves: it is always possible to ask further questions and descend deeper into the world of meanings that these simple, routine activities have for the people who carry them out. Indeed, we have to ask if we can only go up from the context-specific level—maintained in human factors as the most atomistic, basic, low-level data set. In Ross's data, researchers should still question the common sense behind the otherwise taken-for-granted entering of a pilot report: what does a pilot report mean for the controller in a particular context (e.g., weather related), what does entering this report mean for the controller's ability to manage other traffic issues in the near future (e.g., avoiding sending aircraft into severe turbulence)?

While alluding to even more fine-grained details and questions later, these types of activities also point to an intentional strategy at a higher level of analysis: that of the "transformation or translation of information for entry into the system," which, at an even higher level of analysis, could be grouped under a label coding, together with other such strategies (Ross 1995, p. 27). Part of this coding is symbolic, in that it uses highly condensed markings on flight strips (underlining, black circles, strikethroughs) to denote and represent for controllers what is going on. The highly intricate nature of even one flight (where it crosses vs. where it had planned to cross a sector boundary, what height it will be leaving when, whether it has yet contacted another frequency, etc.) can be collapsed or amortized by simple symbolic notation—one line or circle around a code on the strip that stands for a complex, multidimensional problematic that other controllers can easily recognize. Unable to keep all the details of what a flight would do stable in the head, the controller compresses complexity, or amortizes it, as Ed Hutchins might say, by letting one symbol stand for complex concepts and interrelationships, some even temporal.

Similarly, "recognizing a symbol for a handoff" (on a flight strip), though allowing further unpacking (e.g., what do you mean "recognize"?), is an instance of a tactic that "transforms or translates information received," which in turn represents a larger controller competency of "decoding," which in its turn is also part of a strategy to use symbolic notation to collapse or amortize complexity (Ross 1995, p. 27). From recognizing a symbol for a handoff to the collapsing of complexity, there are four steps, each more abstract and less in domain terms than the one before. Not only do these steps allow others to assess the analytical work for its worth, but the destination of such induction is actually a description of work that can be used for guiding the evaluation of a future system. Inspired by Ross's analysis, we can summarize that controllers rely on flight strips for

- Amortizing or collapsing complexity (what symbolic notation conveys)
- Supporting coordination (who gets which flight strip next from whom)
- Anticipating dynamics (how much is to come, from where, when, in what order)

These (no longer so large) jumps to the highest level of abstraction can now be made—identifying the role the flight strip has in making sense of workplace and task complexity. Although not so much a leap of faith any longer (because there are various layers of abstraction in between), the final step, up to the highest level of conceptual description, still appears to hold a certain amount of creative magic. Ross (1995) revealed little of the mechanisms that actually drive his analysis. There is no extensive record that tracks the transformation of survey data into conceptual understandings of work. Perhaps these transformations are taken for granted too: the mystery is left unpacked because it is assumed to be no mystery. The very process by which the researcher manages to migrate from user-language descriptions of daily activities to conceptual languages less anchored in the present remains largely hidden from view. No ethnographic literature guides specifically the kinds of inferences that can be drawn up to the highest level of conceptual understanding. At this point, a lot of leeway is given (and reliance placed on) the researcher and his or her (keenly) developed insight into what activities in the field really mean or do for people who carry them out. The problems of this final step are known and acknowledged in the qualitative research community. Vaughan (1996) and other sociologists referred to it as making the macro–micro connection: locating general meaning systems (e.g., symbolic notation, off-loading) in local contexts (placing a circle around a set of digits on the flight strip). Geertz (1973) noted how inferences that try to make the macro–micro connection often resemble "perfected impressionism" in which "much has been touched but little grasped" (p. 312). Such inferences tend to be evocative, resting on suggestion and insinuation more than on analysis (Vaughan 1996).

In qualitative research, lower levels of analysis or understanding always underconstrain the inferences that can be drawn further on the way to higher levels. At each step, alternative interpretations are possible. Qualitative work does not arrive at a finite description of the system or phenomenon studied (nor does quantitative research, really). But qualitative work does not even aim or pretend to do so (Batteau 2001). Results are forever open to further interpretation, forever subject to increased problematization. The main criterion, therefore, to which we should hold the inferences drawn is not accuracy (Golden-Biddle and Locke 1993), but plausibility: does the conceptual description make sense—especially to the informants, to the people who actually do the work? This also motivates the continuous, circular nature of qualitative analysis: reinterpreting results that have been interpreted once already, gradually developing a theory—a theory of why flight strips help controllers know what is going on that is anchored in the researcher's continually evolving understanding of the informants' work and their world.

The three high-level categories of controller (flight-strip) work tell certifiers that air-traffic controllers have developed strategies for dealing with the communication of complexity to other controllers, for predicting workload and planning future work. Flight strips play a central, but not necessarily exclusive, role. The research account is written up in such a way that the status quo does not get the prerogative: tools other than flight strips could conceivably help controllers deal with complexity, dynamics, and coordination issues. Complexity and dynamics, as well as coordination, are critical to what makes air-traffic control what it is, including difficult. Whatever designers or developers or regulators will want to brand as safe to use, they would do well to take into account that controllers use their artifact(s) to help them deal with complexity, to help them anticipate dynamic futures, and to support their coordination with other controllers. This resembles some kind of human factors requirements that could provide a certifier with meaningful input.

HELPING THE DESIGNER LOOK INTO THE FUTURE

One role of human factors is to help developers and certifiers judge whether a technology is safe for future use. But quantitative and qualitative human factors communities both risk taking the authority of their findings for granted and regarding the translation to future, and claims about the future being either safe or unsafe, as essentially nonproblematic. At least the literature (both literatures) are relatively silent on this fundamental issue. Yet neither the legitimacy of findings nor the translation to claims about the future is in fact easily achieved, or should be taken for granted. More work needs to be done to produce findings that make sense for those who have to certify a system as safe to use. Experimental human factors research can claim empirical legitimacy by virtue of the authority vested in the laboratory researcher and the control over the method used to get data. Such research can speak meaningfully to future use because it tests micro-versions of a future system. Researchers, however, should explicitly indicate where the versions of the future they tested are impoverished, and what subtle effects of context on their experimental settings could produce findings that diverge from what future users will encounter.

Qualitative research in human factors can claim legitimacy and relevance to those who need to certify the next system, because of its authentic encounters with the field where people actually carry out the work. Validation emerges from the literature (what others have said about the same and similar contexts) and from interpretation (how theory and evidence make sense of this particular context). Such research can speak meaningfully to certification issues because it allows users to express their preferences, choices, and apprehensions. Qualitative human factors research, however, must not stop at recording and replaying informant statements. It must deconfound informant understandings with understandings informed by concepts, theory, analysis, and literature. Human factors work, of whatever kind, can help bridge the gap from research findings to future systems. Research accounts need to be both convincing as science and cast in a language that allows an engineer, a developer, a designer, to look ahead to the future: looking ahead to work and a coevolution of people and technology in a system that does not yet exist.

SUPPOSED QUALITATIVE–QUANTITATIVE DIVIDE

Whether quantitative or qualitative research can make more valid claims about the future (thereby helping in the certification of a system as safe to use) is contested. At first sight, qualitative, or field studies are about the present (otherwise, there is no field to study). Quantitative research may test actual future systems, but the setting is typically so contrived and limited that its relationship to a real future is tenuous. As many have pointed out, the difference between quantitative and qualitative research is actually not so great (e.g., Woods 1993; Xiao and Vicente 2000). Claims of epistemological privilege by either are counterproductive and difficult to substantiate. A method becomes superior only if it better helps researchers answer the question they are pursuing, and in this sense, of course, the differences between qualitative and quantitative research can be real. But dismissing qualitative work as subjective misses the point of quantitative work. Squeezing numbers out of an experimental encounter with reality, and then closing the gap to a concept-dependent conclusion on what you just saw, requires generous helpings of interpretation. As we see in the following discussion, there is a great deal of subjectivism in endowing numbers with meaning. Moreover, seeing qualitative inquiry as a mere protoscientific prelude to real quantitative research misconstrues the relationship and overestimates quantitative work. A common notion is that qualitative work should precede quantitative research by generating hypotheses that can then be tested in more restricted settings. This may be one relationship. But often quantitative work only reveals the how or what (or how much) of a particular phenomenon. Numbers in themselves can have a hard time revealing the why of the phenomenon. In this case, quantitative work is the prelude to real qualitative research: it is experimental number crunching that precedes and triggers the study of meaning.

Finally, a common claim is that qualitative work is high in external validity and low in internal validity. Quantitative research, on the other hand, is thought to be low in external validity and high in internal validity. This is often used as justification for either approach and it must rank among the most misconstrued arguments in scientific method. The idea is that internal validity is high because experimental laboratory research allows an investigator almost full control over the conditions in which data are gathered. If the experimenter did not make it happen, either it did not happen, or the experimenter knows about it, so that it can be dealt with as a confound. But the degree of control in research is often overestimated. Laboratory settings are simply another kind of contextualized setting, in which all kinds of subtle influences (social expectations, people's life histories) enter and influence performance just like they would in any other contextualized setting. The degree of control in qualitative research, on the other hand, is often simply assumed to be low. And much qualitative work indeed adds to that image. But rigor and control are definitely possible in qualitative work: there are many ways in which a researcher can become confident about systematic relationships between different factors. Subjectivism in interpretation is not more necessary in qualitative than in quantitative research. Qualitative work, on the other hand, is not automatically externally valid simply because it takes place in a field (applied) setting. Each encounter with empirical reality, whether qualitative or quantitative, generates context-specific data—data from that time and place, from those people, in that language—that are by definition nonexportable to other settings. The researcher has to engage in analysis of those data in order to bring them up to a concept-dependent level, from which terms and conclusions can be taken to other settings.

ONTOLOGICAL ALCHEMY

Living, as it does, on the threshold between psychology and engineering, human factors can have a hard time remaining faithful, or credible, to both. But it tries. The example above, about flight strips, shows that not all statements by operators or practitioners about what they experience (or what goes on "in their minds") can be accommodated pragmatically or analytically within the confines of a typical human factors study. And indeed, such introspection as research method was rejected as untrustworthy and replaced by behaviorism a long time ago. But if we ask people to introspect using a standardized survey, and turn their answers into numbers, will we be believed then? A workload index (the NASA TLX) that does pretty much this, is popular, and its use has gone largely unreflected and unexamined for decades now. One reason might be that it fits the experimentalist approach that lends credibility to many human factors research projects. It fits the empiricist and quantitativist biases too. Again, this was first championed in psychology by Wilhelm Wundt in his Leipzig laboratory. Wundt quickly had to admit that a chronometry of mind was too bold a research goal. But human factors research still reflects pale versions of his ambition. The measurement of workload is in fact the pursuit of a kind of chronometry of mind. Typical workload measurement scales will inquire after the temporal experiences and mental demand. They offer respondent ordinal scales to mark the experienced extent of each.

This is a kind of ontological alchemy: it transforms a mental figment (an invention, a creation, a fabrication, indeed—a construct like workload) into a set of measurable figures. Ontological alchemy is the transformation of an introspective judgment about a mental experience into a number: turning it into something useful. Once workload has become a figure, our analytic machinery stops. We do not concern ourselves with other ontological claims that our informants, our subjects, our participants, might want to put before us, because that really would be introspection and we cannot trust that. Indeed, qualitative inquiries are often suspected as subjectivist and unscientific. Whatever "reality testing" has occurred, has been taken care of by the measurement, the tool, the index, and the numbers are there to prove it. That such work can lead to quantifiable conclusions is an operation that seems to yield objective statements from which human factors derives additional epistemological confidence. At least in contrast to, say, ethnographic studies of human performance (e.g., Hutchins 1995), human factors researchers may not feel as comfortable with language as a device for relating the results of their observations. Quantification of human factors and safety research results reinforces the belief in the existence of an observer-independent reality; the quantified results are an objective window into it:

The use of figures and graphs not only embodies numbers, but gives the reader the sense of "seeing the phenomenon." By using figures and graphs the scientist implicitly says, "You don't have to take my word for it, look for yourself" (Gergen 1999, p. 56).

But what do the numbers "stand for" to begin with? How were they arrived at (recall Albright et al.'s ruler applied to an ordinal scale in the flight-strip example) and, most significantly, what might they possibly be hiding? Validation and certainty, and thus epistemological confidence, are reduced to a kind of ranking or numerical strength, a sort of democracy of numbers (Dekker et al. 2010). In effect, the operationalization of a research design is equated with counting, inventorying, or classifying. In turn, these operations are equated with the construction of strong epistemological statements about the world, the production of facts and science.

This is not unique to experimental or laboratory studies. The literature on ethnographic methods also typically pursues authenticity and some form of objectivity in the description of its empirical encounters with the world. Ethnographic texts ask the reader to accept that the researcher was indeed present in the field and grasped how insiders understood their world (Golden-Biddle and Locke 1993). Ways to achieve this authenticity include particularizing everyday situations in that world (capturing a level of empirical detail that would not have been possible without that field work) and the delineation of the relationship between researcher and informant (i.e., where the perspective of one ends and that of the other begins), as well as qualifying any personal biases that the researcher may have brought to the field. These strategies appeal to some achievement of objectivity, to a separation between the observer and what was observed, and thereby invoke the notion of a world apart from the researcher (even if the perception onto that world by other people—the insiders—all that the researcher can gain access to).

Perhaps another clear example of this comes from the study of decision making, and in particular the evaluation of performance by studying a kind of hit rate of final judgments or decisions (Dekker et al. 2010). One example of a hit rate in human factors, of course, is the tabulation of errors. Recall from earlier in the book how this was described as researchers having perfected new ways way to monitor pilot performance (Croft 2001). Hit rates of observed behavior has also been included in more ethnographic studies of practitioners' adaptations to new technology in, for example, cockpits and operating theaters, which increased the fit of those studies' results with human factors' epistemological preoccupations (Björklund et al. 2006). The research literature on judgment and decision making is built on this. But we must ask what is missing if all that we study about decision making is the hit rate of final judgments? This is largely blind to the richness of knowledge and reasoning and reduces human agency to the outcome of only cognitive processes—ignoring emotive or embodied notions of agency.

FOLK MODELS

And thus we constitute objects that can be measured, quantified, and "proven." Many of our objects can be used interchangeably to refer to the same basic phenomenon. Take, for example, group situation awareness, shared problem models, team situation awareness, mutual knowledge, shared mental models, joint situation awareness, and shared understanding. At the same time, results about what constitutes the phenomenon are fragmented and ideas on how to measure it remain divided. Methods to gain empirical access range from modified measures of practitioner expertise, to questionnaires interjected into suddenly frozen simulation scenarios, to implicit probes embedded in unfolding simulations of natural task behavior. This renders empirical demonstrations of the phenomenon, particularly by our own epistemological standards, unverifiable and inconclusive. After all, how can a researcher claim that he or she saw something if that something was not defined? Perhaps there is no need to define the phenomenon, because everybody knows what it means. Indeed, something like situation awareness is a *folk model*. It has come up from the practitioner community (fighter pilots in this case). Folk models are useful because they can collapse complex, multidimensional problems into simple labels that everybody can relate to. But this is also where the risks lie, certainly when researchers pick up on a folk label and attempt to investigate and model it scientifically.

Situation awareness is not alone in this. Human factors today has more concepts that aim to provide insight into the human performance issues that underlie complex behavioral sequences. It is often tempting to mistake the labels themselves for deeper insight—something that is becoming increasingly common in, for example, accident analyses. Thus, loss of situation awareness, automation complacency and loss of effective crew resource management (CRM) can now be found among the causal factors and conclusions in accident reports. This happens without further specification of the psychological mechanisms responsible for the observed behavior—much less how such mechanisms or behavior could have forced the sequence of events toward its eventual outcome. The labels (modernist replacements of the old *pilot error*) are used to refer to concepts that are intuitively meaningful. Everyone is assumed to understand or implicitly agree on them, yet no effort is usually made to explicate or reach agreement on the underlying mechanisms or precise definitions. People may no longer dare to ask what these labels mean, lest others suspect they are not really initiated in the particulars of their business.

Indeed, large labels that correspond roughly to mental phenomena we know from daily life are deemed sufficient—they need no further explanation. This is often accepted practice for psychological phenomena because as humans we all have privileged knowledge about how the mind works (because we all have one). However, a verifiable and detailed mapping between the context-specific (and measurable) particulars of a behavior on the one hand and a concept-dependent model on the other is not achieved—the jump from context-specifics (somebody flying into a mountainside) to concept dependence (the operator must have lost SA) is immune to critique or verification.

Folk models are not necessarily incorrect, but compared to articulated models, they focus on descriptions rather than explanations, and they are very hard to prove wrong. Folk models are pervasive in the history of science. One well-known example of a folk model from modern times is Freud's psychodynamic model, which links observable behavior and emotions to nonobservable structures (id, ego, superego) and their interactions. One feature of folk models is that nonobservable constructs are endowed with the necessary causal power without much specification of the mechanism responsible for such causation. According to Kern (1998), for example, complacency can cause a loss of situation awareness. In other words, one folk problem causes another folk problem. Such assertions leave few people any wiser. Because both folk problems are constructs postulated by outside observers (and mostly post hoc), they cannot logically cause anything in the empirical world. Yet this is precisely what they are assumed to be capable of. Recall how, in wrapping up a conference on situation awareness, Charles Billings warned against this danger in 1996:

> The most serious shortcoming of the situation awareness construct as we have thought about it to date, however, is that it's too neat, too holistic and too seductive. We heard here that deficient SA was a causal factor in many airline accidents associated with human error. We must avoid this trap: deficient situation awareness doesn't "cause" anything. Faulty spatial perception, diverted attention, inability to acquire data in the time available, deficient decision-making, perhaps, but not a deficient abstraction (Billings 1996, p. 3)!

Situation awareness is too "neat" and "holistic" in the sense that it lacks a level of detail and thus fails to account for a psychological mechanism that would connect features of the sequence of events to the outcome. The folk model, however, was coined precisely because practitioners (pilots) wanted something "neat" and "holistic" that could capture critical but inexplicit aspects of their performance in complex, dynamic situations. We have to see their use of a folk model as legitimate. It can fulfill a useful function with respect to the concerns and goals of a user community.

Folk models can seem like a convenient bridge between basic and applied worlds, between scientific and practitioner communities. Terms like *situation awareness* allow both camps to speak the same language. But such conceptual sharing risks selling out to superficial validity. It may not do human factors a lot of good in the long run, nor may it really benefit the practitioner consumers of research results.

Another folk concept is complacency. Why does people's vigilance decline over time, especially when confronted with repetitive stimuli? Vigilance decrements have formed an interesting research problem ever since the birth of human factors during and just after the Second World War. The idea of complacency has always been related to vigilance problems. Although complacency connotes something motivational (people must ensure that they watch the process carefully), the human factors literature actually has little in the way of explanation or definition. What is complacency? Why does it occur? If you want answers to these questions, do not turn to the human factors literature. You will not find answers there. Complacency is one of those constructs, whose meaning is assumed to be known by everyone. This justifies taking it up in scientific discourse as something that can be manipulated or studied as an independent or dependent variable, without having to go through the bother of defining what it actually is or how it works. In other words, complacency makes a "neat" and "holistic" case for studying folk models.

DEFINITION BY SUBSTITUTION

The most evident characteristic of folk models is that they define their central constructs by substitution rather than decomposition. A folk concept is explained simply by referring to another phenomenon or construct that itself is in equal need of explanation. Substitution is not the same as decomposition: substituting replaces one high-level label with another, whereas decomposition takes the analysis down into subsequent levels of greater detail, which transform the high-level concept into increasingly measurable context specifics. A good example of definition by substitution is the label complacency, in relation to the problems observed on automated flight decks. Most textbooks on aviation human factors talk about complacency and even endow it with causal power, but none really define (i.e., decompose) it:

- According to Wiener (1988, p. 452), "boredom and complacency are often mentioned" in connection with the out-of-the-loop issue in automated cockpits. But whether complacency causes an out-of-the-loop condition or whether it is the other way around is left unanswered.
- O'Hare and Roscoe (1990, p. 117) stated that "because autopilots have proved extremely reliable, pilots tend to become complacent and fail to monitor them." Complacency, in other words, is invoked to explain monitor failures.

- Kern (1998, p. 240) maintained that "as pilots perform duties as system monitors they will be lulled into complacency, lose situational awareness, and not be prepared to react in a timely manner when the system fails." Thus, complacency can cause a loss of situational awareness. But how this occurs is left to the imagination.
- On the same page in their textbook, Campbell and Bagshaw (1991, p. 126) said that complacency is both a "*trait* that can lead to a reduced awareness of danger" and a "*state* of confidence plus contentment" (emphasis added). In other words, complacency is at the same time a long-lasting, enduring feature of personality (a trait) and a shorter-lived, transient phase in performance (a state).
- For the purpose of categorizing incident reports, Parasuraman et al. (1993, p. 3) defined complacency as: "self-satisfaction which may result in nonvigilance based on an unjustified assumption of satisfactory system state." This is part definition but also part substitution: self-satisfaction takes the place of complacency and is assumed to speak for itself. There is no need to make explicit by which psychological mechanism self-satisfaction arises or how it produces nonvigilance.

It is in fact difficult to find much substance on complacency in the human factors literature. The phenomenon is often described or mentioned in relation to some deviation or diversion from official guidance (people should coordinate, double-check, look—but they do not), which is both normativist and judgmental. The "unjustified assumption of satisfactory system state" in Parasuraman et al.'s (1993) definition is emblematic for human factors' understanding of work by reference to externally dictated norms. If we want to understand complacency, the whole point is to analyze why the assumption of satisfactory system state is justified (not unjustified) by those who are making that assumption. If it were unjustified, and they knew that, they would not make the assumption and would consequently not become complacent. Saying that an assumption of satisfactory system state is unjustified (but people still keep making it—they must be motivationally deficient) does not explain much at all.

None of the above examples really provide a definition of complacency. Instead, complacency is treated as self-evident (everybody knows what it means, right?) and thus it can be defined by substituting one label for another. The human factors literature equates complacency with many different labels, including boredom, overconfidence, contentment, unwarranted faith, overreliance, self-satisfaction, and even a low index of suspicion. So if we would ask, "What do you mean by 'complacency'?," and the reply is, "Well, it is self-satisfaction," we can be expected to say, "Oh, of course, now I understand what you mean." But do we really? Explanation by substitution actually raises more questions than it answers. By failing to propose an articulated psychological mechanism responsible for the behavior observed, we are left to wonder. How is it that complacency produces vigilance decrements or how is it that complacency leads to a loss of situation awareness? The explanation could be a decay of neurological connections, fluctuations in learning and motivation, or a conscious trade-off between competing goals in a changing environment. Such definitions, which begin to operationalize the large concept of complacency, suggest possible probes that a researcher could use to monitor for some target effect. But because none of the descriptions of complacency available today offer any such roads to insight, claims that complacency was at the heart of a sequence of events are immune to critique and falsification.

IMMUNITY AGAINST FALSIFICATION

Most philosophies of science rely on the empirical world as touchstone or ultimate arbiter (a reality check) for postulated theories. Following Popper's rejection of the inductive method in the empirical sciences, theories and hypotheses can only be deductively validated by means of falsifiability. This usually involves some form of empirical testing to look for exceptions to the postulated hypothesis, where the absence of contradictory evidence becomes corroboration of the theory. Falsification deals with the central weakness of the inductive method of verification, which, as pointed out by David Hume, requires an infinite number of confirming empirical demonstrations. Falsification, on the other hand, can work on the basis of only one empirical instance, which proves the theory wrong. As seen in Chapter 3, this is of course a highly idealized, almost clinical conceptualization of the scientific enterprise. Yet, regardless, theories that do not permit falsification at all are highly suspect.

The resistance of folk models against falsification is known as *immunization*. Folk models leave assertions about empirical reality underspecified, without a trace for others to follow or critique. For example, a senior training captain once asserted that cockpit discipline is compromised when any of the following attitudes are prevalent: arrogance, complacency, and overconfidence. Nobody can disagree because the assertion is underspecified and therefore immune against falsification. This is similar to psychoanalysts claiming that obsessive–compulsive disorders are the result of unduly harsh toilet training that fixated the individual in the anal stage. In the same vein, if the question of "Where are we headed?" from one pilot to the other is interpreted as a loss of situation awareness, this claim is immune against falsification. The journey from context-specific behavior (people asking questions) to the postulated psychological mechanism (loss of situation awareness) is made in one big leap, leaving no trace for others to follow or critique.

Current theories of situation awareness are not sufficiently articulated to be able to explain why asking questions about direction represents a loss of situation awareness. Some theories may superficially appear to have the characteristics of good scientific models, yet just below the surface they lack an articulated mechanism that is amenable to falsification. Although falsifiability may at first seem like a self-defeating criterion for scientific progress, the opposite is true: the most falsifiable models are usually also the most informative ones, in the sense that they make stronger and more demonstrable claims about reality. In other words, falsifiability and informativeness are two sides of the same coin.

FOLK MODELS VERSUS YOUNG AND PROMISING MODELS

One risk in rejecting folk models is that the baby is thrown out with the bath water. In other words, there is the risk of rejecting even those models that may be able to generate useful empirical results, if only given the time and opportunity to do so. Indeed, the more articulated human factors constructs (such as decision making, diagnosis) are distinguished from the less articulated ones (situation awareness, complacency) in part by their maturity, by how long they have been around in the discipline. This is the pre-paradigmatic answer to any methodological critique. Human factors, it says, is still a young science. And there is nothing, in principle, to prevent it from becoming a normal science in the future. It has not been persuasively demonstrated, after all, that it is impossible for human factors to achieve the rigor and objectivity of the so-called hard sciences. It just takes more time and more evidence. In the pre-paradigmatic view, it is natural—or hard—science that lies at the core of any field's scientific identity. Hard science functions as a kind of index against which any other field's epistemological confidence gets ranked. But, of course, even hard science has been relativized through what has been called the *universality of hermeneutics* (e.g., Feyerabend 1993). Hermeneutics, the study of interpretation, is no longer linked only to the study of humans but to all sciences. Even natural sciences are now seen as historically conditioned and human constructed (Wallerstein 1996). They too must constantly reassess what constitutes relevant facts, methods, and theories and what counts as "nature." Interpretation in human factors involves tacit practical skills and conventions acquired through training and reinforced through the possibilities afforded by the field's professional categories, constructs, and publication outlets.

What opportunity should the younger ones receive before being rejected as unproductive? The answer to this question hinges, once again, on falsifiability. Ideal progress in science is described as the succession of theories, each of which is more falsifiable (and thus more informative) than the one before it. Yet when we assess loss of situation awareness or complacency as more novel explanations of phenomena that were previously covered by other explanations, it is easy to see that falsifiability has actually decreased, rather than increased. Take as an example an automation-related accident that occurred when situation awareness or automation-induced complacency did not yet exist—in 1973. The aircraft in question was on approach in rapidly changing weather conditions. It was equipped with a slightly deficient *flight director* (a device on the central instrument panel showing the pilot where to go, based on an unseen variety of sensory inputs), which the captain of the airplane distrusted. The airplane struck a seawall bounding Boston's Logan Airport about one kilometer short of the runway and slightly to the side of it, killing all 89 people onboard. In its comment on the crash, the National Transportation Safety Board explained how an accumulation of discrepancies, none critical in themselves, can rapidly deteriorate into a high-risk situation without positive flight management. The first officer, who was flying, was preoccupied with the information presented by his flight-director systems, to the detriment of his attention to altitude, heading, and airspeed control (Billings 1997).

Today, both automation-induced complacency of the first officer and a loss of situation awareness of the entire crew could likely be cited under the causes of this crash. (Actually, that the same set of empirical phenomena can comfortably be grouped under either label—complacency or loss of situation awareness—is additional testimony to the undifferentiated and underspecified nature of these concepts.) These supposed explanations (complacency, loss of situation awareness) were obviously not needed in 1974 to deal with this accident. The analysis left us instead with more detailed, more falsifiable, and more traceable assertions that linked features of the situation (e.g., an accumulation of discrepancies) with measurable or demonstrable aspects of human performance (diversion of attention to the flight director vs. other sources of data). The decrease of falsifiability represented by complacency and situation awareness as hypothetical contenders in explaining this crash represents the inverse of scientific progress, and therefore in itself already argues for the rejection of such novel concepts.

NORMATIVISM

Of course, is there such a thing as complacency at all? Can we ever arrive at a satisfactory definition, independent of how much work we might do and data we might gather? The problem, as with situation awareness, is that any definition is inherently (if tacitly) normative. Recall from earlier chapters that we can only define situation awareness relative to some target world that the observer should or could be aware of. This might be the experimenter's understanding of the situation, a collation of process data available to the operator (but perhaps not all observed) or some other "ground truth." The same is the case for complacency. A subtle, nuanced, and thoughtful consideration of complacency is that of Moray and Inagaki (2000). When we speak of complacency, they say, we are concerned with monitoring, with sampling visual displays or other representations or indications of process behavior, particularly in automated systems. "The notion of complacency arises from the suggestion that, particularly when the automation being monitored is highly reliable, operators may not merely trust it, but trust it too much, so that they fail to sample (monitor) the variables often enough" (p. 355). Complacency refers to an incorrect strategy that leads to suboptimal monitoring. Important signals may be missed because of operator complacency, because they have too great a trust in their systems doing the right thing. Such trust tends to grow in the operation of increasingly reliable systems (Billings 1997).

To be complacent, Moray and Inagaki explain, an operator must be shown to sample a process indicator less often than is optimal, given the dynamics of the source. But how often and when should a display be visually sampled in order that monitoring is "optimal"? And who gets to say? Studies that claim to show the existence of complacency is defective, because none have ever rigorously defined optimal behavior in supervisory monitoring. The existence of complacency cannot be proved unless optimal behavior is specified as a benchmark. If signals are missed, then it cannot be claimed that missed signals imply, or are caused by, complacency. More insidiously, the supposed "benchmark" against which actual monitoring behavior is sampled is often defined in hindsight, with knowledge of outcome. Of course with hindsight, it is really easy to show what an optimal sampling rate would have been, because a bad outcome easily shows the suboptimal. If only people had seen *this* piece of data, *then* in the unfolding process, the outcome could have been avoided. This of course means nothing: it is a mere retrospective judgment using a folk model, and not an explanation of anything. As Moray and Inagaki explain:

This is not just a matter of semantics. To claim that an operator missed a signal because of complacency is to have recourse to the classic tactic which all claim to eschew, namely, put the blame on the operator and say that "human error" was the cause of the problem. A term like "complacency" implies, in everyday language, a character trait that the operator has acquired and which could be changed. To say that a person's "complacency" was the cause of an error is not the same as saying, for example, that a fault was due to limited working memory, a characteristic of the nervous system that is not under voluntary control (Moray and Inagaki 2000, p. 362).

Even if optimal monitoring strategies are achievable, would it guarantee that no signals will be missed? Moray and Inagaki demonstrate how even optimal monitoring will often (or almost always) lead to critical signals being missed. In real-world systems, some sources of information are associated with high hazard (e.g., airspeed) and optimal monitoring strategies are typically developed relative to that. But they are not the only sources that need to be monitored. There are always trade-offs between sampling different information sources. These are driven, as signal detection theory explains, by payoffs and probabilities. And as signal detection theory can also explain (see Chapter 7): there is no such thing as an optimal sampling rate to make sure no signals are missed. If that optimum is developed relative to one channel, or one process, or one kind of signal, then immediate sacrifices are made with respect to all others. As Moray and Inagaki dryly observe: "The only way to guarantee to detect all signals is devote attention entirely and continuously to the one process on which the critical signals are expected to appear" (Moray and Inagaki 2000, p. 360). This is indeed what Parasuraman et al. (1993) instructed their operators to do in one experiment, and complacency all but disappeared entirely. But, as Moray and Inagaki point out, in real worlds, this is entirely impractical.

OVERGENERALIZATION

The lack of specificity of folk models and the inability to falsify them contribute to their overgeneralization. One famous example of overgeneralization in psychology is the inverted U-curve, also known as the Yerkes–Dodson law. Ubiquitous in human factors textbooks, the inverted U-curve couples arousal with performance (without clearly stating any units of either arousal or performance), where a person's best performance is claimed to occur between too much arousal (or stress) and too little, tracing a sort of hyperbole. The original experiments were, however, neither about performance nor about arousal (Yerkes and Dodson 1908). They were not even about humans. Examining "the relation between stimulus strength and habit formation," the researchers subjected laboratory rats to electrical shocks to see how quickly they decided to take a particular pathway versus another. The conclusion was that rats learn best (i.e., they form habits most rapidly) at any but the highest or lowest shock. The results approximated an inverted U only with a most generous curve fitting, the x axis was never defined in psychological terms but in terms of shock strength, and even this was confounded: Yerkes and Dodson used different levels of shock that were too poorly calibrated to know how different they really were. The subsequent overgeneralization of the Yerkes–Dodson results (to no fault of their own, incidentally) has confounded stress and arousal, and after a century, there is still little evidence that any kind of inverted-U relationship holds for stress (or arousal) and human performance. Overgeneralizations take narrow laboratory findings and apply them uncritically to any broad situation where behavioral particulars bear some prima facie resemblance to the phenomenon that was investigated under controlled circumstances.

Other examples of overgeneralization and overapplication include "perceptual tunneling" (putatively championed by the crew of an airliner that descended into the Everglades after its autopilot was inadvertently switched off) and the loss of effective CRM as major explanations of accidents (e.g., Aeronautica Civil 1996). A most frequently quoted sequence of events with respect to CRM is the flight of an iced-up airliner from Washington National Airport in the winter of 1982 that ended shortly after takeoff on the 14th Street Bridge and in the Potomac River. The basic cause of the accident is said to be the copilot's unassertive remarks about an irregular engine instrument reading (despite the fact that the copilot was known for his assertiveness). This supposed explanation hides many other factors that might be more relevant, including air-traffic control pressures, the controversy surrounding rejected takeoffs close to decision speed, the sensitivity of the aircraft type to icing and its pitch-up tendency with even little ice on the slats (devices on the wing's leading edge that help it fly at slow speeds), and ambiguous engineering language in the airplane manual to describe the conditions for use of engine anti-ice.

Does It Matter That We Have Folk Models?

It could be argued that none of this matters. After all, human factors is a field born out of pragmatic concerns, not some preoccupation with science or philosophy. As long as human factors is able to produce tangible results, better displays, better work environments, we should perhaps not care about the extent to which it is "science," by whatever criteria. This argument has been made for situation awareness, about which there is now a "substantial body of research pointing to its utility" (Parasuraman et al. 2008, p. 144). However, utility should not be confused with science. Kuhn (1962) warned against assuming that our constructions represent the most accurate or most useful representations of what is going on. In fact, there could be something silently repressive, something exclusionary, about practices that valorize the accumulation of empirical evidence for existing constructs, about practices that prioritize the quantification of research results, which reassert the existence of an observer-independent reality. This, at worst, could blind us to the need to reflect on the nature of our own practices and beliefs. Critical review and self-reflection is necessary because, in the end, it is what distinguishes scientific knowledge from folk modeling (Dekker et al. 2010). Inquiry about the epistemological assumptions of human factors creates additional questions, especially about the ways in which researchers entering the field have been educated. In human factors education, and psychology education more broadly, there seems to have emerged in recent decades a general lack of coverage in the areas of the history of psychology and philosophy of science. In conversations with colleagues and students, Degree programs are becoming less likely to require courses in history, philosophy, or related topics. This might lead to reinvented wheels, as illustrated by the lack of historical scholarship in research on situation awareness.

METHODS IN SAFETY AND HUMAN FACTORS—WHERE TO GO

Counting, measuring, categorizing, and statistically analyzing have been chief tools of the trade for a long time. This trade has been supported by constructing the sorts of objects (SA, complacency, human error) that allow counting and categorizing. Human factors has a realist orientation, assuming that empirical facts are stable, objective aspects of reality that exist independent of the observer or his or her theory. Human errors are among those facts that some researchers think they can see out there, in some objective reality. But the facts researchers see would not exist without them or their method or their theory. The use of such experimental apparatus helps fix their legitimacy as practices that produce knowledge and keep knowledge in circulation. Foucault called this an epistemé: a set of rules and conceptual tools for what counts as evidence. Such practices are of necessity exclusionary. They function in part to establish distinctions between those statements that will be considered true (or factual) and those that will be considered false (or speculative, folk). For example, "lost situation awareness" is a language we accept as scientific to denote a mental state, whereas "losing the bubble" is a statement we leave to practitioners. We explicitly and implicitly agree that we can operationalize (and thereby measure) one and not the other. With one, we can derive some degree of epistemological confidence because we believe that we are not just reporting a subjective experience. Operationalization and our attempts to ensure the accuracy and validity of a measurement affirm that our scientific constructs are truer pictures of the world than what practitioners could arrive at. The true statement is circulated through the literature, reproduced in publications that underpin what is taken to be valid knowledge in our field.

Of course, none of this makes the facts generated through experiments less real to those who observe them, or publish them, or read about them. Heeding Thomas Kuhn (1962), however, this reality should be seen for what it is: an implicitly negotiated settlement among like-minded researchers, rather than a common denominator accessible to all. There is no final arbiter here. It is possible that a componential, experimentalist approach could eventually enjoy an epistemological privilege. There is a hope, for example, that what phrenology once did for psychology, neuroscience will do for human factors: it will come to the empiricist rescue. With more time, and more evidence, particularly more biological or physical evidence that correlates with mental states, we can prove that what we are about really is real. It will allow us to start measuring "real" things; "objectively" real things, visible things, just like real science. The physics envy and empiricist knee jerk of human factors are almost too obvious in this: that which cannot be seen, or cannot be turned into a number, cannot be believed to be "real" (Wilkin 2009). As a result, that which cannot be seen must be turned into something that can (like information processing has always tried, and mental workload scales do as well). Because then it is real. But, of course, even the measurable objects of hard science are relative because of what has been called the "universality of hermeneutics" (Feyerabend 1993). Just like the measurable bumps on crania that once "correlated" with criminal proclivity or mathematical aptitude, the cranial blood flows of neuroimaging are objective facts and correlates of mental states because they have been historically conditioned and human constructed as such (Wallerstein 1996). Whatever the science, it must never overestimate the ontological status of its own constructs, and it must constantly reassess how it itself constitutes its measurable facts.

This also means that there is no automatic imperative for the experimental approach to uniquely stand for legitimate research, as it sometimes seems to do in mainstream human factors. Ways of getting access to empirical reality are infinitely negotiable, and their acceptability is a function of how well they conform to the worldview of those to whom the researcher makes his appeal. The persistent quantitativist supremacy (particularly in North American human factors) seems saddled with this type of consensus authority (it must be good because everybody is doing it). Such methodological hysteresis could have more to do with primeval fears of being branded "unscientific" (the fears shared by Wundt and Watson) than with a steady return of significant knowledge increments generated by the research.

DECLINE OF ESSENTIALISM

Methods, their rules, the objects they study—these all help produce the reality that they seek to understand. Recall that the 1974 accident discussed by Billings above was not an instance of a loss of SA, because the construct simply was not around to make it so. Take memory as another example. Constructs of what memory is and how it works have historically been inscribed on metaphors derived from the technology of the period, ranging from wax tablets and books to photography, telephone switchboards, computers, and even holograms. As the metaphor changed, so has thought about memory. Similarly, the information-processing metaphor, dominant in human factors in the 1970s, turned attention into a phenomenon of storage and retrieval. As technological developments (wax tablets, transistors, computers) influence our models and language, we change the way we think about and therefore how we study memory or attention. This means that the constructs we talk about and do research into are not essential: they have no properties that are immutable or constant across all space and time. Constructs, and the words we use to express them, are choices—consensual agreements—for how to see the world. A technical vocabulary of constructs creates a particular empirical world and would exist differently with other words. It is consensus that cements word to world (Gergen 1999). This robs words (or constructs) of any inherent claim to truthfulness in the sense of describing the world as it is. Their ontological status is only as stable as our agreement about their meaning.

This kind of critical inquiry, a critique of what is assumed to be natural, has emerged as a central preoccupation in late 20th-century science. Knowledge does not come from an increasingly fine-tuned discovery of the underlying structure of the world; rather, knowledge is seen as fabricated, or constructed, in multiple versions. This is seen as something that occurs through and reflects the resources and mechanisms of a given social order. Such constructions especially come into play when we ask how we can mitigate risk, reduce errors, or control people's violations. What this interpretive program stresses is that these things do not simply exist out there "in the world" for us to discover and excise. Instead, its focus is on how and why we bring them into being, how and why we ourselves produce them, through our act of inquiry, through the way we conduct our interrogation(s) of the world (Bader and Nyce 1995; Lützhoft et al. 2010). This requires safety research to assume a reflexivity that it never had before: it is necessary to consider how conceptual and empirical focus that defines the domain of safety science may say more about us as a research community and our fellow practitioners than it does about the world we attempt to describe. This means we will have to investigate how we carry out and legitimate risk assessment and analysis methodologies, error counts, or other audits and interventions. The work of the safety community tends to rely on and reinforce the role "scientific" measurements of human performance play in assessing safety in an organization. The scientific voice takes primacy because of its putatively privileged (i.e., untainted, objective, value-free) access to empirical reality and because its results are represented as numbers that fit right in with contemporary business models for running an organization.

The classical "scientific" approach to safety research may once have assumed that its categories of interest ("risk" to begin with, but also incidents, hazards, errors, violations) can be unproblematically transformed into empirical units of analysis and tabulation, as if they possess immutable identities independent of the observer and her or his context, language, or interests. But today, it is expected that we consistently interrogate traditional conceptions of the real: risk no longer exists in any kind of objective space. Nor does it represent any immutable feature in the physical world. In short, any definition of risk, no matter how rigorous the method of arriving at it, can be reduced to the mind of human observers who impose a particular language, interest, imaginations, and background onto the world around them. Safety and human factors research is no neutral arbiter or producer of knowledge. It, like all knowledge practices, creates descriptions that have certain consequences and end points. As such, it validates and legitimizes certain practices and interpretations at the expense of others.

COMPLEXITY AND METHODOLOGICAL OPEN-MINDEDNESS

Giddens (1984) has argued that the study of human activity, like we do in safety and human factors, is necessarily based on people's situational self-interpretation and the words used to construct what we see. Human factors is an activity in which humans study humans. Humans are self-reflecting actors, not objects in the natural world that do not answer back. For Giddens, this involves a double hermeneutic. First, there are self-interpretations among those people who are studied in human factors research. Recall the Hollnagel and Amalberti (2001) study on human error and that people who were the object of that study disagreed with that interpretation of their performance by some of the researchers. The second hermeneutic applies to the human researchers themselves, who are, of course, humans too. They are constituted in a particular context that offers a particular set of constructs, methods, and techniques. In many of the studies of human factors (error counting and tabulation studies, or research into situation awareness or workload), interpretation governs both the participants' reflections on their own performance and the researchers' subsequent interpretations of those reflections. Any findings reported in such research, says Giddens (1984), are only as stable as those two interpretations. They have no transhistoric truth values that can progress toward a greater accumulation of facts or "science." This is not at all bad in itself. It is only disingenuous to claim that objective, transhistoric facts have been, or can be, generated by such research.

There is great seduction in the progressivist perspective; in the idea that safety and human factors research can and will produce ever more innovative and ever more accurate representations of the world it seeks to understand and control. But no science of any kind today is believed to be capable of producing the facts that 19th-century science thought were possible. An example of this from safety science is that qualitative values, not accurate numbers (e.g., *indicators* of resilience rather than *measures*), are now pursued (Hollnagel et al. 2006). Perhaps the social, or sociotechnical, is not to be taken as so definite—not as it once was in the Wundtian laboratory in any case. With a world that is complex, diffuse, and messy, methods that create order and linearity only do so because they distort things into clarity (Law 2004).

It is not that there should be no room for conventional research methods in human factors and safety. In many cases, such methods may deliver some answers or generate workable questions that, in the words of Levi-Strauss, are "good to think with." But if we accept that the real-world phenomena we are interested in are not necessarily definite, repeatable, or stable, that constancy and accumulation are not achievable with the "facts" we "find," then that is exactly what we should be looking for in our methods too. At least there is a need for heterogeneity and variation in our methods, for open-mindedness and diversity, for generosity and breadth. That way, our methods can bring out the unique, the extraordinary, the revelatory, not just the repetitive or canonical. That is the only way to begin to match the diversity and richness of our world. Indeed, if the systems we want to study, understand, and influence are complex, then let us learn to live with that. Let us take that vision of a complex world, rather than a merely complicated Newtonian one, directly to questions of new technology and automation. We can see how both traditional simulated world experiments and a new open-mindedness might help us understand the relationship between technology, humans, and safety.

STUDY QUESTIONS

1. In what ways are human factors and safety research methods still dominated by a Cartesian–Newtonian worldview? That is, how and where can you recognize Newtonian assumptions (e.g., linearity and proportionality of cause and effect, knowledge as correspondence, reductionism).
2. Which aspects of the Cartesian–Newtonian worldview help and hamper our ability to anticipate the impact of safety or human factors interventions in complex operating worlds?
3. In what sense is the qualitative–quantitative opposition in research methods spurious?
4. By what features can you recognize a folk model? What distinguishes it from a young, promising pre-paradigmatic model?

5. Why does normativism so easily slip into folk models, and how can this pose a risk to practitioners whose performance is described by them?
6. What is essentialism and why does it often get assumed for the constructs (e.g., "complacency") that are the topic of our research?
7. In human factors research, what does the double hermeneutic mean? Does it have consequences for how confident we can be in the validity of our results?

7 New Technology and Automation

CONTENTS

KEY POINTS

- The idea that we can replace human work by automation without any consequences (other than greater safety or efficiency) is based on Tayloristic assumptions about decomposing work into components that can be divided up between human and machine. This is also known as the substitution myth.
- Automation, or any new technology, changes the tasks that it is designed to support or replace. It creates new human work, new pathways to success and failure, and new capabilities and complexities.
- Data overload is a generic human performance problem that often attracts automation. Depending on how the problem is framed, however, automation and new technology can help or hinder the human capacity to manage anomalies in their monitored process.
- New technology is not the manipulation of a single variable in an otherwise stable system. New technology instead triggers transformations in human and social practices, and human adaptation of, and to, the new technology.

- Envisioning practice recognizes that future is not given, that humans and technology coevolve and that new capabilities and complexities can emerge. By building or imagining envisioned worlds, designers can explore the reverberations of technology change before committing resources to fixing particular solutions.

CAN WE AUTOMATE HUMAN ERROR OUT OF THE SYSTEM?

The possibility for automating human part-tasks is a by-product of the scientific revolution, industrialization, and the microprocessor revolution. Conceptually, it was given a boost by Taylor's idea about redundancy of parts. His approach to understanding work and making it more efficient was by decomposing, or reducing it, to its most elemental parts. These, he suggested, could then be reorganized, planned better, or redistributed across those who could do those tasks in the most reliable, quickest way. The parts of the system carrying out those tasks were seen as redundant, or reducible to one another. If things were planned, organized, and supervised well, it did not matter whether one worker or another did a particular job on the production line. So when technology made it possible, such part-tasks could be automated as well, as long as the goal of greater efficiency was achieved. That is the legacy of Taylorist thinking: where technically possible, automation can replace human work without any consequences for the human–machine ensemble, except on some output measure (like, indeed, efficiency). Taylor's legacy makes the expectation that automation will help reduce human error a sensible one. If we automate part of a task, then the human does not carry out that part. And if the human does not carry out that part, there is no possibility of human error. As a result of this logic, there was a time (and in some quarters there perhaps still is) that automating everything we technically could was considered the best idea. The Air Transport Association of America (ATA) observed, for example, that "during the 1970's and early 1980's... the concept of automating as much as possible was considered appropriate" (Billings 1997, p. xi). It would lead to greater safety, greater capabilities, and other benefits. Automation today not only replaces human work but also indeed extends human capabilities in many safety-critical fields. In fact, automation is often presented and implemented precisely because it helps systems and people perform better and cheaper. It may even make operational lives easier: reducing task load, increasing access to information, helping the prioritization of attention, providing reminders, and doing work for us where we cannot.

FUNCTION ALLOCATION AND THE SUBSTITUTION MYTH

But can automation, in the Taylorist sense, replace human work, thereby reducing human error? Or is there a more complex coevolution of people and technology? Engineers and others involved in automation development are sometimes led to believe that there is a simple answer, and in fact a simple way of getting the answer. MABA-MABA lists, or "Men-Are-Better-At, Machines-Are-Better-At" lists have appeared over the decades in various guises. What these lists basically do is try to enumerate the areas of machine and human strengths and weaknesses, in order to provide engineers with some guidance on which functions to automate and which ones to give to the human. The process of function allocation as guided by such lists sounds straightforward but is actually fraught with difficulty and often unexamined assumptions.

MABA-MABA OR ABRACADABRA

One problem is that the level of granularity of functions to be considered for function allocation is arbitrary. For example, it depends on the model of information processing on which the MABA-MABA method is based (Hollnagel 1999). Others offer, for example, four stages of information processing (acquisition, analysis, selection, response) that form the guiding principle to which functions should be kept or given away (Parasuraman et al. 2000). This, too, is an essentially arbitrary decomposition based on a notion of a human–machine ensemble that resembles a linear input–output device. In cases where it is not a model of information processing that determines the categories of functions to be swapped between human and machine, the technology itself often determines it. MABA-MABA attributes are then cast in mechanistic terms, derived from technological metaphors. Even Paul Fitts applied terms such as information capacity and computation in his list of attributes for both the human and the machine (Fitts 1951). If the technology gets to pick the battlefield (i.e., determine the language of attributes), it will win most of them back for itself. This results in human-uncentered systems where typically heuristic and adaptive human abilities such as not focusing on irrelevant data, scheduling and reallocating activities to meet current constraints, anticipating events, making generalizations and inferences, learning from past experience, and collaborating easily fall by the wayside.

Moreover, MABA-MABA lists rely on a presumption of fixed human and machine strengths and weaknesses. The idea is that, if you get rid of the (human) weaknesses and capitalize on the (machine) strengths, you will end up with a safer system. This is what Hollnagel (1999) called "function allocation by substitution." The idea is that automation can be introduced as a straightforward substitution of machines for people—preserving the basic system while improving some of its output measures (lower workload, better economy, fewer errors, higher accuracy, etc.). Indeed, Parasuraman et al. (2000) recently defined automation in this sense: "Automation refers to the full or partial replacement of a function previously carried out by the human operator" (p. 287). But automation is more than replacement (although perhaps automation is about replacement from the perspective of the engineer). The interesting issues from a human performance standpoint emerge after such replacement has taken place.

Behind the idea of substitution lies the idea that people and computers (or any other machines) have fixed strengths and weaknesses and that the point of automation is to capitalize on the strengths while eliminating or compensating for the weaknesses. The problem is that capitalizing on some strength of computers does not replace a human weakness. It creates new human strengths and weaknesses—often in unanticipated ways (Bainbridge 1987). For instance, the automation strength to carry out long sequences of action in predetermined ways without performance degradation amplifies classic human vigilance problems. It also exacerbates the system's reliance on the human strength to deal with the parameterization problem, or literalism (automation does not have access to all relevant world parameters for accurate problem solving in all possible contexts). As we have seen, however, human efforts to deal with automation literalism, by bridging the context gap, may be difficult because computer systems can be hard to direct (How do I get it to understand? How do I get it to do what I want?). In addition, allocating a particular function does not absorb this function into the system without further consequences. It creates new functions for the other partner in the human–machine equation—functions that did not exist before, for example, typing, or searching for the right display page, or remembering entry codes. The quest for a priori function allocation, in other words, is intractable. And such new kinds of work create new error opportunities (What was that code again? Why can't I find the right page?).

NEW CAPABILITIES, NEW COMPLEXITIES

With new technological capabilities come new sociotechnical complexities. We cannot just automate part of a task and assume that the human–machine relationship remains unchanged. Though it may have shifted (with the human doing less and the machine doing more), there is still an interface between humans and technology. And the work that goes on at that interface has likely changed—sometimes drastically. Increasing automation transforms hands-on operators into supervisory controllers, into managers of a suite of automated and other human resources. With their new work come new vulnerabilities, new error opportunities. With new interfaces (from pointers to pictures, from single parameter gauges to computer displays) come new pathways to human–machine coordination breakdown. Many operational fields have witnessed the transformation of work by automation first-hand and documented its consequences widely. Automation does not do away with what we typically call human error, just as (or precisely because) it does not do away with human work. There is still work to do for people. It is not that the same kinds of errors occur in automated systems as in manual systems. Automation changes the expression of expertise and error; it changes how people can perform well and changes how their performance breaks down, if and when it does. Automation also changes opportunities for error recovery (often not for the better) and in many cases delays the visible consequences of errors. New forms of coordination breakdowns and accidents have emerged as a result.

AUTOMATION SURPRISE

During a normal approach to landing at Nagoya runway 34 in visual meteorological conditions, the captain indicated that he was going around but did not indicate why. Within the next 30 seconds, witnesses saw the aircraft in a nose-up attitude, rolling to its right before crashing tail-first 300 feet to the right of the approach end of the runway.

During the approach, the copilot flying apparently triggered the autopilot TOGA (takeoff-go-around) switch, whereupon the automation added power and commanded a pitch-up. The captain warned the copilot of the mode change, but the copilot continued to attempt to guide the aircraft down the glide slope while the automation countered his inputs with nose-up elevator trim. Ultimately, with stabilizer trim in an extreme nose-up position, the copilot was unable to counteract the trim with the nose-down elevator. The aircraft nosed up to an attitude in excess of 50°, stalled, and slid backward to the ground. Two hundred and sixty-four people were killed in the crash.

...It is thought that the pilots failed to realize that their decision [to continue to approach] contradicted the logic of the airplane's automated safety systems. In February 1991, an Interflug A310 at Moscow experienced a sudden, steep pitch-up similar to the one observed in this accident.

On August 31, 1994, the NTSB issued Safety Recommendations A-94-164 through 166 to the FAA. Its recommendation stated, "the Safety Board is concerned that the possibility still exists for a pilot-induced 'runaway trim' situation at low altitude and that...such as situation could result in a stall or the airplane landing in a nose-down attitude" (p. 5). Referring to other transport category aircraft autopilot systems, the Board said

> It is noted that the (autopilot) disconnect and warning systems are fully functional, regardless of altitude, and with or without the autopilot in the land or go-around modes. The Safety Board believes that the autopilot disconnect systems in Airbus A-300 and A-310 are significantly different...additionally, the lack of a stabilizer-in-motion warning appears to be unique to (these aircraft). The accident in Nagoya and the incident in Moscow indicate that pilots may not be aware that under some circumstances the autopilot will work against them if they try to manually control the airplane (p. 5).

The Board recommended that these autopilot systems be modified to ensure that the autopilot would disconnect if the pilot applies a specified input to the flight controls or trim system, regardless of the altitude or operating mode of the autopilot, and also to provide a sufficient perceptual alert when the trimmable horizontal stabilizers is in motion, irrespective of the source of the trim command (Billings 1997).

MODE AWARENESS

One important issue in automated systems is knowing what mode the automation is in. Mode confusion can lie at the root of automation surprises, with people thinking that they told the automation to do one thing whereas it was actually doing another. The formal instrument for tracking and checking mode changes and status in a modern cockpit is the FMA, or Flight Mode Annunciator. It is a small strip that displays contractions or abbreviations of modes (e.g., Heading Select mode is shown as HDG or HDG SEL) in various colors, depending on whether the mode is armed (i.e., about to become engaged) or engaged. Most airline procedures require pilots to call out the mode changes they see on the FMA.

SEEING AND CALLING OUT MODE CHANGES

Even if the mode annunciations in the case above had been less confusing, it is not clear that they would have been of much help in a busy operational phase like a transition to landing. One study monitored flight crews during a dozen return flights between Amsterdam and London on a full flight simulator (Björklund et al. 2006). Where both pilots were looking and how long was measured by EPOG (eye-point-of-gaze) equipment, which uses different kinds of techniques ranging from laser beams to measuring and calibrating saccades, or eye jumps that can track the exact focal point of a pilot's eyes in a defined visual field (see Figure 7.1). Like other studies, this one found that pilots do not look at the FMA much at all (Mumaw et al. 2001). And they talk about it even less. Very few callouts are made the way they should be (according to the procedures). Yet this does not seem to have much of an effect on automation-mode awareness or on the airplane's flight path. Without looking or talking, most pilots apparently still know what is going on inside the automation. In this one study, 521 mode changes occurred during the 12 flights. About 60% of these were pilot induced (i.e., because of the pilot changing a setting in the automation); the rest were automation induced. Two out of five mode changes were never visually verified (meaning neither pilot looked at his FMA during 40% of all mode changes). The pilot flying checked a little less than the pilots not flying, which could be a natural reflection of the role division. Pilots who are flying the aircraft have other sources of flight-related data they need to look at, whereas the pilot not flying can oversee the entire process, thereby engaging more often in checks of what the automation modes are. There are also differences between captains and first officers (even after correcting for pilot-flying vs. pilot-not-flying roles). Captains visually

FIGURE 7.1 The eye-point-of-gaze fixations during a one-hour flight between Amsterdam and London. The mode annunciation panel (on top of the primary flight display) does not attract a large number of looks. (Data taken from Björklund C. et al., *International Journal of Aviation Psychology*, *16*(3), 257–269, 2006.)

verified the transitions in 72% of the cases, versus 47% for first officers. This may mirror the ultimate responsibility that captains have for safety of flight, yet there was no expectation that this would translate into such concrete differences in automation monitoring. Amount of experience on automated aircraft types was ruled out as being responsible for the difference.

Of 512 mode changes, 146 were called out. If that does not seem like much, consider this: only 32 mode changes (that is about 6%) were called out after the pilot looked at the FMA. The remaining callouts came either before looking at the FMA or without looking at the FMA at all. Such a disconnect between seeing and saying suggests that there are other cues that pilots use to establish what the automation is doing. The FMA might not serve as a major trigger for getting pilots to call out modes. Two out of five mode transitions on the FMA are never even seen by entire flight crews. In contrast to instrument monitoring in non-glass-cockpit aircraft, monitoring for mode transitions is based more on a pilot's mental model of the automation (which drives expectations of where and when to look) and an understanding of what the current situation calls for. Such models are often incomplete and buggy and it is not surprising that many mode transitions are neither visually nor verbally verified by flight crews. At the same time, a substantial number of mode transitions are actually anticipated correctly by flight crews. In those cases where pilots do call out a mode change, four out of five visual identifications of those mode changes are accompanied or preceded by a verbalization of their occurrence. This suggests that there are multiple, underinvestigated resources that pilots rely on for anticipating and tracking automation-mode behavior (including pilot mental models). The FMA, designed as the main source of knowledge about automation status, actually does not provide a lot of that knowledge. It triggers a mere one out of five callouts and gets ignored altogether by entire crews for a whole 40% of all mode transitions. Proposals for new regulations are unfortunately taking shape around the same old display concepts. For example, a Joint Aviation Authorities Advisory Circular (ACJ 25.1329) from 2003 said that: "The transition from an armed mode to an engaged mode should provide an additional attention-getting feature, such as boxing and flashing on an electronic display (per AMJ25-11) for a suitable, but brief, period (e.g., ten seconds) to assist in flight crew awareness" (p. 28). But flight-mode annunciators are not good at attention getting at all, whether there is boxing or flashing or not. Indeed, empirical data show that the FMA does not "assist in flight crew awareness" in a dominant or relevant way. If design really is to capture the crew's attention about automation status and behavior, it will have to do radically better than annunciating abstruse codes in various hues and boxing or flashing times.

The callout procedure appears to be miscalibrated with respect to real work in a real cockpit, because pilots basically do not follow formal verification and callout procedures at all. Forcing pilots to visually verify the FMA first and then call out what they see bears no similarity to how actual work is done, nor does it have much sensitivity to the conditions under which such work occurs. Callouts may well be the first task to go out the window when workload goes up, which is also confirmed by this type of research. In addition to the few formal callouts that do occur, pilots communicate implicitly and informally about mode changes. Implicit communication surrounding altitude capture could for example be "Coming up to one-three-zero, (capture)" (referring to flight level 130). There appear to be many different strategies to support mode awareness, and very few of them actually overlap with formal procedures for visual verification and callouts. Even during the 12 flights of the Björklund et al. (2003) study, there were at least 18 different strategies that mixed checks, timing, and participation. These strategies seem to work as well as, or even better than, the official procedure, as crew communications on the 12 flights revealed no automation surprises that could be traced to a lack of mode awareness. Perhaps mode awareness does not matter that much for safety after all.

There is an interesting experimental side effect here: if mode awareness is measured mainly by visual verification and verbal callouts, and crews neither look nor talk, then are they unaware of modes, or are the researchers unaware of pilots' awareness? This poses a puzzle: crews who neither talk nor look can still be aware of the mode their automation is in, and this, indeed seems to be the case. But how, in that case, is the researcher (or your company, or line-check pilot) to know? The situation is one answer. By looking at where the aircraft is going, and whether this overlaps with the pilots' intentions, an observer can get to know something about apparent pilot awareness. It will show whether pilots missed something or not. In the research reported here, however, pilots missed nothing: there were no unexpected aircraft behaviors from their perspective.

DATA AVAILABILITY AND AUTOMATION CERTIFICATION

Human factors has a vital role to play here. Certifying a system on this kind of basis (and assuming that data availability will mean that such data will be observed) is short-sighted and risky. Data that are available, after all, do not mean data that are observable. This is patently obvious from the studies on mode awareness and the flight mode annunciator discussed above. They show that there can be a significant gap between data availability and data observability, depending not only on the physical and temporal features of any display, but on the interests, goals, workload, and attentional direction of the observer(s) (Woods and Hollnagel 2006). The same goes for procedural compliance. Certifying a system on assumptions of 100% compliance with work-as-imagined is risky too. Work-as-imagined is not necessarily work-as-done. There is a gap between those as well (Dekker 2003; McDonald et al. 2002). The world is too complex for there to be complete overlap: the modernist ideal that nature and society can be ordered in advance, administratively, is at odds with operational complexity. It is easy for us to become miscalibrated. It is easy for us to become overconfident that if our envisioned system can be realized, the predicted consequences and only the predicted consequence will occur. We lose sight of the fact that our views of the future are tentative hypotheses and that we would actually need to remain open to revision, that we need to continually subject these hypotheses to empirical jeopardy.

REPRESENTATIONS OF AUTOMATION BEHAVIOR

One way to fool ourselves into thinking that only the predicted consequences will occur when we introduce automation is to stick with substitutional practice of function allocation. Remember that this assumes a fundamentally uncooperative system architecture in which the interface between human and machine has been reduced to a straightforward "you do this, I do that" trade. If that is what it is, of course we should be able to predict the consequences. But it is not that simple. The question for successful automation is not who has control over what or how much. That only looks at the first parts, the engineering parts. We need to look beyond this and start asking humans and automation the question: "How do we get along together?" Indeed, where we really need guidance today is in how to support the coordination between people and automation. In complex, dynamic, nondeterministic worlds, people will continue to be involved in the operation of highly automated systems. The key to a successful future of these systems lies in how they support cooperation with their human operators—not only in foreseeable standard situations but also under novel, unexpected circumstances.

One way to reframe the question of how to involve people in what the automation is doing is to ask how to turn automated systems into effective team players (Sarter and Woods 1997). Good team players make their activities observable to fellow team players, and they are easy to direct. To be observable, automation activities must be presented in ways that capitalize on well-documented human strengths (our perceptual system's acuity to contrast, change, and events, and our ability to recognize patterns and know how to act on the basis of this recognition). For example (Woods 1996):

- Event based: Representations need to highlight changes and events in ways that the current generation of state-oriented displays do not.
- Future oriented: In addition to historical information, human operators in dynamic systems need support for anticipating changes and knowing what to expect and where to look next.
- Pattern based: Operators must be able to quickly scan displays and pick up possible abnormalities without having to engage in difficult cognitive work (calculations, integrations, extrapolations of disparate pieces of data). By relying on pattern- or form-based representations, automation has an enormous potential to convert arduous mental tasks into straightforward perceptual ones.

Team players are directable when the human operator can easily and efficiently tell them what to do. Designers could borrow inspiration from how practitioners successfully direct other practitioners to take over work. These are intermediate, cooperative modes of system operation that allow human supervisors to delegate suitable subproblems to the automation, just as they would be delegated to human crew members. The point is not to make automation into a passive adjunct to the human operator who then needs to micromanage the system each step of the way. This would be a waste of resources, both human and automatic. Human operators must be allowed to preserve their strategic role in managing system resources as they see fit, given the circumstances.

DATA OVERLOAD

Automation does not replace human work. Instead, it changes the work it is designed to support. And with these changes come new burdens. Take system monitoring, for example. There are concerns that automation can create data overload. Rather than taking away cognitive burdens from people, automation introduces new ones, creating new types of monitoring and memory tasks. Because automation does so much, it also can show much (and indeed, there is much to show). If there is much to show, data overload can occur, especially in pressurized, high-workload, or unusual situations. Our ability to make sense of all the data generated by automation has not kept pace with systems' ability to collect, transmit, transform, and present data.

But data overload is a complex phenomenon, and there are different ways of looking at it (Woods et al. 2002). For example, we can see it as a workload bottleneck problem. When people experience data overload, it is because of fundamental limits in their internal information-processing capabilities. If this is the characterization, then the solution lies in even more automation. More automation, after all, will take work away from people. And taking work away will reduce workload.

DATA OVERLOAD AND WARNING SYSTEMS

One area where the workload-reduction solution to the data-overload problem has been applied is in the design of warning systems. It is there that fears of data overload are often most prominent. Incidents in aviation and other transportation modes keep stressing the need for better support of human problem solving during dynamic fault scenarios. People complain of too much data, of illogical presentations, of warnings that interfere with other work, of a lack of order, and of no rhyme or reason to the way in which warnings are presented. Workload reduction during dynamic fault management is so important because problem solvers in dynamic domains need to diagnose malfunctions while maintaining process integrity. Not only must failures be managed while keeping the process running (e.g., keeping the aircraft flying); their implications for the ability to keep the process running in the first place need to be understood and acted on. Keeping the process intact and diagnosing failures are interwoven cognitive demands in which timely understanding and intervention are often crucial.

A fault in dynamic processes typically produces a cascade of disturbances or failures. Modern airliners and high-speed vessels have their systems tightly packed together because there is not much room onboard. Systems are also cross-linked in many intricate ways, with electronic interconnections increasingly common as a result of automation and computerization. This means that failures in one system quickly affect other systems, perhaps even along nonfunctional propagation paths. Failure crossover can occur simply because systems are located next to one another, not because they have anything functional in common. This may defy operator logic or knowledge. The status of single components or systems, then, may not be that interesting for an operator. In fact, it may be highly confusing. Rather, the operator must see, through a forest of seemingly disconnected failures, the structure of the problem so that a solution or countermeasure becomes evident. Also, given the dynamic process managed, these are among the questions that practitioners might well be pursuing:

- Which issue should be addressed first?
- What are the postconditions of these failures for the remainder of operations (i.e., what is still operational, how far can I go, what do I need to reconfigure)?

- Is there any trend?
- Are there noteworthy events and changes in the monitored process right now?
- Will any of this get worse?

Current warning systems in commercial aircraft do not go far in answering these questions, something that is confirmed by pilots' assessments of these systems. For example, pilots comment on too much data, particularly all kinds of secondary and tertiary failures, with no logical order, and primary faults (root causes) that are rarely, if ever, highlighted. The representation is limited to message lists, something that we know hampers operators' visualization of the state of their system during dynamic failure scenarios. Yet not all warning systems are the same. Current warning systems show a range of automated support, from not doing much at all, through prioritizing and sorting warnings, to doing something about the failures, to doing most of the fault management and not showing much at all anymore. Which works best? Is there any merit to seeing data overload as a workload bottleneck problem, and do automated solutions help?

- An example of a warning system that basically shows everything that goes wrong inside an aircraft's systems, much in order of appearance, is that of the Boeing 767. Messages are presented chronologically (which may mean the primary fault appears somewhere in the middle or even at the bottom of the list) and failure severity is coded through color.
- A warning system that departs slightly from this baseline is, for example, the Saab 2000, which sorts the warnings by inhibiting messages that do not require pilot actions. It displays the remaining warnings chronologically. The primary fault (if known) is placed at the top, however, and if a failure results in an automatic system reconfiguration, then this is shown too. The result is a shorter list than the Boeing's, with a primary fault at the top.
- The Airbus A320 family has a fully defined logic for warning-message prioritization. Only one failure is shown at a time, together with immediate action items required of the pilot. Subsystem information can be displayed on demand. Primary faults are thus highlighted, together with guidance on how to deal with them.
- The MD-11 has the highest degree of autonomy and can respond to failures without asking the pilot to do so. The only exceptions are nonreversible actions (e.g., an engine shutdown). For most failures, the system informs the pilot of system reconfiguration and presents system status. In addition, the system recognizes combinations of failures and gives a common name to these higher-order failures (e.g., Dual Generator).

As might be expected, response latency on the Boeing 767-type warning system is the longest (Singer and Dekker 2000). It takes a while for pilots to sort through the messages and figure out what to do. Interestingly, they also get it wrong more often on this type of system. That is, they misdiagnose the primary failure more often than on any of the other systems. A nonprioritized list of chronological messages about failures seems to defeat even the speed–accuracy trade-off: longer dwell times on the display do not help people get it right. This is because the production of speed and accuracy are cognitive: making sense of what is going wrong inside an aircraft's systems is a demanding cognitive task, where problem representation has a profound influence on people's ability to do it successfully (meaning fast and correct). Modest performance gains (faster responses and fewer misdiagnoses) can be seen on a system like that of the Saab 2000, but the Airbus A320 and MD-11 solutions to the workload bottleneck problem really seem to pay off. Performance benefits really accrue with a system that sorts through the failures, shows them selectively, and guides the pilot in what to do next. In our study, pilots were quickest to identify the primary fault in the failure scenario with such a system and made no misdiagnoses in assessing what it was (Singer and Dekker 2001). Similarly, a warning system that itself contains or counteracts many of the failures and shows mainly what is left to the pilot seems to help people in quickly identifying the primary fault.

These results, however, should not be seen as justification for simply automating more of the failure-management task. Human performance difficulties associated with high-automation participation in difficult or novel circumstances are well known, such as brittle procedure following where operators follow heuristic cues from the automation rather than actively seeking and dealing with information related to the disturbance chain. Instead, these results indicate how progress can be made by changing the representational quality of warning systems altogether, not just by automating more of the human task portion. If guidance is beneficial, and if knowing what is left is useful, then the results of this study tell designers of warning systems to shift to another view of referents (the thing in the process that the symbol on the display refers to). Warning-system designers would have to get away from relying on single systems and their status as referents to show on the display and move toward referents that fix on higher-order variables that carry more meaning relative to the dynamic fault-management task. Referents could integrate current status with future predictions, for example, or cut across single parameters and individual systems to reveal the structure behind individual failures and show consequences in terms that are operationally immediately meaningful (e.g., loss of pressure, loss of thrust).

DATA OVERLOAD AS A CLUTTER PROBLEM

Another way of looking at data overload is as a clutter problem—there is simply too much on the display for people to cope with. The solution to data overload as a clutter problem is to remove stuff from the display. In warning-system design, for example, this may result in guidelines that stress how no more than a certain number of lines must be filled up on a warning screen. Seeing data overload as clutter, however, is completely insensitive of context. What seems clutter in one situation may be highly valuable, or even crucial, in another situation.

DATA OVERLOAD AND DE-CLUTTER

The crash of an Airbus A330 during a test flight at the factory field in Toulouse, France in 1994 provides a good demonstration of this (see Billings 1997). The aircraft was on a certification test flight to study various pitch-transition control laws and how they worked during an engine failure at low altitude, in a lightweight aircraft with a rearward center of gravity (CG). The flight crew included a highly experienced test pilot, a copilot, a flight-test engineer, and three passengers. Given the lightweight and rearward CG, the aircraft got off the runway quickly and easily and climbed rapidly, with a pitch angle of almost 25° nose-up. The autopilot was engaged 6 seconds after takeoff. Immediately after a short climb, the left engine was brought to idle power and one hydraulic system was shut down in preparation for the flight test. Now the autopilot had to simultaneously manage a very low speed, an extremely high angle of attack, and asymmetrical engine thrust. After the captain disconnected the autopilot (this was only 19 seconds after takeoff) and reduced power on the right engine to regain control of the aircraft, even more airspeed was lost. The aircraft stalled, lost altitude rapidly, and crashed 36 seconds after takeoff.

When the airplane reached a 25° pitch angle, autopilot and flight-director mode information were automatically removed from the primary flight display in front of the pilots. This is a sort of de-clutter mode. It was found that, because of the high rate of ascent, the autopilot had gone into altitude-acquisition mode (called ALT* or "Altitude Star" in the Airbus) shortly after takeoff. In this mode, there is no maximum pitch protection in the autoflight system software (the nose can go as high as the autopilot commands it to go, until the laws of aerodynamics intervene). In this case, at low speed, the autopilot was still trying to acquire the altitude commanded (2000 feet), pitching up to it, and sacrificing airspeed in the process. But ALT* was not shown to the pilots because of the de-clutter function. So the lack of pitch protection was not announced, and may not have been known to them. De-clutter has not been a fruitful or successful way of trying to solve data overload, precisely because of the context problem. Reducing data elements on one display calls for that knowledge to be represented or retrieved elsewhere (people may need to pull it from memory instead), lest it be altogether unavailable.

DATA OVERLOAD AS A PROBLEM OF MEANING

Seeing data overload as a workload or clutter problem is based on false assumptions about how human perception and cognition work. Questions about maximum human data-processing rates are misguided because this maximum, if there is one at all, is highly dependent on many factors, including people's experience, goals, history, and directed attention. As alluded to earlier in the book, people are not passive recipients of observed data; they are active participants in the intertwined processes of observation, action, and sensemaking. People employ all kinds of strategies to help manage data and impose meaning on it. For example, they redistribute cognitive work (to other people, to artifacts in the world), they re-represent problems themselves so that solutions or countermeasures become more obvious. Clutter and workload characterizations treat data as a unitary input phenomenon, but people are not interested in data, they are interested in meaning. And what is meaningful in one situation may not be meaningful in the next. De-clutter functions are context insensitive, as are workload-reduction measures. What is interesting, or meaningful, depends on context. This makes designing a warning or display system challenging. How can a designer know what the interesting, meaningful or relevant pieces of data will be in a particular context? This takes a deep understanding of the work as it is done, and especially as it will be done once the new technology has been implemented. Recent advances in cognitive work analysis (Vicente 1999) and cognitive task design (Hollnagel 2003) present ways forward.

In addition to knowing what (automated) systems are doing, humans are also required to provide the automation with data about the world. They need to input things. In fact, one role for people in automated systems is to bridge the context gap. Computers are dumb and dutiful: they will do what they are programmed to do, but their access to context, to a wider environment, is limited—limited, in fact, to what has been predesigned or preprogrammed into them. They are literalist in how they work. This means that people have to jump in to fill a gap: they have to bridge the gulf between what the automation knows (or can know) and what really is happening or relevant out there in the world. The automation, for example, will calculate an optimal descent profile in order to save as much fuel as possible. But the resulting descent may be too steep for crew (and passenger) taste, so pilots program in an extra tailwind, tricking the computers into descending earlier and eventually shallower (because the tailwind is fictitious). The automation does not know about this context (preference for certain descent rates over others), so the human has to bridge the gap. Such tailoring of tools is a very human thing to do: people will shape tools to fit the exact task they must fulfill. But tailoring is not risk- or problem-free. It can create additional memory burdens, impose cognitive load when people cannot afford it, and open up new error opportunities and pathways to coordination breakdowns between human and machine.

Automation changes the task for which it was designed. Automation, though introducing new capabilities, can increase task demands and create new complexities. Many of these effects are in fact unintended by the designers. Also, many of these side effects remain buried in actual practice and are hardly visible to those who only look for the successes of new machinery. Operators who are responsible for (safe) outcomes of their work are known to adapt technology so that it fits their actual task demands. Operators are known to tailor their working strategies so as to insulate themselves from the potential hazards associated with using the technology. This means that the real effects of technology change can remain hidden beneath a smooth layer of adaptive performance. Operational people will make it work, no matter how recalcitrant or ill suited to the domain the automation, and its operating procedures, really may be. Of course, the occasional breakthroughs in the form of surprising accidents provide a window onto the nature of automation and its operational consequences. But, as is often the case, it is easy to blame the human for not staring hard enough, or not intervening quickly enough.

LIMITS OF SAFETY CERTIFICATION

A rigorous certification process will typically have different people look at color coding, data organization, character size and legibility, issues of human–computer interaction, software reliability and stability, seating arrangements, button sensitivities, and so forth. Certification can spend a long time following the footsteps of a design process to probe and poke it with methods and forms and questionnaires and tests and checklists and tools and guidelines—all in an effort to ensure that local human factors or ergonomics standards have been met. But such static snapshots may mean little. A lineup of microcertificates of usability does not guarantee safety. As soon as they hit the field of practice, systems start to drift. A year (or a month) after its inception, no sociotechnical system is the same as it was in the beginning. As soon as a new technology is introduced, the human, operational, organizational system that is supposed to make the technology work forces it into locally practical adaptations. Practices (procedures, rules) adapt around the new technology, and the technology in turn is reworked, revised, and amended in response to the emergence of practical experience.

SAFETY AS MORE THAN THE SUM OF CERTIFIED PARTS

So safety is more than the sum of the certified parts. An example from *Drift into Failure* (Dekker 2011b) concerns an MD-80 jet whose trim system failed on Alaska 261, leading to a loss of control and plunge into the ocean, killing all aboard. The jackscrew in the tail of the aircraft (that makes it go up or down) cannot do much to maintain the integrity of an MD-80/DC-9 trim system without a maintenance program that guarantees continued operability. But ensuring the existence of such a maintenance system is nothing like understanding how the local rationality of such a system can be sustained while safety standards are in fact continually being eroded (in this case, these jets went from a 350- to a 2550-hour lubrication interval over a number of years, approved by the regulator). The redundant components may have been built and certified. The maintenance program (with 2550-hour lubrication intervals—certified) may be in place. But safe parts do not guarantee system safety.

Certification processes do not typically take lifetime wear of parts into account when judging an aircraft airworthy, even if such wear will render an aircraft, like Alaska 261, quite unworthy of flying. Certification processes certainly do not know how to take sociotechnical adaptation of new equipment, and the consequent potential for drift into failure, into account when looking at nascent technologies. Systemic adaptation or wear is not a criterion in certification decisions, nor is there a requirement to put in place an organization to prevent or cover for anticipated wear rates or pragmatic adaptation, or fine-tuning. As a certification engineer from the regulator testified, "Wear is not considered as a mode of failure for either a system safety analysis or for structural considerations" (NTSB 2002, p. 24). Because how do you take wear into account? How can you even predict with any accuracy how much wear will occur? McDonnell-Douglas surely had it wrong when it anticipated wear rates on the trim jackscrew assembly of its DC-9 (the predecessor of the MD-8- series aircraft).

Originally, the assembly was designed for a service life of 30,000 flight hours without any periodic inspections for wear. But within a year, excessive wear had been discovered nonetheless, prompting a reconsideration.

The problem of certifying a system as safe to use can become even more complex if the system to be certified is sociotechnical and thereby even less calculable. What does wear mean when the system is sociotechnical rather than consisting of pieces of hardware? In both cases, safety certification should be a lifetime effort, not a still assessment of decomposed system status at the dawn of a nascent technology. Safety certification should be sensitive to the coevolution of technology and its use, its adaptation. Using the growing knowledge base on technology and organizational failure, safety certification could aim for a better understanding of the ecology in which technology is released—the pressures, resource constraints, uncertainties, emerging uses, fine-tuning, and indeed lifetime wear.

Safety certification is not just about seeing whether components meet criteria, even if that is what it often practically boils down to. Safety certification is about anticipating the future. Safety certification is about bridging the gap between a piece of gleaming new technology in the hand now, and its adapted, coevolved, grimy, greased-down wear and use further down the line. But we are not very good at anticipating the future. Certification practices and techniques oriented toward assessing the standard of current components do not translate well into understanding total system behavior in the future. Making claims about the future, then, often hangs on things other than proving the worthiness of individual parts.

Take the trim system of the DC-9 again. The jackscrew in the trim assembly had been classified as a "structure" in the 1960s, leading to different certification requirements from when it would have been seen as a system. The same piece of hardware, in other words, could be looked at as two entirely different things: a system, or a structure. In being judged a structure, it did not have to undergo the required system safety analysis (which may, in the end, still not have picked up on the problem of wear and the risks it implied). The distinction, this partition of a single piece of hardware into different lexical labels, however, shows that airworthiness is not a rational product of engineering calculation. Certification can have much more to do with localized engineering judgments, with argument and persuasion, with discourse and renaming, with the translation of numbers into opinion, and opinion into numbers—all of it based on uncertain knowledge.

As a result, airworthiness is an artificially binary black-or-white verdict (a jet is either airworthy or it is not) that gets imposed on a very gray, vague, uncertain, and evolving world—a world where the effects of releasing a new technology into actual operational life are surprisingly unpredictable and incalculable. Dichotomous, hard yes or no meets squishy reality and never quite gets a genuine grip. A jet that was judged airworthy, or certified as safe, may or may not be in actual fact. It may be a little bit unairworthy. Is it still airworthy with an end-play check of 0.0042 inches, the set limit? But "set" on the basis of what? Engineering judgment? Argument? Best guess? Calculations? What if a following end-play check is more favorable? The end-play check itself is not very reliable. The jet may be airworthy today, but no longer tomorrow (when the jackscrew snaps). But who would know?

Promises versus Problems of New Technology

Ergonomics has historically documented a number of disconnects between the promises of new technology and its actual problems (Meister and Farr 1967; Wiener 1988; Woods and Dekker 2001; Woods and Hollnagel 2006). Vendors, in many of these cases, promise better results through the replacement of human work with computerized aids (components can be substituted with predictable results), promise to offload humans (there is a fixed amount of knowable work the system needs to do: if the machine does more, the human does less), and claim they can focus human attention (meaning there is a right answer that they already know). Of course, designing new equipment is largely a technical and problem-solving activity, and deeper social or complex questions about the nature of workplaces may not make it to the development agenda (Lützhoft et al. 2010). As Wilkin (2009, p. 4) sums up, such an "approach assumes an atomistic social world composed of variables... that [it] is a closed system of study where the individual parts (variables) can be separated, measured and controlled." Design sometimes assumes it can meaningfully work with the units of a closed system that it has identified and can control but this is inconsistent with complexity. The components that make up a complex system are not simply exchangeable; not all are even fully known. As explained in the beginning of the book, a complex system is open and keeps changing in interaction with its environment, and its boundaries are fuzzy. Rather than a fixed, known amount of work, complex systems entail various ways of working (i.e., order) that emerge from the multitude of relationships and interactions between its parts. This diversity gives complex systems the resilience that allows it to cope with a changing environment.

INTERVENTION DECISIONS AND MONITORING TECHNOLOGY

A good example of the effects of emergence of complexity comes from the introduction of new monitoring technology in obstetrics. For a variety of anatomical and physiological reasons, labor is still dangerous for both baby and mother, even though in the West it has never been safer (Amer-Wåhlin and Dekker 2008). Hypoxic injury to the child is one important risk, and fetal monitoring aims to capture its early signs. Traditionally, fetal monitoring is done through cardiotocography (CTG) via an abdominal Doppler on the mother or fetal scalp electrode on the baby. A historical–graphical representation of the baby's heart rate and uterine contractions, CTG is both highly sensitive (good at detecting true positives) and unspecific (it also generates many false positives). It can get complemented with fetal blood sampling to identify metabolic acidosis in the baby, reflected by a low fetal blood pH that is in turn an indication of inadequate fetal oxygenation. Interventions in the labor process may be necessary on a variety of clinical indications, of which fetal hypoxia is a very important one. Intervention can range from offering drugs to speed up labor to removing the fetus by emergency Caesarian section.

Intervention decisions (when to increase inputs to, or take over from, an autonomous process, for example) are fundamental to many ergonomic problems (Kerstholt et al. 1996). They are also problematic because, particularly in escalating situations, they come either too early or too late. An early intervention may delete the evidence of a problem that warranted one, whereas a late intervention may lag behind any meaningful ability to retain or reestablish process integrity (Dekker and Woods 1999). The difficulty in timing interventions right may be reflected, for instance, in the increase of C-sections in a number of US States, despite postoperative risks to the mother and a lack of convincing clinical indications (Zaccaria 2002).

Obstetric technology is now available that can analyze a fetal electrocardiogram where particular waveforms can warn of compromised fetal adaptation to hypoxia and is an earlier sign of trouble than what can be obtained through fetal blood sampling. It is also continuous and thought to be more specific than the CTG curve. Consistent with the introduction of new technology in other safety-critical worlds (Wiener 1989), vendors purvey this technology with promises to make intervention decisions easier and better, focus clinicians' attention on the right answer, avoid unnecessary Caesarian sections or other difficult instrumental deliveries, and provide reassurance to clinicians when interventions are not necessary.

Systems that monitor selected process data (like a CTG curve) for potentially abnormal conditions (e.g., indications of fetal hypoxia) can be modeled as signal detection devices. These aim to discriminate between noise and signal + noise. Signal detectors take measurements (e.g., fetal CTG and other indicators of patient condition) that together form a multidimensional input vector called X, which is compared against expected characteristics of noise versus signal + noise events in the monitored process (Sorkin and Woods 1985). In obstetrics, for example, multidimensional inputs of CTG quality, patient condition, and history are compared against heart rate, heart rate variations, and fetal blood pH (measures whose limit values are often contained in clinical guidelines).

Signal detection theory (see Figure 7.2) separates the monitoring problem into a sensitivity parameter, d', and a response criterion, β. d' refers to the ability of the system to discriminate signal from noise on the input channel X and is a function of the normalized distance between the means of the two distributions ($d' = (\mu_n - \mu_{s+n})/\sigma$). In obstetrics, d' can be influenced by the resolution of a CTG readout consulted by the human or machine monitor. β, in contrast, specifies how much evidence is required to decide on the presence of a signal (impaired fetal compensation for hypoxia). β moves with payoffs. With any high-consequence payoff (e.g., malpractice risk, β is set low, demanding only some evidence before intervention) and prior knowledge of probabilities (a product of training and experience). With partially overlapping distributions, β is always a compromise: a response to the costs and benefits as well as historical probabilities of seeing or missing evidence.

The insertion of a machine monitor (such as ST waveform technology) between human and monitored process doubles the signal detection problem (Sorkin and Woods 1985). The automated subsystem (ST waveform technology)

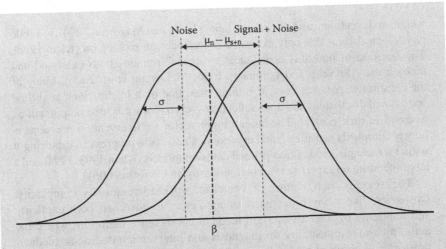

FIGURE 7.2 Noise and Signal + Noise distributions in Signal Detection Theory.

monitors a noisy input channel (fetal ECG) for occasional signal events, which compares its input vector, X_a, with prespecified thresholds. The automated monitor itself has a particular sensitivity (or d_a') and a decision criterion (β_a). Upon detecting a signal, the machine will create an output: an ST event alarm (see Figure 7.3). If the probability of a machine output is high (through its high d_a' or low β_a), β_h' will increase. Given that β_a is often set to minimize the number of missed signals (and can be manipulated without much cost or computing capacity, unlike d_a'), the overall performance of the monitoring ensemble is likely suboptimal (Moray and Inagaki 2000; Sorkin and Woods 1985).

X_a in ST-waveform technology relies on a relatively coarse division of T through QRS in a fetal ECG. X_h, in contrast, is a rich and constantly shifting amalgam of inputs. For the work described here, all kinds of patient parameters enter into it, ranging from obvious ones such as numbers of previous births,

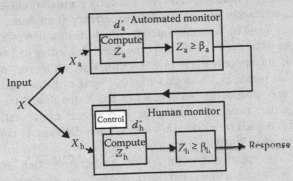

FIGURE 7.3 Automated and human monitor in sequence, yet both take their own input vector X from the monitored process. (Adapted from Sorkin R. D. and Woods, D. D., *Human–Computer Interaction*, 1(1), 49–75, 1985.)

weight and condition, and duration of labor (Drife and Magowan 2004), to subtle physiological signs that only extensive experience can pick up on (Klein 1998). The question of how this constitutes "evidence" remains both contested and problematic (De Vries and Lemmens 2006; McDonald et al. 2006). Many of the constitutive parameters of X_h remain inexplicit both in clinicians' real-time operational discussions and upon reflection—even if any machine output will be assessed in their context. The human input vector for intervention decisions is hugely complexly sensitized, and ergonomics has made progress in capturing it with, for example, recognition-primed decision models (Klein 1993, 1998) and a large literature on expertise (Farrington-Darby and Wilson 2006).

The socially rich nature of obstetric work increases its complexity. Obstetrics, like many other parts of medicine, is governed both explicitly and implicitly by a relatively rigid medical competence hierarchy, where the authority and responsibility for diagnosis and intervention decisions, medication orders, control of medical technology, and continuation of care decisions rest at the top (Ödegård 2007). This is populated by doctors (often male, even in obstetrics), recruited from a limited socioeconomic slice. Underneath lies nursing, which monitors patient condition, carries out medication orders, and offers patient continuity of care (doctors often only "visit" a patient) (Benner, Malloch, and Sheets 2010; Ehrenreich and English 1973). Below that is caring, which handles physiological (if not psychological) needs of feeding, cleaning, and rehabilitation. And below that is the patient, who is generally assumed not to know much of value about his or her own disease or condition other than possessing the privileged experience to describe its surface features to clinicians (Ödegård 2007).

This strict hierarchy makes each layer subordinate to the one above, which can lead to intriguing divergences between expertise and decision authority. In Swedish obstetrics in particular (but elsewhere too), midwives occupy an important swath of clinical experience and judgment (Sibley, Sipe, and Koblinsky 2004). Nowadays, midwives are fully registered nurses with extra training and education, and accumulate experience from hundreds or thousands of hours spent by bedsides. Intervention decisions, however, belong formally to those who haven't spent such time there. According to Swedish praxis and protocol, doctors only come in when things are no longer "normal." But what does that mean, and who gets to say? Signs of "abnormality" are the interpretive and often contested product of the X, the β, and the d' of those present when they occur: the midwives. A physician's intervention decision is thus often preceded by a midwife's intervention decision: to call the doctor (Amer-Wåhlin et al. 2010).

But midwives typically know the doctors, and doctors often know the midwives. The midwife's setting of β, then, depends on which physician is on duty—estimates of individual physician age, experience, competence, and sometimes even gender affect how much evidence midwives need to call them. In our study, this seemed quite independent of the number of machine alarms.

Doctors, in turn, accumulate their own experience with midwives' βs. Some are known to call some physicians earlier (this may itself depend on a variety of factors, ranging from time of day to bed load and estimates of physician workload, or, more specifically, estimates of that particular physician's ability to handle workload in a socially and professionally acceptable way), which in turn leads to different physician βs for those midwives' calls for help.

It gets more intricate still, because a doctor on duty can call a doctor on call (who may be elsewhere in the hospital, or at home). Midwives, under the rules in the hospitals we studied, cannot call the backup physician themselves. In our observations, midwives made estimates among themselves about the likelihood that a duty doctor would call the backup vis-à-vis their desire that he or she do so (which in turn depended on the duty doctor's perceived attributes) and adjusted the construction of their message to the duty doctor on this basis. In other words, the duty doctor's assumed β for calling the backup, and his or her assumed d' (sensitivity to evidence), was used to adjust both the timing and the content of the message delivered. Remarks such as "we hope he'll be smart enough to call the backup if we put it this way" seemed to create an identity of the physician (McDonald et al. 2006) that linked directly to that doctor's assumed β and d'.

Further technological developments in one of the hospitals we studied led to summary monitors (of CTG curves from the various delivery rooms) being placed in the obstetric break room. The arrangement was akin to Foucault's descriptions of the panopticon prison where the very possibility of being observed any moment changed inmates' behavior every moment (Foucault 1977). In the obstetric ward, the silent knowledge that others could, unbeknownst to oneself, be watching the very evidence trace on which you would be taking action or not was enough to affect clinicians' β again. The boundaries of the complex system were made fuzzier—where exactly did the delivery room end now? Even if physicians might have liked to take action on some CTG traces shown in the break room, they knew better than to enter some midwives' delivery rooms without a call. They had their β set very high. It seemed here that the very medical competence hierarchy could become renegotiated and, in some pockets, inverted through processes of social learning and adaptation, fueled largely through assumptions about colleagues' β and d', or even those of colleagues of colleagues.

In the case described above, it was in this world, this complex system, that the new monitoring (ST waveform) technology was introduced:

- X_{1i} is hugely complexly sensitized (which ergonomics has recognized for a long time).
- There are predictable interactions between $β_h$ and $β_a$ (also not new), but there are also multiple interdependent $β_h$ values in the delivery room as clinicians of multiple rankings interact in the consideration and construction of evidence from the data traces they monitor.

- This system, in turn, is open. A part of nurses' and midwives' β_h is adjusted on the basis of assumptions that they make and share about the various β_h of other clinicians present somewhere in the system.
- d' is actively tuned by midwives (through the modulation of tone and message content) depending on where they estimate a physician's β to lie in that context (patient, business on the ward) and on the imagined d' of the receiver (how clear do we need to be? How thick is this doc?).
- The boundaries of the system that thus calls in physicians are fuzzy (potentially including any other part of the hospital or people's own homes), and can even include the assumed β_h of others about others' β_h (e.g., duty doctor to doctor-on-call).
- The placement of displays of data traces throughout the workplace (and beyond the walls of delivery rooms) once again influences clinicians' β_h, and does so contextually (who might be watching "over my shoulder" here?).

Understanding such a complex system is about understanding the intricate web of relationships they weave, their interconnections and interdependencies, and the constantly changing nature of those as people come and go and technologies get adapted in use. A formal extension of the signal detection paradigm to capture this is beyond the scope of this book, but the outlines are sketched in the points above: multiple, changeable interconnections between agents who come and go onto the clinical scene (sometimes unbeknownst to those in the delivery room), whose d' and β are known, estimated or the subject of discussion, negotiation, and manipulation.

TRANSFORMATION AND ADAPTATION

As the example above shows, the introduction of new technology changes, in many subtle and less subtle ways, the relationships and interconnections that hold a complex system together. Introducing a new piece of technology or automating a part-task does not come down to the manipulation of a single variable. Automation and new technology transform people's practices and force them to adapt in novel ways: "It alters what is already going on—the everyday practices and concerns of a community of people—and leads to a resettling into new practices" (Flores et al. 1988, p. 154). Unanticipated consequences are the result of these much more profound, qualitative shifts. For example, during the Gulf War in the early 1990s, "almost without exception, technology did not meet the goal of unencumbering the personnel operating the equipment. Systems often required exceptional human expertise, commitment, and endurance" (Cordesman and Wagner 1996, p. 25).

The introduction of new technology thus often creates reverberations typical of complexity (Cilliers 2002). New human roles emerge, and relationships between people and artifacts get transformed (Woods and Dekker 2000). New kinds of human work are produced that call on new sorts of expertise. Interconnections between people and departments and artifacts proliferate as computerized networks tighten the couplings between new players both intentionally and unintentionally (Perry et al. 2005). The example above shows how even colleagues on a break can become unwitting players in others' intervention decisions, given the possibility of even casual glances at obstetric monitors in their break room. The knowledge that they are watching, or might watch (and depending on who those colleagues are in the medical competence hierarchy and the social dynamics of the ward), likely influences the decision criterion of the clinician(s) by the bedside. As a result, new technology opens up new possibilities for success that are not necessarily aligned with designer intention, and pathways to failure that can be altogether hard to foresee. While complexity theory does not provide the exact tools to solve such complex problems, it can provide rigorous accounts of the challenges our field needs to meet (Cilliers 2005, p. 258):

> ...because complex systems are open systems, we need to understand the system's complete environment before we can understand the system, and, of course, the environment is complex in itself. There is no human way of doing this. The knowledge we have of complex systems is based on the models we make of these systems, but in order to function as models—and not merely as a repetition of the system—they have to reduce the complexity of the system. This means that some aspects of the system are always left out of consideration. The problem is confounded by the fact that that which is left out, interacts with the rest of the system in a non-linear way and we can therefore not predict what the effects of our reduction of the complexity will be, especially not as the system and its environment develops and transforms in time.

Engineers, given their professional focus, may believe that automation transforms the tools available to people, who will then have to adapt to these new tools. But people's practices and relationships get transformed by the introduction of new tools. New technology, in turn, gets adapted by people in locally pragmatic ways so that it will fit the constraints and demands of actual practice. For example, controlling without flight-progress strips (relying more on the indications presented on the radar screen) asks controllers to develop and refine new ways of managing airspace complexity and dynamics. In other words, it is not the technology that gets transformed and the people who adapt. Rather, people's practice gets transformed and they in turn adapt the technology to fit their local demands and constraints.

The key is to accept that automation and new technology will transform people's practice and relationships. As human factors and safety researchers, designers, and practitioners, we need to be prepared to learn from these transformations as they happen. This is by now a common (but not often successful) starting point in contextual design. Here, the main focus of system design is not the creation of artifacts per se, but getting to understand the nature of human practice in a particular domain, and changing those work practices rather than just adding new technology or replacing human work with machine work. This recognizes that

- Design concepts represent hypotheses or beliefs about the relationship between technology and human cognition and collaboration.

- They need to subject these beliefs to empirical jeopardy by a search for disconfirming and confirming evidence.
- These beliefs about what would be useful have to be tentative and open to revision as they learn more about the mutual shaping that goes on between artifacts and actors in a field of practice.

Subjecting design concepts to such scrutiny can be difficult. Traditional validation and verification techniques applied to design prototypes may turn up nothing, but not necessarily because there is nothing that could turn up. Validation and verification studies typically try to capture small, narrow outcomes by subjecting a limited version of a system to a limited test. The results can be informative, but hardly about the processes of transformation (different work, new cognitive, and coordination demands) and adaptation (novel work strategies, tailoring of the technology) that will determine the sources of a system's success and potential for failure once it has been fielded. Another problem is that validation and verification studies need a reasonably ready design in order to carry any meaning. This presents a dilemma: by the time results are available, so much commitment (financial, psychological, organizational, political) has been sunk into the particular design that any changes quickly become unfeasible.

Such constraints through commitment can be avoided if human factors can say meaningful things early on in a design process. What if the system of interest has not been designed or fielded yet? Are there ways in which we can anticipate whether automation, and the human role changes it implies, will create new error problems rather than simply solving old ones? This has been described as Newell's catch: in order for human factors to say meaningful things about a new design, the design needs to be all but finished. Although data can then be generated, they are no longer of use, because the design is basically locked. No changes as a result of the insight created by human factors data are possible anymore. Are there ways around this catch? Can human factors say meaningful things about a design that is nowhere near finished? One way that has been developed is future incident studies, and the concept they have been tested on is exception management.

ENVISIONING PRACTICE

Envisioning future practice or anticipating how automation and technology will create new human roles is really hard. What are the performance consequences of those roles? Introducing automation to turn people into exception managers can sound like a good idea and responsive to the unpredictabilities of complexity. Of course, the obstetric example above shows that this involves a host of sociotechnical issues that go beyond a mere decision to intervene on the part of one individual clinician. The example raises significant questions about social order, about power, about hierarchy, even history and tradition, about gender and the distribution of resources across a workplace. These are all factors that can significantly influence the success or failure of new technologies in a workplace, and finding ways to account for them in our models should be one important aim. Just saying "manager of exceptions," then, is insufficient: it does not make explicit what it means to practice. What work does an exception manager do? What cues does he or she base decisions on? The downside of underspecification is the risk of remaining trapped in a disconnected, shallow, unrealistic view of work. And when our view of (future) practice is disconnected from many of the pressures, challenges, and constraints operating in that world, our view of practice is distorted from the beginning. It misses how operational people's strategies are often intricately adapted to deal effectively with these constraints and pressures.

AUTOMATION AND HUMAN EXCEPTION MANAGEMENT

In ever-busier systems, where operators are vulnerable to problems of data overload, turning humans into exception managers is a powerfully seductive concept. It has, for example, been practiced in the dark cockpit design that essentially keeps the human operator out of the loop (all the annunciator lights are out in normal operating conditions) until something interesting happens, which may then be the time for the human to intervene. This same envisioned role, of exception manager, dominates recent ideas about how to effectively let humans control ever increasing air-traffic loads. Recognize how this goes far beyond the question of the last chapter—whether and how air-traffic controllers can anticipate and handle their traffic without access to manipulable paper flight strips. Perhaps, the thought goes, controllers should no longer be in charge of all the parameters of every flight in their sector. A core argument is that the human controller is a limiting factor in traffic growth. Too many aircraft under one single controller leads to memory overload and the risk of human error. Decoupling controllers from all individual flights in their sectors, through greater computerization and automation on the ground and greater autonomy in the air, is assumed to be the way around this limit.

The reason we may think that human controllers will make good exception managers is that humans can handle the unpredictable situations that machines cannot. In fact, this is often a reason why humans are still to be found in automated systems in the first place. Following this logic, controllers would be very useful in the role of traffic manager, waiting for problems to occur in a kind of standby mode. The view of controller practice is one of passive observer, ready to act when necessary. But intervening effectively from a position of disinvolvement has proven to be difficult—particularly in air-traffic control. For example, Endsley and colleagues pointed out, in a study of direct routings that allowed aircraft deviations without negotiations, that with more freedom of action being granted to individual aircraft, it became more difficult for controllers to keep up with traffic (Endsley et al. 1997). Controllers were less able to predict how traffic patterns would evolve over a foreseeable time frame. In other studies, too, passive monitors of traffic seemed to have trouble maintaining a sufficient understanding of the traffic under their control and were more likely to overlook separation infringements (Galster et al. 1999). In one study, controllers effectively gave up control over an aircraft with communication problems, leaving it to other aircraft and their collision-avoidance systems to sort it out among themselves (Dekker and Woods 1999). This turned out to be the controllers' only route out of a fundamental double bind: if they intervened early, they would create a lot of workload problems for themselves (suddenly a large number of previously autonomous aircraft would be under their control). Yet if they waited on intervention (in order to gather more evidence on the aircraft's intentions), they would also end up with an unmanageable workload and very little time to solve anything in. Controller disinvolvement can create more work rather than less and produce a greater error potential.

UPSIDE OF UNDERSPECIFICATION

There is an upside to underspecification, however, and that is the freedom to explore new possibilities and new ways to relax and recombine the multiple constraints, all in order to innovate and improve. Will automation help you get rid of human error? With air-traffic controllers as exception managers, it is interesting to think about how the various designable objects would be able to support them in exception management. For example, visions of future air-traffic control systems typically include data linking as an advance that avoids the narrow bandwidth problem of voice communications—thus enhancing system capacity. In one of our studies mentioned before (Dekker and Woods 1999), a communications failure affected an aircraft that had also suffered problems with its altitude reporting (equipment that tells controllers how high it is and whether it is climbing or descending). At the same time, this aircraft was headed for streams of crossing air traffic. Nobody knew exactly how datalink, another piece of technology not connected to altitude-encoding equipment, would be implemented (its envisioned use was, and is, to an extent underspecified). One controller, involved in the study, had the freedom to suggest that air-traffic control should contact the airline's dispatch or maintenance office to see whether the aircraft was climbing or descending or level. After all, datalink could be used by maintenance and dispatch personnel to monitor the operational and mechanical status of an aircraft, so "if dispatch monitors power settings, they could tell us," the controller suggested. Others objected because of the coordination overheads this would create. The ensuing discussion showed that, in thinking about future systems and their consequences for human error, we can capitalize on underspecification if we look for the so-called leverage points (in the example: datalink and other resources in the system) and a sensitivity to the fact that envisioned objects only become tools through use—imagined or real (datalinks to dispatch become a backup air-traffic control tool).

Anticipating the consequences of automation on human roles is also difficult because—without a concrete system to test—there are always multiple versions of how the proposed changes will affect the field of practice in the future. Different stakeholders (in air-traffic control this would be air carriers, pilots, dispatchers, air-traffic controllers, supervisors, flow controllers) have different perspectives on the impact of new technology on the nature of practice. The downside of this plurality is a kind of parochialism where people mistake their partial, narrow view for the dominant view of the future of practice, and are unaware of the plurality of views across stakeholders. For example, one pilot claimed that greater autonomy for airspace users is "safe, period" (Baiada 1995). The upside of plurality is the triangulation that is possible when the multiple views are brought together. In examining the relationships, overlaps, and gaps across multiple perspectives, we are better able to cope with the inherent uncertainty built into looking into the future.

A number of future incident studies examined controllers' anomaly response in future air-traffic control worlds precisely by capitalizing on this plurality (Dekker and Woods 1999). To study anomaly response under envisioned conditions, groups of practitioners (controllers, pilots, and dispatchers) were trained on proposed future rules. They were brought together to try to apply these rules in solving difficult future airspace problems that were presented to them in several scenarios. These included aircraft decompression and emergency descents, clear air turbulence, frontal thunderstorms, corner-post overloading (too many aircraft going to one entry point for airport area), and priority air-to-air refueling and consequent airspace restrictions and communication failures. These challenges, interestingly, were largely rule or technology independent: they can happen in airspace systems of any generation. The point was not to test the anomaly response performance of one group against that of another, but to use triangulation of multiple stakeholder viewpoints—anchored in the task details of a concrete problem—to discover where the envisioned system would crack, where it would break down. Validity in such studies derives from (a) the extent to which problems to be solved in the test situation represent the vulnerabilities and challenges that exist in the target world and (b) the way in which real problem-solving expertise is brought to bear by the study participants.

Developers of future air-traffic control architectures have been envisioning a number of predefined situations that call for controller intervention, a kind of reasoning that is typical for engineering-driven decisions about automated systems. In air-traffic management, for example, potentially dangerous aircraft maneuvers, local traffic density (which would require some density index), or other conditions that compromise safety would make it necessary for a controller to intervene. Such rules, however, do not reduce uncertainty about whether to intervene. They are all a form of threshold crossing—intervention is called for when a certain dynamic density has been reached or a number of separation miles have been transgressed. But threshold-crossing alarms are very hard to get right—they come either too early or too late. If too early, a controller will lose interest in them: the alarm will be deemed alarmist. If the alarm comes too late, its contribution to flagging or solving the problem will be useless and it will be deemed incompetent. The way in which problems in complex, dynamic worlds grow and escalate and the nature of collaborative interactions indicate that recognizing exceptions in how others (either machines or people) are handling anomalies is complex. The disappointing history of automating problem diagnosis inspires little further hope. Threshold-crossing alarms cannot make up for a disinvolvement—they can only make a controller acutely aware of those situations in which it would have been nice to have been involved from the start.

Future incident studies allow us to extend the empirical and theoretical base on automation and human performance. For example, supervisory-control literature makes no distinction between anomalies and exceptions. This indistinction results from the source of supervisory-control work: how do people control processes over physical distances (time lag, lack of access, etc.). However, air-traffic control augments the issue of supervisory control with a cognitive distance: airspace participants have some system knowledge and operational perspective, as do controllers, but there are only partial overlaps and many gaps. Studies on exception management in future air-traffic control force us to make a distinction between anomalies in the process, and exceptions from the point of view of the supervisor (controller). Exceptions can arise in cases where airspace participants are dealing with anomalies (e.g., an aircraft with pressurization or communications problems) in a way that forces the controller to intervene. An exception is a judgment about how well others are handling or going to handle disturbances in the process. Are airspace participants handling things well? Are they going to get themselves in trouble in the future? Judging whether airspace users are going to get in trouble in their dealings with a process disturbance would require a controller to recognize and trace a situation over time—contradicting arguments that human controllers make good standby interveners.

Future, as Prigogine said, is not given. Things are not predetermined; new behaviors and effects can emerge. This is where human factors can help. If it plays its capabilities smartly, it can rescue the discourse of prediction, verification, and validation from the grasp of Newton's and Descartes', where individually certified components and systems supposedly live in time-reversible, stable worlds where nothing new can appear. It can introduce a language and logic of complexity, shift the goal to a prediction of probabilities, not certainties, and orient engineers' conversations toward imagined futures, toward possible worlds and envisioned systems. It can help lift the gaze from components to systems, where safe and unsafe outcomes emerge from interactions between components—not from malfunctioning components, not from a few final frontline operators who do not stare hard enough or intervene aggressively enough.

STUDY QUESTIONS

1. Can human error be automated out of a system?
2. What is Tayloristic about function allocation methods that divide tasks up between human and machine? What is the substitution myth? Do you have examples from your own world where automation changed the tasks it was supposed to support or replace?
3. What are different ways to characterize data overload, and which differing assumptions do these characterizations make about the nature of human cognition?
4. Complexity is more about relationships than components. Can you think of an example from your own world (like the obstetric one in this chapter) where the introduction of new technology demonstrably changed the relationships between people (and people and the technology)?
5. If the future is not "given," is safety certification of complex systems possible at all? How can we still say meaningful things about future systems without overextending our epistemological reach?
6. How can data overload be characterized? Which characterization is suited best for capturing the complexities of the introduction of technologies to help humans manage data overload?
7. If design concepts are hypotheses about what might be useful in the collaboration between humans and machines, then how can the underspecification of early design ideas help? Why could it actually hamper progress on usability and safety if design concepts are fixed early on?

8 New Era in Safety

CONTENTS

KEY POINTS

- To create the conditions for a new era in safety, we should not see people as a problem to control but as a solution to harness. We should not define safety as an absence of negatives, but the presence of positive capacities. And we should dare to trust people more than we trust bureaucracy, protocol, and process.
- Zero vision on safety is a modernist, moral goal. It can also have negative practical consequences, including the suppression of negative events and bad news, the stigmatization of people involved in incidents, and an inefficient use of investigative resources.
- There is an inverse correlation between low numbers of incidents and fatality risk, something that has been empirically established in fields from construction to aviation. Having only few incidents (or only reporting few) thus creates greater serious safety risks.
- Safety culture is typically seen as a functionalist object, which can be studied from the outside-in, decomposed, measured, and manipulated from the

top-down. In contrast, culture has traditionally been seen as a bottom-up emergent property of social interactions, and as neither "good" nor "bad."
- The safety literature regards deference to expertise as a key to preventing a drift into failure. This sees expert practitioners as a solution to harness. Yet expertise can lie locked up in what managers and others see as "prima donnas": operators whose special status gives them informal power.
- Resilience engineering provides a new frame to understand people as a solution to harness. It studies, among other things, the way individuals, teams, and organizations handle sacrificing decisions, problem fragmentation, evidence of past success, the apparent absence of risk, and opportunities for vicarious learning.

FROM MODERNIST SAFETY TO NEW ERA SAFETY

If we want a new era in safety, what would it look like? A new era would certainly reconsider, once again, the role of human beings in the creation of safety. Our governance of safety is often organized around bureaucratic process, driven by high-modernist ideas, and held up by Cartesian–Newtonian assumptions about how things go right and wrong. It supports and legitimates systems of counting and tabulating, and largely relies on vocabularies of control, constraint, and human deficit. The new era, instead, calls for a form of governance that sends power over many safety decisions back to the floor, to the projects; a form of governance that sees people there as the source of diversity, insight, creativity, and wisdom about safety, not as sources of risk that undermine an otherwise safe system. It calls for governance that trusts people and mistrusts bureaucracy. A governance that is once again committed to preventing injury rather than insurance claims. The contrast is laid out in Table 8.1.

A continued pursuit of existing safety strategies is not going to lead to different outcomes, and it is unlikely that we can break through the asymptote on safety progress with them. Perhaps it is time for entirely different indicators, or measures. That said, we should probably not discontinue what we are doing all at once, and some

TABLE 8.1
Transitioning from Modernist Safety Thinking to the New Era

Modernist Safety	New Era Safety
People are a problem to control	People are a solution to harness
Safety is defined as the absence of negatives (injuries, incidents) that show where things go wrong	Safety is defined as the presence of capabilities, capacities, and competencies that make things go right
Safety is a bureaucratic accountability directed upward in the organization	Safety is an ethical responsibility directed downward in the organization
Cause–effect relationships are linear and unproblematic	Cause–effect relationships are complex and nonlinear
Vocabularies of control, constraint, and human deficit	Vocabularies of empowerment, diversity, and human opportunity

things we should not ever discontinue—much of what we have been doing in safety is quite worthwhile. It has led to significant reductions in harm and damage. But we should not have the expectation that it will help us do much more than maintaining current levels of safety in many industries. Further progress instead hinges on a number of key transitions:

- We need to transition from seeing people as a problem to control to seeing people as a solution to harness.
- We need to transition from seeing safety as a bureaucratic accountability *up* to seeing it as an ethical responsibility *down*.
- We need to transition from seeing safety as an absence of negatives to seeing it as the presence of a positive capacity to make things go right.

This chapter runs through some of the research that can help with some of the conditions for such a new era. These conditions are certainly not in place yet, and in fact, in many instances, the research points more to what is keeping conditions for the old, or current, era in place instead. But even that can be helpful as we move to sketch the outlines of a new era.

MODERNISM AND ZERO VISION

OPTIMISM AND PROGRESS TOWARD ZERO

Remember from the first chapter that the modernist vision of work is essentially an optimistic one. The Enlightenment suggested that if we use our reason, if we think harder about a problem, then we can make the world better, we can constantly improve it. Modernism says that technical–scientific rationality can create that better, safer, more predictable, more controllable world. A promise offered by modernism is that we can eradicate diseases like polio or cholera, or get them under control; and that we might achieve workplaces without injuries, incidents, or accidents. If, for example, we plan the work carefully, if we design well and train, discipline, supervise, and monitor the people who are going to execute the work (just like Taylor recommended), we can eventually live in a world without human error.

This commitment to constant improvement is reflected in the zero visions of many industries and organizations around the world, from road traffic to construction. In Finland, for example, more than 280 companies have joined in a Zero Accident Forum, representing some 10% of the country's workforce (Zwetsloot et al. 2013). Similar networks or forums exist in countries around the world (Donaldson 2013). Such membership, and the commitment it implies, gets organizations to realize safety improvements because they need to back up their commitment with resources. But these were already very safe and committed companies. Being a high achiever partly explains one's membership in such a group. But what does "being a high achiever" mean? The US Government Accountability Office, or GAO, recently studied these issues in the United States and asked whether some safety incentive programs and other workplace safety policies may actually discourage workers' reporting of injuries and illnesses (GAO 2012). It found the following:

Little research exists on the effect of workplace safety incentive programs and other workplace safety policies on workers' reporting of injuries and illnesses, but several experts identified a link between certain types of programs and policies and reporting. Researchers distinguish between rate-based safety incentive programs, which reward workers for achieving low rates of reported injuries or illnesses, and behavior-based programs, which reward workers for certain behaviors. Experts and industry officials suggest that rate-based programs may discourage reporting of injuries and illnesses and reported that certain workplace polices, such as post-incident drug and alcohol testing, may discourage workers from reporting injuries and illnesses. Researchers and workplace safety experts also noted that how safety is managed in the workplace, including employer practices such as fostering open communication about safety issues, may encourage reporting of injuries and illnesses (GAO 2012, p. 2).

Thus, zero vision can sometimes lead to a suppression of evidence about incidents, injuries, or other safety issues, as well as to the numerical gymnastics and relabeling that happened in the example above. The unethical behavior it can incentivize might sometimes even be judged as illegal or criminal:

IN JAIL FOR ZERO

A Louisiana man is spending time in prison for lying about worker injuries at a local power utility, which allowed his company to collect $2.5 million in safety bonuses. A federal court news release says that the 55-year-old was sentenced to serve 6.5 years in prison followed by two years of supervised release.

He was the safety manager for a construction contractor. He was convicted in November of not reporting injuries at two different plants in Tennessee and Alabama between 2004 and 2006. At his federal trial, jurors heard evidence of more than 80 injuries that were not promptly recorded, including broken bones, torn ligaments, hernias, lacerations, and injuries to shoulders, backs and knees. The construction contractor paid back double the bonuses (Anon. 2013).

ZERO UPSIDE DOWN

The aim to reduce harm to zero is consistent with a modernist, Enlightenment inspiration. As GAO observed, however, little is known about the sorts of activities and mechanisms that lie underneath the reductions in harm that committed companies have witnessed, and not much research has been conducted into this (Zwetsloot et al. 2013). One important reason is that the goal, the zero vision, is defined by its dependent variable, not its manipulated variables. In typical scientific work, the experimenter gets to manipulate one or a number of variables (called the independent or manipulated variables). These are in turn presumed to have an effect on one or a number of dependent variables. In this, safety is always the dependent variable—it is influenced by a lot of other things (independent or manipulated variables). Increases in production pressure and resource shortages (independent variables), for example, pushes the operating state closer to the marginal boundary, leading to a reduction in safety margins (the dependent variable) (Rasmussen 1997). A decrease in the transparency of interactions and interconnections (the independent variable) can increase the likelihood of a systems accident (the dependent variable) (Perrow 1984). Structural secrecy and communication failures associated with bureaucratic organization (independent variables) can drive the accumulation of unnoticed safety problems (the dependent variable) (Turner 1978; Vaughan 1996). Managerial visibility on work sites (an independent variable) can have an impact on worker procedural compliance rates (the dependent variable).

Zero vision has got this upside-down. It tells managers to manipulate a dependent variable. Zero vision was never driven by safety theory or research. It has grown out of a practical commitment and a faith in its morality. It is a logical continuation, completion even, of the modernist project. Safety theory, in contrast, is mostly about manipulated variables, even though it often considers which dependent variables to look for (e.g., are incident counts meaningful dependent variables to measure?). But mostly, theories tend to specify the kinds of things that engineers, experts, managers, directors, supervisors, and workers need to do to organize work, communicate about it, and write standards for it. What they need to manipulate, in other words. Outcomes (measured in terms of incidents or accidents, or in terms of indicators of resilience) then are what they are. In retrospect (and the study of past accidents is often what drives theorizing on safety), outcomes can be traced back to manipulated variables (whether validly or not). Zero vision turns all of this on its head. Managers are expected to manipulate a dependent variable—an oxymoron. Manipulating a dependent variable is something that science considers to be either experimentally impossible or professionally unethical. With a focus on the dependent variable—in terms of how bonuses are paid, contracts are awarded, promotions are earned—manipulation of the dependent variable (which is, after all, a variable that literally depends on a lot of things not under one's control) becomes a logical response. Honesty can suffer, as can learning. And indeed, in the longer run, safety itself can be the victim.

GETTING ZERO BY DEMANDING IT FROM YOUR WORKERS

Countering the reformation halfway through the 20th century, behavior-based safety programs and practices focus on worker behavior rather than on workplace hazards as the cause of injuries and incidents. These practices might include the following (Frederick and Lessin 2000):

- Safety incentive programs, where workers receive prizes or rewards when they don't report work-related injuries.
- Injury discipline policies, where workers are threatened with or receive discipline (including termination) when they do report injuries.
- Post-injury drug testing, where workers are automatically drug tested when they report an injury.
- Workplace signs that track the number of hours or days without a lost-time or recordable injury, which encourages numbers games.

- Other posters or signs, such as those stuck to washroom mirrors stating, "You are looking at the person most responsible for your safety."
- Programs where workers observe coworkers and record their "safe behaviors" or "unsafe acts." This focuses attention away from hazards and reinforces the idea that injuries result from workers' bad behavior rather than hazardous conditions.

Behavioral safety programs can harm social cohesion and worker solidarity. When workers lose prizes if a coworker reports an injury, peer pressure comes into play. I learned of an oil installation where the "prize" for a particularly long injury-free period had gone from a six-pack of beers to a cooler to hold them, up to a leisure boat to hold the cooler. A few days before the period was about to expire and the boat awarded, a new worker suffered an injury that could not be hidden. She reportedly was not popular with her coworkers, to put it mildly.

Whereas it is possible to reclassify incidents and hide injuries, it is more difficult to do that with fatalities. In 2005, BP, for example, was touting an injury rate many times below the national average at its Texas City facility, but then an explosion there took the lives of 15 workers and injured 180 (Baker 2007). More recently, BP (whose oil rig Deepwater Horizon exploded in the Gulf of Mexico on April 20, killing 11 and causing untold damage in the Gulf) had been a recipient of safety awards for having low recorded injury rates in their facilities (Elkind and Whitford 2011).

The concerns expressed in the case description(s) above are borne out by research as well. A study of Finnish construction and manufacturing from 1977 to 1991 showed a strong correlation between incident rate and fatalities, but reversed ($r = -0.82$, $p < 0.001$). In other words, the fewer incidents a construction site reported, the higher its fatality rate was. Figure 8.1 shows this. The horizontal (x) axis shows the fatality rate for the hours worked in a given year, the vertical (y) axis shows incident frequency for that same year. As the incident rate increases, the fatality rate declines. Efforts to reduce incident rates (which may involve discouraging or suppressing their reporting, see above) are strongly correlated with a higher fatality rate (Saloniemi and Oksanen 1998).

Perhaps not so surprisingly, there is no evidence that zero vision has an impact on safety that is any greater than the next safety intervention (Donaldson 2013). This may not matter, however, as zero visions are a strong instrument of what is known as bureaucratic entrepreneurialism. In perhaps a cynical reading, zero vision allows people involved in safety to say two things simultaneously: they can claim that great things have been accomplished already because of their work, but that more work is necessary because zero has not yet been reached. And because it never will, or because the organization fears backsliding away from zero, safety people will stay relevant, employed, contracted, and funded. Whether people in these positions genuinely believe that injuries and accidents can be fully expunged is hard to know. But perhaps they have to be seen to believe it—in order to attract investments, work, federal grants, contracts, regulatory approval, and affordable insurance.

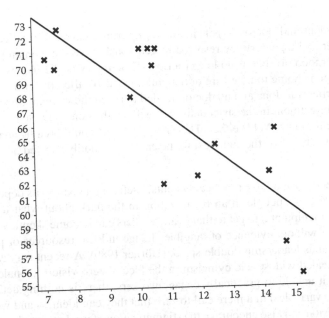

FIGURE 8.1 As the incident rate declines, the fatality rate increases. These data come from the Finnish construction industry between 1977 and 1991 (Regression figure made on the basis of data from Saloniemi and Oksanen 1998). The horizontal (x) axis shows the fatality rate for all the manhours worked in a given year (represented by the data points in the plot). The vertical (y) axis shows incident frequency for that same year. Efforts to reduce incident rates (which may involve suppressing their reporting) are strongly correlated with a higher fatality rate.

ZERO BY ANY MEANS

A few years ago, I learnt of a woman who was slightly injured at work. She told her supervisor, showed the injury, and went to see the doctor that same afternoon. While in the waiting room, she got a call from school. Her son had fallen ill and been sent home. After her appointment and having her gash cleaned and glued by a nurse, she rushed home to take care of her boy. She later informed her supervisor.

News of the injury made its way to the company's safety manager. He was horrified. Not necessarily because of the injury or the employee's fate, but because he had been on a "winning streak." Next to the entrance of the plant, a sign announced that the company had been without injury for 297 days. 300 had been within reach! The number would have looked so good. It would have made him look so good.

The day after the incident, the safety manager went to see the supervisor before committing, with lead in his shoes, to change the sign to 0. Zero days since the last injury. Then he learnt that the employee had gone home after the incident and doctor's visit. It was a gift from heaven. He called the HR manager and together they resolved to generously give the woman the previous afternoon off. It was no longer a Loss-Time Injury (LTI). The woman had simply gone home to take care of her child. He also called the clinic. Because no suturing was done and no doctor needed to touch the woman, this was not a medical treatment in the strict definition held by the company. So no Medical-Treatment Injury (MTI) either. The safety manager could breathe again.

A few days later, the sign next to the entrance proudly showed 300.

Does a zero vision have practical benefits? Defining a goal by its dependent variable tends to leave people in an organization in the dark about what to do (which variables to manipulate) to get to that goal. Workers can become skeptical about zero sloganeering without evidence of tangible change in local resources or practices. It is easily seen as leadership double-speak (Dörner 1989). A recent survey of 16,000 workers revealed widespread cynicism in the face of zero vision (Donaldson 2013). Not only is it unable to practically engage workers, there is nothing actionable (no manipulable variables) in a mere call to zero that they can identify and work with.

A zero vision can also encourage the stigmatization of workers involved in an incident. As such, it can become a technology or means to control problematic workers.

WEARING THE ORANGE VEST

Consider the example of a food warehouse, where 150 workers load and unload trucks, lift boxes, drive fork trucks, and move pallets. Each month that no one reports an injury, all workers receive prizes, such as $50 gift certificates. If someone reports an injury, no prizes are given that month. Management then added a new element to this "safety incentive" program: if a worker reported an injury, not only would coworkers forgo monthly prizes but the injured worker had to wear a fluorescent orange vest for a week. The vest identified the worker as a safety problem, and alerted coworkers: he or she lost you your prizes (Frederick and Lessin 2000).

To ensure that progress toward zero is maintained—or to be able to identify where it wasn't—organizations are increasingly able to monitor and track compliance to their stated values through a variety of "panoptisms" (or "see-all" technologies). This ranges from cockpit voice recorders, to intelligent vehicle monitoring systems, to video recorders in operating theaters in some hospitals. Safety audits, safe work observations, local supervision, and safety culture measurements and safety climate assessments add to this mix. Together, these things constitute a form of social control that connects individual conduct to organizational norms and expectations. And they also exert a kind of disciplinary effect. Behavior and transgressions get recorded in files, reports, case notes, and more. It can even become a means of extracting confessions and publicly humiliating and pillorying people (see the orange vest example above).

One of the more enduring instances of this can be found in medicine, where many are still battling the very idea that errors don't occur (Vincent 2006). Many in that world are faced daily with a world where errors are considered to be shameful lapses, moral failures, or failures of character in a practice that should aim to be perfect (Bosk 2003; Cook and Nemeth 2010). Errors are not seen as the systematic by-product of the complexity and organization and machinery of care, but as caused by human ineptitude (Gawande 2010); as a result of some people lacking the "strength of character to be virtuous" (Pellegrino 2004, p. 94). The conviction is that if we all pay attention and apply our human reasoning like our Enlightenment forebears, we too can make the world a better place (Gawande 2008). The 2000 Institute of Medicine report (IOM 2003) was accompanied by a political call to action to obtain a 50% reduction in medical mistakes over five years. This was not quite a zero vision, but halfway there.

Investigative resources can be misspent as a result of a zero vision too. If zero is assumed to be achievable, then everything is preventable. And if everything is preventable, everything needs to be investigated, including minor sprains and paper cuts. And if an organization doesn't investigate, it can even have direct legal implications. A documented organizational commitment to zero harm can lead a prosecutor to claim that if the organization and its managers and directors really believed that all harm was preventable, then such prevention was reasonably practicable (Donaldson 2013). They are liable if harm occurs after all, since they or their workers must have failed to take all reasonably practicable steps to prevent it.

ARE ACCIDENT-FREE ORGANIZATIONS POSSIBLE?

A zero vision is a commitment. It is a modernist commitment, inspired by Enlightenment thinking, that is driven by the moral appeal of not wanting to do harm and making the world a better place. It is also driven by the modernist belief that progress is always possible, that we can continually improve, can always make things better. Past successes of modernism are taken as a reason for such confidence in progress. After all, it has helped us achieve remarkable increases in life expectancy and reduce all kinds of injuries and illnesses. With even more of the same efforts and commitments, we should be able to achieve more of the same results, ever better. But a commitment should never be mistaken for a statistical probability. The statistical probability of failure in a complex, resource-constrained world—both empirically, and in terms of the predictions made by the theory—simply rules out zero. In fact, safety theorizing of almost any pedigree is too pessimistic to allow for an incident- and accident-free organization.

Look at man-made disaster theory, for example. On the basis of empirical research on a number of high-visibility disasters, it has concluded that "despite the best intentions of all involved, the objective of safely operating technological systems could be subverted by some very familiar and 'normal' processes of organizational life" (Pidgeon and O'Leary 2000, p. 16). Such "subversion" occurs through usual organizational phenomena such as information not being fully appreciated, information not correctly assembled, or information conflicting with prior understandings of risk. Turner noted that people were prone to discount, neglect, or not take into discussion relevant information. So no matter what vision managers, directors, workers, or other organization members commit to, there will always be erroneous assumptions and misunderstandings, rigidities of human belief and perception, disregard of complaints or warning signals from outsiders, and a reluctance to imagine worst outcomes—as the normal products of bureaucratically organizing work (Turner 1978).

Not long after, Perrow suggested that accident risk is a structural property of the systems we operate (Perrow 1984). The extent of their interactive complexity and coupling is directly related to the possibility of a systems accident. Interactive complexity makes it difficult for humans to trace and understand how failures propagate, proliferate, and interact, and tight coupling means that the effects of single failures reverberate through a system—sometimes so rapidly or on such a massive scale that intervention is impossible, too late, or futile. The only way to achieve a zero vision in such a system is to dismantle it, and not use it altogether, which is what Perrow essentially recommended societies do with nuclear power generation. Some would argue that Perrow's prediction has not been borne out quantitatively since the theory was first publicized in 1984. Perrow's epitome of extremely complex and highly coupled systems—nuclear power generation—has produced only a few accidents, for example. Yet the 2011 earthquake-related disaster at Fukushima closely followed a Perrowian script. The resulting tsunami flooded low-lying rooms at the Japanese nuclear plant, which contained its emergency generators. This cut power to the coolant water pumps, resulting in reactor overheating and hydrogen–air chemical explosions and the spread of radiation. Also, increasingly coupled and complex systems like military operations (Snook 2000), spaceflight (CAIB 2003), and air-traffic control (BFU 2004) have all produced Perrowian accidents since 1984.

Vaughan's analysis of the 1986 Space Shuttle Challenger launch decision reified what is known as the banality-of-accidents thesis (see Chapter 5). Similar to manmade disaster theory, it says that the potential for having an accident grows as a normal by-product of doing business under normal pressures of resource scarcity and competition. Telling people not to have accidents, to try to get them to behave in ways that make having one less likely, is not a very promising remedy. The potential for mistake and disaster is socially organized: it comes from the very structures and processes that organizations implement to make them less likely. Through cultures of production, through the structural secrecy associated with bureaucratic organizations, and a gradual acceptance of risk as bad consequences are kept at bay, the potential for an accident actually grows underneath the very activities an organization engages in to model risk and get it under control. Even high-reliability organization theory is so ambitious in its requirements for leadership and organizational design that a reduction of accidents to zero is all but out of reach. Leadership safety objectives, maintenance of relatively closed operational systems, functional decentralization, the creation of a safety culture, redundancy of equipment and personnel, and systematic learning are all on the required menu for achieving high-reliability organization status (Rochlin et al. 1987). While some organizations may hew more closely to some of these ideals than others, there is none that has closed the gap perfectly, and there are no guarantees that manipulating these attributes will keep an organization at zero (Sagan 1993).

SAFETY CULTURE AND PEOPLE AS A PROBLEM TO CONTROL

But what about safety culture? Even if becoming a high-reliability organization may be out of reach for many, can they not strive for a better safety culture? To judge the value of that question, let's first look at the concept itself a bit more. Classical–mechanical descriptions of the sorts of organizational failures like Macondo, Challenger, and Columbia have increasingly been acknowledged to fall short (e.g., CAIB 2003). These are not linear cause–effect breaches of defined, fixed-order layers of defense. And if we insist on seeing them that way, we do violence to the complexity of the events, and shortchange ourselves by eschewing more compelling narratives and change initiatives. This goes for smaller organizational failures as well—"down" even to a single medication misadministration at a patient's bedside (Dekker 2011c). Acknowledging the overriding importance of "softer," social aspects for creating the organizational conditions for such failures, the safety literature has embraced the concept of "safety culture" over the last 10 years. Safety culture has been increasingly deployed to fill the social–organizational complexity vacuum left by sequence-of-events models.

"Safety culture" seems to have responded to the organizational and political need to explain accidents that befall basically safe and mechanically sound systems, or cases where the sorts of failures and problems that turned up in the accident are widely shared in the industry without leading to similar problems elsewhere. To understand this a bit better, consider two quite different ways to understand culture. As anthropologist Clifford Geertz may have put it, we can approach culture as an interpretive effort in search of meaning, or as an experimental effort in search of law. This has important consequences for what we believe culture is, and what we can do with (or to) it. Roughly, these two ways to consider safety culture work out like this (Henriqson 2013):

- An interpretivist approach of anthropology and sociology, where a safety culture is something that an organization *does*. Culture is bottom-up; it emerges from the multitude of interactions between members of a system. It cannot be controlled, only influenced—and likely with all kinds of hard-to-predict side effects. From this perspective, saying there is a "good" versus "bad" safety culture is either nonsensical or arrogantly moralizing. Culture can be studied through qualitative methods, taking an emic or inside-out approach—that is, seeing the world through the eyes of those whose outlook you are trying to understand.
- A functionalist approach of psychology, management science, and engineering. Here, culture is something an organization *has*. It is a property, or a possession, and as such it can be taken away and replaced (if partially). It can also be controlled and manipulated. Culture can be studied through quantitative methods, where surveys go in with predetermined questions that ask participants about their experiences, beliefs, attitudes, and values. This is an etic, or outside-in approach. It is, from this perspective, legitimate to speak about "good" versus "bad" safety cultures, as aggregates of

beliefs and attitudes can be argued to be more harmful to safety in some context than others.

In safety, the functionalist approach dominates. It is a view of safety culture that renders humanity decomposable and redesignable. As Batteau summarized, "safety culture results from a decomposition of culture ... a set of homilies toward prudence, following rules, and communicating openly. By decomposing habits of the heart into elements sized to a designer's checklist, one eliminates from view the ineffable impusles [sic] that distinguish cultures from commands. In a list focusing on specific items, such as reporting systems (check), open communication (check), non-punitive fact-finding (check) and adherence to standard operating procedures (double-check), there is no formatted space for a dialogue," a dialogue or conversation that might construct other versions or perceptions of operational risks (Batteau 2001, p. 203). It turns culture away from the emergent phenomena and subtle connections within human contexts, and into a branch of engineering. It uses engineering strategies and engineering discipline to solve an engineering problem. "When social sciences support this strategy, they strike a pose of positivism, with any insights of interpretation, of social construction, of discourse, and of historical conflict carried along as contraband" (p. 204).

It is a construction of culture with consequences. First, safety culture implies a normative homogeneity of values, beliefs, and behaviors, something that organizations can demand by policy and incentive. Its homogenizing effects can stand in the way of diversity, creativity, and improvisation (or at least push them out of view). Yet it is that kind of diversity that is linked to resilient systems; systems that are able to adapt to change and absorb disruptions that they were not directly designed for. Second, safety culture initiatives can go hand-in-glove with neoliberal and behavioral safety initiatives toward worker responsibilization (Gray 2009). This is what Foucault referred to as governmentality: a complex form of power that links individual conduct and administrative practices, in this case extending responsibility for safety from the state to organizations, and from organizations to individuals, expecting self-responsibilization and self-discipline. People are seen as a problem to control, and through subtle and less subtle processes and technologies, organizations exercise such control. Workers are expected to commit to safety, adopt accountabilities and responsibilities, participate and comply, communicate openly, and actively look out for each other (Henriqson 2013). Safety, through all of this, is once again seen as a moral commitment (remember "hearts and minds" from not so long ago, and the human factor as the "moral or mental shortcomings" of individual people from the turn of the last century). Safety again becomes a normative choice, a supposedly "free" choice, even though it can be in great conflict with other stated and unstated organizational goals on production and efficiency, or with the design provisions offered in the equipment people need to work with. The sorts of behavioral encouragements that come from, and with, a safety culture, however, are based on faith more than on evidence—much like zero vision. Their embrace seems to be driven by trust, conviction, and moral commitment more than by science.

SAFETY AND FAITH: POSTERS TO ENGAGE THE WORKFORCE

I was at a construction site not long ago, accompanying the site's production director. Posters with a photograph of him, wearing the requisite protective equipment, were plastered all over the place. On the photograph, he had a stern look, his arms folded across his chest. Next to his head were the company's slogans about its commitment to zero harm. I asked him what evidence the company had that any of this was working to prevent injury and incident. He had no such evidence, nor had his company, anywhere. The posters seemed based entirely on faith. His company was not alone of course. I recall working with an air-traffic service provider, whose latest anti-incident campaign was based on posters throughout the control room that goaded air-traffic controllers into greater "situation awareness."

Leaving the construction site, I asked the production director whether the company would do the same when facing a decision to engage in a project, in say, Tanzania. Whether they would print posters with him, or the CEO on it, and a slogan next to his head saying "Tanzania is a great place to invest in a project!" and then paste them in the lifts, and board room, and hallways and toilets. Would he truly have the hope that in a few weeks' time, most people in the company would be committed to investing in a project in Tanzania? He looked at me blankly. Yet are we not asking people to embrace a zero vision of safety on the same sort of faith and absence of evidence?

Some science has actually been done on this. To find out what worked and what didn't, Hallowell and Gambatese (2009) analyzed the relative contributions of a number of safety interventions in the construction industry. Their Delphi analysis revealed some surprising things (or perhaps not). Some programs and initiatives that safety managers or entire companies cherish did not actually generate much safety yield. In descending order of importance for their effects on safety (with 1 the most, 13 the least), their results were as follows (Hallowell and Gambatese 2009):

1. Upper management support. This involves the explicit consideration of safety as a primary goal of the firm. Upper management can demonstrate this commitment by, for example, participating in regular safety meetings, serving on committees, and providing funding for safety equipment and initiatives.
2. Subcontractor selection and management. This involves the consideration of safety performance during the selection of subcontractors. That is, only subcontractors with demonstrated ability to work safely should be considered during the bidding or negotiating process. This should involve skepticism about mere low numbers of negatives, of course.
3. Employee involvement in safety and work evaluation. This may include activities such as performing job hazard analyses, participating in toolbox talks, or performing inspections. For this to work well, the focus of such activities needs to be driven bottom-up, not imposed top-down.

4. Job hazard analyses. This includes reviewing the activities associated with a work process and identifying potential hazardous exposures that may lead to an injury.

5. Training and regular safety meetings. These aim to establish and communicate project-specific or work-specific safety goals, plans, and policies before the start of the project or work.

6. Frequent worksite inspections. These help identify uncontrolled hazardous exposures to workers, violations of safety standards or regulations, or behavior of workers.

7. Safety manager on site. The primary responsibility of a safety manger is to perform, direct, and monitor the implementation of safety program elements and serve as a resource for employees.

8. Substance abuse programs. These target the identification and prevention of substance abuse of the workforce. Testing is a crucial component of this safety program element. Methods of testing and consequences of failure may differ between organizations and industries.

9. Safety committees made up of supervisors, practitioners, workers, representatives of contractors, owner representatives, and safety consultants can be formed with the sole purpose of addressing the safety of work.

10. Safety orientation. This involves the organization- or site-specific, but not necessarily project-specific, orientation, and training of all new hires and contractors.

11. Written safety plan serves as the foundation for an effective safety program. The plan must include documentation of project- or work-specific safety objectives, goals, and methods for achieving success.

12. Record-keeping and incident/accident analyses involve documenting and reporting the specifics of incidents and accidents including information such as time, location, work-site conditions, or cause. It includes the analyses of accident data to reveal trends or weak spots.

13. Emergency response planning. This may be required by owners, insurers, or regulators, and involves the creation of a plan in case of a serious incident. Planning for emergencies can define the difference between an accident and a catastrophic event.

While these results were generated out of data from the construction industry, parallels with other industries and safety-critical settings are not hard to find. Frequent work inspections, for example, are common in aviation, where airlines themselves do line checks (where line-check captains fly with other crews to monitor and discuss their performance). This is true for a safety manager on site as well: one of the airlines I flew for had the safety manager sit in an office directly off the crew room at the airline's hub airport. Pilots and others could easily walk in and discuss anything related to safety. The airline also had a safety committee, on which I participated, for example, when a change was proposed to the pre-takeoff procedures to ensure the correct wingflap setting before departure. What is known as job hazard analyses or regular pre-work safety meetings in construction and other project work is known as pre-flight briefings in aviation. These happen not only in the crew room to discuss the entire duty day but also in the cockpit before every critical phase of flight.

It is interesting, for any industry, to reflect on the order of importance of safety initiatives as identified in Hallowell's and Gambatese's research. Does this order, and what it says about the role of the human factor, apply to them as well? The critical importance of upper management support has been identified in high-reliability organization research across multiple industries, for example. Maintaining the goal of avoiding serious operational failures in those organizations has been said to nurture an organizational perspective in which short-term efficiency gains take a second seat to high-reliability operations. This has been found in air-traffic control, navy operations, and electricity generation (LaPorte and Consolini 1991). In industries such as those, there are obvious reasons why organizational leaders must place a high priority on safety:

> High reliability organizations require both significant levels of redundancy and constant operational training, and both of these factors cost a great deal of money... If political authorities and organizational leaders are not willing to devote considerable resources to safety, accidents will become more likely (Sagan 1993, p. 18).

Many organizations might say they have safety as their highest priority, but such talk is cheap. It needs to be backed up by visible action and particularly with resources. Whether high-reliability organizations have really accomplished this or whether other factors account for their apparent success is still a matter for debate (Sagan 1994). The selection of subcontractors or partners who can demonstrate acceptable safety performance is also something that is widely shared beyond construction. In aviation, for example, airlines are not allowed to join a global alliance without undergoing a whole series of audits. The problem with these initiatives, of course, is that it easily reproduces the practices that we saw in zero visions above. Where safety is mainly assessed by an absence of negatives, incentives exist to make negative events invisible. In industries such as construction and energy, it is unfortunately still canonical for contractors and subcontractors to demonstrate low numbers of negatives; otherwise, they may not even be considered for tender at all. If the client is a government (city, state, federal), the demonstration of low numbers of incident and injury rates is often obligatory. Selecting a contractor or subcontractor on such low numbers may say more about their creativity in making the numbers look good than about their safety or resilience. It moved David Capers, a long-time expert in energy processing, to quip that the numbers that matter are not LTIs (Loss Time Injuries) or MTIs (Medical Treatment Injuries), but an LGI—a Looking Good Index.

Let's turn to the bottom of the list now. Safety inductions or orientations have developed, in many industries, a negative connotation. They can be seen as a waste of time and money. Some sites or companies have orientations that take days, not hours. And even then, they may be ill-connected to the work people will do or the risks they will face. Rather, they are seen as yet another exercise in liability management and reduction on part of the company (so that they can say, "see, told you so, warned you!" if things do go wrong). Many inductions or orientations are also top-down in an almost Tayloristic sense ("we are smart, you are dumb, so listen to us for how to be safe"). Success stories do exist, however, and it is not difficult to see what it takes. One chemical production site, for example, made sure that its safety orientation was very short in the classroom but much longer on the site. It had developed a buddy system, in which new hires were mentored for weeks or more by somebody specifically assigned to them (and trained as a "buddy"). This allowed the experienced person and the new hire to actively engage with the work and jointly explore what was dodgy about it. Learning about safety was a co-production, set in context—not a one-way dictation in a vacuum. The relationship between those already on site or in the company and those new to it was, as a result, more organic, more equal.

The low score of record-keeping and incident and accident analysis may come as a damper to those who make it their livelihood. Record-keeping is no aim in itself, of course, and they won't speak to safety people like some oracle. Records and data are only any good if you know what to ask of them. There is probably no organization in the world that has a perfect balance between all the data it gathers about things going wrong (or right) and the analytic yield it gets from it. In project-driven industries, like construction, it is interesting to see how little data are mined on previous projects to understand what accounted for the successes and the foundering. The data are there, but the questions that need to be asked of it in order to be wider before the next project are not consistently reflected on. The position of incident analysis in the list is consistent with Amalberti's findings on safe organizations (recall this from Chapter 5). Analyzing the causal factors of incidents, and sorting and recombining them into an understanding of how accidents are going to happen, has a lower predictive value if your organization is safer. One reason for this seems to be that accidents are preceded by normal work, not by incidents. The daily frustrations and difficulties in getting the job done—which do not get reported—can be a stronger predictor of fatality risk than the reported incidents that have made it into the database and company records.

That written safety plans and emergency response planning have hit the bottom of the list may be no surprise to people familiar with these sorts of things in their own organizations. It is no surprise either to normal accident theory, which has concluded that organizations regularly ignore the bulk of their experience that shows these plans and documents to be inaccurate. This is why Clarke called them "fantasy documents," which are neither wholly believed nor disbelieved. They are tested against reality only rarely and seem to draw from a rather unrealistic or idealistic view of the organization or the environment in which it operates (i.e., every contingency is known and prepared for). The extent to which these plans and documents amount to fantasy was once again obvious from the Fukushima nuclear disaster in 2011 (already alluded to above), the largest since Chernobyl in 1986:

Japan seemed underprepared for the disaster. At the plant itself, for instance, procedures and guidelines were woefully insufficient for a meltdown, forcing operators to respond on an almost entirely ad hoc basis. When the lights went out they had to borrow flashlights from nearby homes to study the plant's gauges. On-site dosimeters maxed out at levels that were far below those that could be expected from a catastrophe, unable to display readings any higher. And, as the plant's former safety manager would later testify, emergency plans "...had no mention of using sea-water to cool the core," an oversight that caused unnecessary, and perhaps critical, delays. The bureaucratic infrastructure beyond the plant evinced similar shortcomings. Official announcements were often ill-considered and characterized by denial, secrecy, and refusal to accept outside help... The pervasive idealization of formal risk assessments, which so many narratives of Fukushima reaffirm, both narrows the democratic discussions around nuclear power, and perverts the processes through which it is governed. The false surety it projects allows the cost–benefit projections that frame nuclear decision-making to silently discount the evidence of past accidents in the tacit understanding that disasters are somehow aberrant and avoidable rather than endemic. The result, as outlined above, is a deep-rooted institutional reluctance to adequately plan for worst-case scenarios (Downer 2013, pp. 2–3).

And finally, back toward the top of the list. For employee involvement in safety and work assessments to score as high as it did is encouraging and consistent with the growing literature on high-performance teams and resilient organizations. The themes that animate these literatures, after all, include participative leadership, coordinative relationships and decision making, common purpose, a preference for bottom-up rather than top-down initiatives, open communication and mutual trust, valuing of diversity of inputs, and openly dealing with conflict between viewpoints. It may read like an ideal list. That it seems more ideal to people in safety than it does to people in, say, sports coaching, perhaps says a lot about people in safety. At the same time, there are undoubtedly both managers and safety people who would consider an uncritical deference to high-performing teams and their expertise to be a mixed blessing. We turn to that now.

DEFERENCE TO EXPERTISE

Deference to operational or engineering expertise is seen as a critical ingredient of safety in a variety of industries. If people are not a problem to control, but a resource to harness, then taking their experience and expertise seriously is a strong first step. After all, operational activities in any industry contain situations whose subtle and infinite variations may mismatch the exact circumstances of training. This includes "fundamental surprises" (Lanir 1986), that is, situations that fall outside the procedures constructed for normal and emergency operations. On these occasions, operators must be able to apply skills and knowledge that no training department was able to foresee or deliver. This leaves a residue of potential problems that crews are not prepared for (i.e., they are not in their inventory [Dismukes et al. 2007]). The same is true for other industries and operators as well. Formal mechanisms of safety regulation (through, for example, design requirements, policies, procedures, training programs, checks of operator competence) will always fall short in foreseeing and meeting the shifting demands posed by uncertainty, limited resources, and multiple conflicting goals. For this residue, one has to count on crews' generic competencies to add to the ability of a system to respond to unexpected and escalating situations. These surprises at the margins of otherwise safe systems stem from limits in the industry's knowledge or, more often, limits on its ability to put together diverse pieces of knowledge, as well as from limits on the understanding of operational environments (Lanir 1986). In other words, the knowledge base for creating safety in complex systems is inherently imperfect (Rochlin 1999). Often the problem is not that the industry lacks the data. The problem is that this accumulation of noise and signals can muddle both the perception and conception of "risk." Pockets of expertise that may have predicted what could go wrong often exist in some corner of the industry long before any accident.

Such signals of potential danger, and of a gradual drift into failure, can be missed by those who are not familiar with the messy details of practice. For a manager, deference to expertise means engaging those who are practiced at recognizing risks and anomalies in operational processes. So-called high-reliability organizations, for example, have been acclaimed for their sensitivity to operations and deference to expertise. They are attentive to their operational front end, the sharp end where the "real" work gets done, where workers are in direct contact with the organization's safety-critical processes. High-reliability organizations push decision making down and around, creating a recognizable "pattern of decisions 'migrating' to expertise" (Weick and Sutcliffe 2007, p. 16). Such engagement must happen even for decisions that have, at the surface, little connection to operations or design. "Budgets," for example, "are often insensitive to operations" (Weick and Sutcliffe 2007, p. 13) but can in the long run very well have operational or safety consequences. Paying attention to the sharp end is generally thought to pay off: recent research has linked leadership involvement in daily work operations with worker competence, role clarity, and safety involvement (Dahl and Olsen 2013). It is echoed in organizational–psychological research into the effects of leadership presence on employee performance, loyalty, and attachment (Dekker and Schaufeli 1995; Schein 1992).

In hindsight, *not* deferring to expertise is often constructed as a major safety shortcoming. Prior to the Texas City refinery explosion in 2005, for example, BP had eliminated several thousand US jobs and outsourced much of its refining technology work, and several hundred engineers who had been centrally located within the company left. Many experienced others retired. It was only years later that BP realized it had lost too much in-house technical expertise. The loss of human expertise in the refining business, both in the line and supporting functions, meant that line managers became miscalibrated about process safety. The focus was on cost-control and rationalization of the US refining business, and managers believed that process safety was adequately tracked and controlled. They tended to simplify operational and technical problems and not defer to expertise or operations. The Baker panel concluded that

> Although many members of BP's technical and process safety staff have the capabilities and expertise needed to support a sophisticated process safety effort, the Panel believes that BP's system for ensuring an appropriate level of process safety awareness, knowledge, and competence in the organization relating to its five U.S. refineries has not been effective in a number of respects (Baker 2007, p. 147).

In a similar vein, the appointment of Sean O'Keefe (Deputy Director of the White House Office of Management and Budget) to lead NASA signaled that the organization's focus should be on management and finances, continuing a trend that had been set years before. O'Keefe's experience was in the management and bureaucratic administration of large government programs—a skill base deemed important by the new Bush administration for the NASA top job. NASA had already vastly reduced its in-house safety-related technical expertise during the 1990s. Its Apollo-era research and development culture once prized deference to the technical expertise of its working engineers (Mindell 2008; Murray and Cox 1989). This had become overridden by bureaucratic accountability—managing upward with an allegiance to hierarchy, procedure, and following the chain of command (Feynman 1988; Vaughan 1996). After the accident, the investigation board concluded that

> Management decisions made during Columbia's final flight reflect missed opportunities, blocked or ineffective communications channels, flawed analysis, and ineffective leadership. Perhaps most striking is the fact that management... displayed no interest in understanding a problem and its implications. Managers failed to avail themselves of the wide range of expertise and opinion necessary to achieve the best answer to the debris strike question—"Was this a safety-of-flight concern?"... In fact, their management techniques unknowingly imposed barriers that kept at bay both engineering concerns and dissenting views, and ultimately helped create "blind spots" that prevented them from seeing the danger the foam strike posed (CAIB 2003, p. 170).

In the wake of the Columbia accident, NASA was told it needed "to restore deference to technical experts, empower engineers to get resources they need, and allow safety concerns to be freely aired" (CAIB 2003, p. 203). The two Space Shuttle accidents—Challenger in 1986 and Columbia in 2003—have led to calls for organizations to take engineering and operational expertise more seriously. This has become well established in the literature on high-reliability organizations and resilience.*

THE "PRIMA DONNA SYNDROME"

There could be another implication of deferring to expertise. When an operation relies on specialized knowledge, "notably because certain decisions are highly technical ones, certain experts attain considerable informal power" (Mintzberg 1979, p. 199). If such key people are a resource to be harnessed, are there any problematic implications of this? Tobacco factories in the 1960s, for example, were ruled by maintenance men because only they could handle the major source of uncertainty: machine stoppage. Everybody relied on them to keep things running, but nobody understood what they did, nor could they check on them. Supervisors lost in the perpetual fight for control. The arrangement in fact challenged the collaboration between experts and any other organization members (Crozier 1964). The popular organizational literature has referred to the "prima donna syndrome" (Girard 2005). Prima donnas typically have privileged contact with (and influence over) the organization's primary or safety-critical process. They often enjoy more authority than the formal organizational structure

---

* Interestingly, the emergence of this appeal has coincided with unprecedented growth in generic management (MBA) programs and a simultaneous rise in corporations retaining external subject-matter expert consultants. If there is a role for expertise, it is not in-house and not in management. See Mintzberg (2004) and also Khurana (2007).

allows and can receive preferential treatment and outsized compensation. They are recognized for their high technical competence, assertiveness, and self-confidence, yet can have an insensitivity to larger organizational goals and difficulty working under direction or as part of a mixed-composition team (Girard 2005). The "prima donna" syndrome has been described in fields as varied as construction (Schultz 1998), nursing (Girard 2005), technology (Dickerson 2001), business (Wright 2009), sports (Dubois 2010), manufacturing (Pollock 1998), and aviation (Bertin 1997).

Not all these fields have safety-critical processes. But when they do, and "when people are entrusted with dangerous technologies, it is easy for them to feel self-important and central since they live on a steady diet of people telling them that they are important" (Weick and Sutcliffe 2007, p. 159). Feeling "self-important and central" can turn into a sense of psychological entitlement (Harvey and Martinko 2008). Specialized, detailed insight into safety-critical technology can amplify the organizational leverage of groups tasked with operating it (Edwards and Jabs 2009; Lovell and Kluger 1994). Some of the observations made about "prima donnas" in this literature are as follows:

- They feel they don't have to play by the rules. They might resent being held accountable for their performance like other employees, because they believe their performance is as good as it will ever get.
- They have unrealistic expectations about what the organization will do for them and are resistant to negative feedback.
- Superior knowledge of frontline activities or technologies is allowed to trump all other organizational concerns.
- Their qualities as operators of frontline or safety-critical technologies (persuasive, self-confident, productive when they want to be) are the same that make them hard to manage and hard to get along with.
- They have an inflated sense of self-importance, a belief that gets justified by praise, attention, and "coddling" from others (Dubois 2010, p. 22). At the same time, they can show deliberate rudeness and have corrosive effects on team and organizational cohesion.
- They thrive in workplaces with high levels of ambiguity. Where credit and blame are easily diffused (because of a lack of documentation and accountability), self-serving attributional biases can solidify.

Popular technical literature sometimes expresses intolerance with Prima donnas, driven in part by emerging economic realities. High-quality operators in IT, for instance, may be easier to find than a few years ago (Dickerson 2001), though not necessarily in the healthcare or energy sectors (Aiken et al. 2002). In addition, research has questioned whether deference to prima donna expertise is critical for safety. Accident potential has been linked to structural factors that supersede expert input into organizational decision making (Perrow 1984). A recent analysis of the Macondo deepwater oil well blowout (BP 2010) shows that deference to expertise underdetermined whether the operation remained safe (Hayes 2012). And like Jensen's (1996) account of the Space Shuttle Challenger launch decision, Vaughan showed in much historical detail how expertise, too, gets absorbed by cultures of production; how subtle normalization of signs of deviance in technical and operational details affects experts too (Vaughan 1996). Others have also shown how technical neutrality in the face of larger organizational goals and pressures is often an illusion (Weingart 1991; Wynne 1988). Is deference to expertise critical for safety, then, or fearmongering? And if such deference is required, how is that voice modulated and balanced against other legitimate organizational concerns?

At least two psychological literatures are available to guide any observations about the prima donna syndrome. The first literature is that of psychological entitlement, broadly organized around cognitive and attributional frameworks. The second is that of organizational narcissism, an originally psychodynamic concept adopted to social–organizational contexts. They can be deployed not in a componential hunt for personality types or clinical predispositions of "prima donnas." Rather, both literatures allow a prima donna syndrome to be constructed as systematically produced by the interplay between environment, organization, cognition, professional expectations, and social and industrial relations. Prima donnas, in other words, are not bad components, but emergent from a complex of organizational, professional, and institutional relationships.

PSYCHOLOGICAL ENTITLEMENT

Psychological entitlement refers to the belief that one should receive preferential treatment that is not matched to actual deservingness. Reference to it in the psychological literature has increased over the last decade (Harvey and Martinko 2008). Entitlement means an inflation of expectations about rewards and compensation above what actual performance deserves. Psychological entitlement is not based on an equitable exchange. The core operational groups in the organizations studied for this book expected more benefits, attention, power and input without seeing the need to reciprocate with high(er) levels of performance or other sacrifices (Naumann et al. 2002).

CASE STUDIES OF PRIMA DONNAS

A chemical process plant and an airline pilot training school, both located in Western countries, were studied to better understand the "prima donna syndrome." Their respective top managers were followed and observed during meetings and decision-making processes—at the plant for nine months, and at the school for a period of two years. Both sites were responsible for safety-critical processes, with its personnel dominated (in staff numbers, organizational status, and compensation) by a core operator group: instructor pilots and process operators, respectively. The chemical plant employed over 100 process operators who worked a 24/7 schedule. The school employed between 20 and 30 instructor pilots during the observed period, active mostly 8 a.m. to 4 p.m. shifts on working days. These groups enjoyed strong union representation (almost 100%) at both sites. Both organizations studied had gone the last three or more years without any major incident or accident—safety performance (if measured by the absence of negatives) was deemed adequate or better by relevant parties (e.g., regulator, mother organization, auditors).

At the two worksites observed, core groups (who would typically be labeled prima donnas by managers) expected their organizations to pay for their time while getting dressed for and after work, and expected free high-standard facilities for doing so. Instructor pilots at one of the organizations had been given not only uniforms and laundry facilities by the school but also a considerable shoe allowance. At the same time, both groups were able to successfully block organizational initiatives at increasing productivity for several years. At the plant, reducing night-shift manning was blocked, and at the school, instructor pilots successfully kept a requirement to fly four (rather than three) sorties per day. In both cases, safety was used as the legitimating device of the positions taken, though threats of industrial action were never far below the surface. Indeed, safety was almost invariably used as the lever by the core group to get the attention or issue it sought. Even a challenge to seemingly trivial perquisites (like a shoe allowance, or having to fly with more than four students per course) was constructed as a threat to safety and quality. At both sites, managers had essentially stopped seeking expertise on safety matters from their core operator groups. Rather, the initiative was inverted, with operators raising as safety problems a host of industrial, work-context and lifestyle issues, ranging from lighting to work hours to desk size.

Injuries attributed to maintenance contractors at the chemical plant were always caused by contractor inexperience or incompetence, even though in some cases operator contributions (e.g., leaving particular piping pressurized) were easy to find. Maintenance had only recently been outsourced in a large reorganization (involving redundancies in the mother company), so these attributions may have been driven by additional motives. And to be sure, existing accountability mechanisms were perhaps perverse as well—contractors were penalized for loss-time injuries (which might hurt their chances to gain additional work in the future), but operators at the mother organization running the plant went scot-free.

Entitlement expectations are not just monetary. The term has been used to describe the extent to which individuals prefer being treated as special or unique in social settings. Entitlement promotes an inaccurate view of the world and oneself (Snow et al. 2001). Such psychological entitlement got members of the core groups at both organizations to react negatively to criticism and challenges to their worldview—and be very public with those reactions. Meetings at both sites were observed on multiple occasions where both core group representatives and managers lost their temper when not enough heed was seen to be given to complaints and grievances of "prima donna" employees.

Harvey and Martinko (2008) have shown that psychological entitlement diminishes the cognitive processing that employees apply to workplace situations. They tend to overlook important situational information. Entitlement is associated with a self-serving attributional tendency, which allows them this cognitive shortcut. Negative events or outcomes tend to get attributed to other people or external circumstances, whereas positive ones are attributed to the self. Individuals with strong entitlement perceptions tend to take credit for positive outcomes and are likely to feel estranged and blame others when negative outcomes occur. Entitlement thus functions as a mental patterning device. It inhibits their desire to engage in an elaborate cognitive evaluation process and will get people to shy away from information that contradicts these attributions:

> An attribution style that biases individuals toward attributing the negative events in their lives, such as receiving a poor performance evaluation at work, to external factors is one way in which this ego-protecting perceptual distortion can manifest itself. When undesirable events are attributed to external factors, such as another persons' incompetence, the individual fails to accept responsibility and the positive self-view is protected. We propose that this attributional tendency is likely among entitled individuals, whose positive self-images would logically bias them toward assuming that they are not to blame for negative outcomes (Harvey and Martinko 2008, p. 463).

Attributional processes (i.e., searches for causes of workplace events) tend to become less effortful and less detailed with psychological entitlement. This is likely to reduce the authenticity or richness of people's attributions, reifying their attributional bias. Job dissatisfaction and poor working relationships result from unmet expectations and a warped view of workplace responsibilities (Naumann et al. 2002). Conflict with supervisors is more likely, as are job turnover intentions (Harvey and Martinko 2008). At both sites, core group turnover was high. Some 30% of instructor pilots left during the observation period, while many plant operators expressed hanging on for a few more years only to maximize their eventual payout.

Narcissism in Groups

Narcissism, originally a psychodynamic concept, refers to a collection of cognitions and behaviors that help in the regulation of self-esteem (Freud 1950). It involves ego-defense mechanisms such as denial, rationalization, attributional egotism, sense of entitlement, and ego aggrandizement. Narcissism has been used in the management literature to understand organizational behavior and collective identity. It does not literally see organizations, groups or teams as narcissistic entities, but their behaviors and social cognitions are analogous to those exhibited by narcissistic individuals (Brown 1997). Groups, after all, have needs for self-esteem too, and these can be regulated in narcissistic ways—displayed as follows (Godkin and Allcorn 2009):

- Exceptional pride in own accomplishments and belief in continued success
- Entitlement that supports exploiting others both inside and outside the organization
- Envy and rage that arise when pride or pursuit of own goals are threatened
- History of banning or driving out nonconformers or resistors

- Management by intimidation
- Suppression of accurate reality testing and creativity
- Filtering of information and magical thinking
- Frequent blaming and scapegoating of others
- Volatile mood swings, from celebrating success one day to despair over not achieving the smallest of goals the next
- Alienation of management and leadership to their "foxholes" (cubicles, offices)
- Destructive internal competition and open organizational warfare

What characterizes narcissistic groups is a sense of anxiety, which stems from a dependence on others to validate self-esteem (Brown 1997). Narcissistic groups or "prima donnas" are in a constant dilemma. For their self-esteem, positive regard, and affirmation, they depend on the very people whom they hold in contempt or even feel threatened by. These may be supervisors, managers or plant leadership, or colleagues who do not interact daily with the organization's safety-critical processes. Organizational change, threats of job insecurity, or industrial uncertainty can exacerbate this anxiety.

ANXIETY AND ORGANIZATIONAL NARCISSISM

The sites studied were remarkably similar in this regard, even though they were not selected on this variable: during the periods of observation, threats to their continued existence emerged for different economic factors (though both are still in existence today). Mother organizations of both sites were considering realignment or even complete closure and made little effort to hide their possible intentions. This could have made core group members more likely to settle on disordered ways to shore up self-esteem and get organizational affirmation of their role and relevance (Dekker and Schaufeli 1995). It exacerbated the managerial dilemma too: while less able to afford perceived indulgences of their core groups (yet getting more demands from them), managers perhaps needed to tap into expertise more than ever before: times of cost-cutting, downsizing, and possible closure can make safety-critical organizations extra vulnerable (CAIB 2003; Reason 1997).

Many of the behaviors and cognitions of narcissistic groups were captured by myth-making. Core operator groups at both sites often told stories of their own mythical past—preceding the current management. Plant operators would point to a new parking lot built on their former sports field and now filled with contractor vehicles (who had "stolen" their colleagues' jobs). Instructor pilots pointed to a past where links with the Air Force resulted in cheap fuel, ample manpower and expertise, well-recognized pilot selection processes, a very favorable instructor-to-student ratio, small class sizes, higher-performance aircraft, a bigger building, and generally more collegiality. Interestingly, at the school, more than the plant, core operator group members increasingly mistook the amount, extent, and status of their expert knowledge for its currency and accuracy. On multiple occasions, managers were able to point out places where developments in the industry or technology had overtaken what was still upheld as the ideal of expertise and competence. Largely unbeknown to the operators, that ideal had become spotty with obsolescence.

SELF-ESTEEM AND MYTH-MAKING

Schwartz (1989) similarly found how myth-making, or the dramatization of its own ideal character, played a role in NASA's first Space Shuttle accident. The institutionalized fiction was of NASA as an organization destined for success and incapable of failure—negating the expertise and budgetary gap that had opened wide since Apollo times. Organizational narcissism and concomitant myth-making can thus interfere with accurate calibration of expert knowledge, creating holes and bugs in expertise and even safety awareness that may go unrecognized (Hall 2003; Schwartz 1989). Myths are patterning devices that cohere and order group members' understandings. As with psychological entitlement, they lead to lower levels of cognitive processing because they offer attributional shortcuts (Harvey and Martinko 2008). They thrive on attributional or causal ambiguity (Wright 2009): when things go wrong, it is easy—natural even—to blame others (Brown 2000). Myths are a vehicle for denying errors and responsibilities when self-esteem is threatened. They immunize groups against contradicting evidence. Actual injuries and incidents at both sites (for some did occur during the observation periods) were almost always attributed to other groups (e.g., students, administrators, contractors).

Denial plays a general role in the preservation of self-esteem. It is first among the characteristics of narcissism as applicable to groups:

- Through *denial*, groups may seek to disavow or disclaim awareness, knowledge, or responsibility for faults that might otherwise attach to them. Denial at the group level is helped by myth-making: myths not only overtly deny that something is the case; they often conceal conflicting or contradictory information and exclude other equally valid interpretations (Brown 1997).
- *Rationalization* is the attempt to justify unacceptable behavior and present it in a form that is tolerable or acceptable. Groups will offer explanations for their activities that secure legitimacy for what they did and preserve their self-esteem (Weick 1995).
- *Self-aggrandizement* refers to the tendency to overestimate abilities and accomplishments. It is accompanied by self-absorption, the seeking of gratification, exhibitionism, claims to uniqueness, and a sense of invulnerability. Groups use myth and humor to exaggerate their sense of self-worth and fantasize about their unlimited abilities during times of stress (Janis 1982). They also engage in social cohesion ceremonies that are unduly exhibitionistic—sometimes highly visible, noisy rituals that make others feel acutely excluded. This can include intentional manipulation of physical

space that is designed to separate, intimidate, or excite admiration, the use of special language and symbols, and the use of power to make others wait or feel worthless by other means (Brown 1997; Schwartz 1989).

- *Attributional egotism* means finding explanations for events that are self-serving. Favorable outcomes are attributed to the group, unfavorable ones to anybody else. As shown above, process or maintenance failures were attributed to inexperienced contractors rather than to inaccurate permitting, for example. Or (by extension) they are blamed on managers who outsourced maintenance in the first place or regulators who allowed it (Campbell et al. 2011).
- *Sense of entitlement*, as explained in an earlier section, is driven by a belief in the right to exploit others and a simultaneous inability to empathize with them (Harvey and Martinko 2008).

These behaviors and cognitions can become persistent, pervasive, and significant (Godkin and Allcorn 2009). Yet for safety-critical organizations, this is not necessarily all bad. Employing "prima donnas" is a trade-off, a mixed blessing (Campbell et al. 2011) of which confidence, charisma, and technical prowess form the bright side. Strong self-esteem can be healthy when faced regularly with uncertain outcomes and the demand to make decisions with incomplete knowledge or information. It also helps in the face of substantial and existential risk to the self or others— whether by conducting surgery, flying jets, or operating the hot and critical parts of petrochemical facilities. Ritual task performance and denial of feelings of attachment are ways to deal with the daily stress, anxiety, and tension of safety-critical work (Aiken et al. 2002; Brown 1997). This can hold true even if narcissism leads to overconfidence, less accurate decision making, and a willingness to court risk (Campbell et al. 2004).

Some groups will obviously be more socially adaptive than others in regulating their self-esteem (just as many individuals are). In safety-critical organizations, the dilemma for colleagues and managers is that they depend, for their own safety and the organization's, on the expertise of their "prima donnas." Yet always deferring to this expertise, being sensitive to their operations and allowing decisions to float down toward them, can feel as unfair and undeserved as the high-reliability organizational literature says it is necessary. In some cases, myth-making, alienation, and self-aggrandizement of expert practitioner groups can strain the relationship with the rest of the organization and its leadership. This becomes visible, for instance, in physical separation of work spaces, higher turnover, industrial strife, and ultimately less safe outcomes for the entire organization (McCartin 2011; Schwartz 1989).

DEFERRING TO EXPERTISE WHILE DEALING WITH "PRIMA DONNAS"

The managerial dilemma, then, consists in this: a push to legitimize experts' concerns about operations and safety, versus a pull to limit their power and influence over organizational decision making and other work groups. As with any dilemma, this wouldn't be one if it were easy to solve. But perhaps there are the following managerial possibilities:

- List the potentially corrosive consequences of prima donna behavior in the organization, including when and where these consequences are most visible and harmful. Recognize that prima donna behavior is often about something else, try to find out what the employees' real interest is, and work with that.
- Recognize that low self-esteem and anxiety may drive the behaviors observed. The literature on job insecurity and anxiety recommends keeping periods of industrial uncertainty as short as possible. Certainty of a bad outcome is generally better than uncertainty (Dekker and Schaufeli 1995; Lazarus 1966). It allows people to start coping with a new reality, rather than relying on the myth and fantasy of exaggerated self-worth during a bygone reality.
- Align treatment with performance and build ways to generate accountability equitable with other employees or contractors. Dispensation from rules applicable to other people is corrosive for morale and organizational cohesion. All employees and contractors are part of something bigger than themselves: no player is bigger than the entire team. There needs to be a place in the organization for good performers, but not necessarily in the role or level of influence they would like.
- Legitimize operational expertise. Encourage the sharing of that knowledge and experience, also by showing a willingness to listen to it and time made in managerial agendas to engage with it. Make the conversation about the expertise, not the person (where it is easy to either knock down or overinflate egos). Recognize that autonomy, mastery, and purpose are the intrinsic motivators that likely attracted them to the job in the first place. If they want to contribute to decisions, it was originally not about grabbing power but because they really *do* have something to contribute. Differences of opinion or style are not a problem of attitude. In fact, it is that sort of diversity that keeps organizations resilient in the face of challenge and surprise.
- Consider making changes to physical space, layouts, and co-location arrangements if necessary to prevent isolation or counterproductive status enhancement of particular groups.

Both literatures here allow, to an extent, for a relational or transactional construction of a "prima donna syndrome." Rather than focusing on essentialist personality characteristics of a bad few, they encourage thinking about *what* is responsible for the creation of the "syndrome." If entitlement and narcissism are the systematic products of cognitions, behaviors, and organizational dynamics, then potential managerial pathways out of the dilemma, like the above, can be developed. The safety literature emphasizes deference to *expertise*, not necessarily to experts (Weick and Sutcliffe 2007). Expertise is seen as relational. Expertise and its effects emerge from people querying each other, supplying data, opinions, and other input to conversations in which it can be rejected, deferred to, modified, delayed, and more. Expertise, in other words, is a co-production: a construct that requires as much social and organizational legitimation from outside as substantive credibility from within. Expertise sometimes emerges spontaneously outside existing organizational structures, when knowledgeable people self-organize into ad hoc networks to solve problems (Murray and Cox 1989; Rochlin et al. 1987). This can only work effectively in organizations that value expertise and experience more than rank and hierarchy, particularly when novel or unexpected situations arise (Rochlin 1999; Schwenk and Cosier 1980). Let us now briefly turn to the literature on resilience and see what it has to offer on this.

NEED FOR RESILIENCE

Resilience is the ability of a system to recognize, absorb, and adapt to disruptions that fall outside a system's design base (Hollnagel et al. 2006), where the design base incorporates soft and hard aspects that went into putting the system together (e.g., equipment, people, training, procedures).

NO INVENTORY FOR THIS...

An example from aviation is the Swissair 111 accident in 1998, where the crew responded promptly to the presence of smoke in the cockpit by adhering to the relevant checklist. However, while following the established and trained procedures, fire engulfed the aircraft (TSB 2003). This accident reminded the industry of the immense difficulty that crews face in making trade-offs when adapting plans and procedures under duress and uncertainty and of industry shortcomings in preparing them for incidents like these (Dekker 2001). In this case, following the procedures turned out to be the problem rather than the solution and this tragedy initiated further research on the use of checklists and procedures (Burian and Barshi 2003). Regarding training for these types of situations, Burian and Barshi (2003, p. 3) concluded that: "the degree to which training truly reflects real life emergency and abnormal situations, with all of their real-world demands, is often limited."

More recent is the accident of Pinnacle Airlines flight 3701 in 2004, where pilots ferrying an empty 50-seat aircraft carried out several nonstandard maneuvers that made the engines shut down in flight and then failed in their attempts to restart them. This accident exposed gaps in crew knowledge of high-altitude operations, handling of low-speed and stall situations, and recovery from double-engine failures and on how the absence of passengers erodes operational margins (NTSB 2007a). The type of engine on the Pinnacle-type aircraft had a history of problems with in-flight restarts during flight tests. But few or no operational crews would have been aware of any of this, in part because of structural industry arrangements that regulate who gets or needs to know what and in what depth. Similar restrictions of access to knowledge have played a role in accidents in other industries (e.g., nuclear industry, the Three Mile Island accident, and Chernobyl).

As a result, some operators will, at some point or other, be left to "fend for themselves" at the edges of an extremely safe industry. It is at these edges that the skills trained for meeting standard threats need transposing to counter threats no one has foreseen. The flight of United Airlines 232 in 1989 is an extreme example (NTSB 1990). The triple engine DC-10 lost total hydraulic power and became seemingly uncontrollable as a result of a mid-flight tail engine rupture, with debris ripping through all hydraulic lines that ran through the tail-plane and subsequent loss of hydraulic fluid. The crew figured out how to use differential power on the two remaining engines and steered the craft toward an extremely difficult high-speed landing at Sioux City, Iowa. Despite their efforts, the plane broke up on the runway, but the majority of the passengers and crew subsequently survived the landing. In simulator reenactments of this scenario, none of 42 crews managed to get the aircraft down on the runway. Both the crew and the investigation concluded that the relatively successful outcome of this impossible situation could largely be attributed to the training of general competencies in the carrier's human factors and crew resource management (CRM) training programs.

Thinking outside the box, taking a system way beyond what it was designed to do (even making use of an adverse design quality such as pitching moments with power changes) is a hallmark of resilience. Resilience is about enhancing people's adaptive capacity so that they can recognize and counter unanticipated threats. Adaptive capacity with regard to a narrow set of challenges can grow when an organization courts exposure to smaller dangers (Rochlin et al. 1987). This allows it to keep learning about the changing nature of the risk it faces—ultimately forestalling or being able to absorb larger dangers. Such adaptation, by the way, could be one explanation behind recent data that suggest that the passenger mortality risk on major airlines that have suffered nonfatal accidents is lower than that on airlines that have been accident-free (Barnett and Wang 2000), just like the data from the construction industry that show that organizations with low incident counts have a higher fatality risk.

TOO AMBIGUOUS FOR FEEDFORWARD, TOO DYNAMIC FOR FEEDBACK

Take the case of a DC-9 that got caught in wind shear while trying to go around from an approach to Charlotte, North Carolina, in 1994 (NTSB 1995). Noting the dynamic weather situation in front of them, the crew invested continually in their awareness of the potential threat ahead. They asked for pilot reports from aircraft on the approach in front of them, and consistently received information (e.g., "smooth ride all the way down") that confirmed that continuing the approach made sense. Given the ambiguity of the situation (the storm was visible ahead of them, yet pilot reports in front of them showed unproblematic, smooth approaches), effective feedforward was difficult. What to do? When to break off the approach and go around, if at all? As an additional investment in safety, the crew planned to turn right in case they would have to go around. They adapted the procedure to deal better with local circumstances: the planned go-around route would have taken them right into the center of the storm and a right turn will take them away from it in case they have to go around.

Once it got close to the runway, the DC-9 encountered rain and then experienced airspeed variations, which prompted the pilots to go around. They turned right. What they flew into at that moment, invisible to them, was a microburst—a highly concentrated bubble of colder air that falls through the layers of air underneath it and hits the ground, exploding in all directions. Because of the right turn, taken to avoid the storm, the crew was now in the worst possible place, giving them the most tailwind and rapidly robbing them of forward flying speed. This was the sort of stochastic element that is an ingredient in many accidents; the sort of bad luck that turns a set of good plans and intentions into a bad outcome. While trying to engineer in more buffer, more margin between themselves and the weather, the crew ended up in a situation where no slack was available at all. Once inside the microburst, the DC-9 rapidly lost speed, started to hit trees and broke apart on a residential street next to the airport.

This crew's training had never covered the encounter of a microburst on approach to an airport, but even if it had, it may not have helped. Theirs was a terrible double bind: a situation that was both hard to foresee (except if one would decide not to fly at all during thunderstorm season, which in the Southern United States can last half a year) and hard to respond to. In other words, the situation was too ambiguous for feedforward and too dynamic for effective feedback. This was a mechanism by which people's investment in safety was outwitted by a rapidly developing situation and adapting effectively was extremely difficult.

Of course, it is easy, in hindsight, to point to defective decisions (or a "loss of SA," but this was not yet popular as causal explanation when this report was written). What matters much more is changing the way we think about safety and how "the human factor" relates to them. If accidents are linked to the complexity and dynamics of the operational worlds in which we try to prevent them, we are not going to get very far with thinking in terms of Newtonian components and cause–effect relationships. Reductionist, linear models always get it wrong when they try to address a world where safety and risk are emergent and nonlinear. Cartesian models always get it wrong in a world where the border between what is in the head and what is in the world does not neatly follow dualist lines once drawn by him. This book has shown in multiple places how this conception becomes untenable when confronted with a world that is characterized by change, complexity, and diversity. It is a conception that becomes untenable when confronted with a world in which the notion of "cause" is not only unstably suspended somewhere between human and machine but tends to fade out of view and slip from our hands altogether as we attempt to track it down ever further into the organized sociotechnical complexity of our systems. Let's see what directions the ideas of resilience can offer for a new era in safety.

ENGINEERING RESILIENCE INTO ORGANIZATIONS

If we want to apply a Rasmussian image, then we could say that all open systems are continually adrift inside their safety envelopes (Rasmussen 1997). Pressures of scarcity and competition, the intransparency and size of complex systems, the patterns of information that surround decision makers, and the incrementalist nature of their decisions over time can cause systems to drift into failure. Recall from Chapter 5 how drift is generated by normal processes of reconciling differential pressures on an organization (efficiency, capacity utilization, safety) against a background of uncertain technology and imperfect knowledge. Drift is about incrementalism contributing to extraordinary events, about the transformation of pressures of scarcity and competition into organizational mandates, and about the normalization of signals of danger so that organizational goals and supposedly normal assessments and decisions become aligned. In safe systems, the very processes that normally guarantee safety and generate organizational success can also be responsible for organizational demise. The same complex, intertwined sociotechnical life that surrounds the operation of successful technology is to a large extent responsible for its potential failure. Because these processes are normal, because they are part and parcel of normal, functional organizational life, they are difficult to identify and disentangle. The role of these invisible and unacknowledged forces can be frightening. Harmful consequences can occur in organizations constructed to prevent them. Harmful consequences can occur even when everybody follows the rules (Vaughan 1996).

Resilience is an emergent property, and its erosion is not about the breakage or lack of quality of single components. This makes the conflation of quality and safety management counterproductive. To illustrate this, let's make a distinction between robust and resilient systems:

- Robust systems are effective at meeting predicted threats that represent finite configurations of what the industry could anticipate. They are capable of maintaining process integrity inside of known design, training, and procedural parameters.
- Resilient systems are effective at meeting threats that represent infinite reconfigurations of—or ones that may lie entirely beyond—what the industry could anticipate. Resilient systems are capable of maintaining process integrity well outside the design base or outside training or procedural provisions.

Many organizations have quality and safety management wrapped up in one function or department. Yet managing quality (or robustness) is about single components or systems, about seeing how they meet particular specifications, about removing or repairing defective components. Managing safety (let alone creating the conditions for resilience) has little to do anymore with single components. A different level of understanding, a different vocabulary is needed to understand safety, in contrast to quality. Perhaps drifting into failure is not so much about breakdowns or malfunctioning of components, as it is about an organization not adapting effectively to cope with the complexity of its own structure and environment. Organizational resilience, then, is not a property. It comes from a set of capabilities:

- A capability to recognize the boundaries of safe operations
- A capability to steer back from them in a controlled manner
- A capability to recover from a loss of control if it does occur

It can even mean a different conceptualization of what we understand a system to be. Recall from Chapter 5 how systems can be seen as dynamic relationships, a notion that opens up different ways of looking at it and designing for its safety.

SHARED SPACES AND A DIFFERENT IDEA OF "SYSTEM"

For an example of a different conceptualization of a system, let's briefly turn to road safety. Here, the realization is that the very idea of a system may need to change—at least for certain circumstances. We may have to let go of the idea of the system as the engineered, mechanical imposition of high-modernist ideas onto individual road users. Hans Monderman's "shared space" model is one way to do that (Hamilton-Baillie 2008). Monderman, a soft-spoken traffic engineer from the Netherlands, proposed a radical change to the management of sizable volumes of car traffic (mixed with bicyclists and pedestrians) through the town of Drachten in the north of his country. He decided to propose removing all barriers, signs, lights, and separators, ending up with just one big, featureless but pleasant-looking square into which roads empty out. Such thinking makes entirely different assumptions about the nature of control and order in a system. Instead of Newtonian cause–effect sequences, where the only way to create order is to control energy by putting engineered and legal barriers in place, order is left to emerge organically. It is not controlled by hampering the movements of single components in a system but influenced by allowing and even encouraging the interaction between many components. The results have been very encouraging. Vehicle speeds, for instance, drop remarkably. Monderman's ideas have been copied the world over, with many towns in the United Kingdom and the Unite States adopting his so-called shared space concept.

The "system" in the ideas of Monderman is allowed to be an ecological, living organism created by a self-organizing social order. Instead of achieving lower speeds by increased regulatory control, enforcement, and conventional urban design, removing such controls appears to allow much more behavioral constraints and social obligations to come back into play. "Shared space, fostering the multiple uses of streets and spaces for every kind of social activity as well as movement, requires the formal abandonment of the principle of segregation in urban traffic engineering" (Hamilton-Baillie 2008, p. 137). This idea is consistent with the physical mechanics and mathematics that try to capture the workings of complex (rather than linear) systems (Cilliers 1998; Dekker 2011b; Dekker et al. 2011). Road users can create and contribute to reading situations, adapting to them, and working out plausible solutions. Solutions literally "emerge" from the complex, nonlinear interactions between the components that make up the system. Resilience is created bottom-up.

This may not work everywhere, of course. But it is the sort of "system" that is capable of inverting the risk compensation or homeostasis so often demonstrated by researchers like John Adams (1995). Traditional interventions that make traffic spaces look and feel safer, after all, can and do paradoxically invite riskier behavior. Making those spaces instead look and feel riskier has empirically been shown to invite safer behavior (Hamilton-Baillie 2008). What it also does is introduce an entirely different vision of humanity into the debate about systems and road safety; a gentler, less misanthropic vision, one where people see each other—are forced to look one another in the eye. This pushes back against the increasing manifestations of "the highway" in urban landscapes and the psychological retreat it has implied for most road users (Engwicht 1999). It is a vision where road users are not the problem to be controlled by somebody who always knows better (and who has the law and engineering interventions to back it up and enforce it), but where people themselves are the potential around and from which solutions can emerge.

Examples like "shared space" mean that human factors and safety should look at new ways of "engineering" resilience into organizations, of equipping organizations with a capability to recognize, absorb, adapt to, and recover from harmful influences that could imply a loss of control:

- How can an organization monitor its own adaptations (and how these bound the rationality of decision makers) to pressures of scarcity and competition, while dealing with imperfect knowledge and uncertain technology?
- How can an organization become aware, and remain aware, of its models of risk and danger?
- How does the organization respond to evidence of things that went wrong? Does it respond with measures that target individual components or does it acknowledge the complex, deeply interactive ways in which both failure and success come about?

But what does that mean for the people who work in the organization? What are some possible indicators of resilience on their part? The literature so far has this to suggest (Hollnagel et al. 2008, 2009):

- How do people handle sacrificing decisions? Pressures for faster, better, and cheaper service are often readily noticeable and can be satisfied in an easily measurable way. But how do people consider and negotiate how much they are willing to borrow from safety in order to achieve those other goals? When faced with sacrificing decisions, resilient teams and organizations are able to take small losses in order to invest in larger margins (e.g., an airliner exiting the takeoff queue in winter conditions to go for another dose of deicing fluid). Organizational resilience is about finding means to invest in safety even under pressures of scarcity and competition, because that may be when such investments are needed most.

- Do people take past success as a guarantee of future safety? Having been in that same situation many times before (and succeeded at it) may have crews believe that acting the same way will once again produce safety. In a dynamic, complex world, this is not automatically so.
- Are the operators keeping a discussion about risk alive even when everything looks safe? Continued operational success is not necessarily evidence of large safety margins, and people who actively engage in a risk analysis of their next plan may be better aware of the edges of the safety envelope and how close they are to it.
- Have people invested in the possibility of role flexibility and role breakout? For example, they may include what to do in case of an unstabilized approach to a runway during the briefing of that approach, with a particular emphasis on giving the subordinate the confidence that challenging or taking over is alright and that disagreements over the interpretation of the situation can be sorted out later, on the ground.
- Do people distance themselves from vicarious learning through differencing? In this process, people would decline to look at other failures and other situations, as they are judged not to be relevant to them and their setting. They discard other events because they appear to be dissimilar or distant. This is unfortunate, as nothing is too distant to contain at least some lessons at some level.
- Is people's problem solving fragmented? With information incomplete, disjointed, and patchy, none of the participants in some work process may be able to recognize a gradual erosion of safety constraints, for example.
- Are people open to generating and accepting fresh perspectives on a problem? Systems that apply fresh perspectives (e.g., people from other backgrounds, diverse viewpoints) on problem-solving activities can be more effective: they generate more hypotheses, cover more contingencies, openly debate rationales for decision making, and reveal hidden assumptions.

The pursuit of resilience should not about a new kind of heroics or individual courage on part of frontline workers. That would turn it into a newly cloaked behavioral-based safety program of its own, a new moral code of conduct. This would do little more than telling or expecting people at the sharp end to try a little harder despite the organizational or design constraints surrounding them. So if it is not this, then how does resilience want to be understood? And is it something we can train?

SAFETY AS THE PRESENCE OF CAPACITIES

Resilience sees safety as a presence of capacities, capabilities, and competencies to make things go right. No longer should we define safety as the absence of things that go wrong, no longer should it be seen as an absence of negatives (such as errors, violations, or incidents). The sections on resilience above (and how to train for it, how to recognize indicators of it) suggest that the absence of negatives tells us very little about the presence of resilience. In fact, the absence of negatives, as shown in the 2000 study by Barnett and Wang, can increase the mortality risk of users and operators of the system. The examples above suggest strongly that we need to dispense with the high-modernist illusion that we are mere custodians of well-designed, well-planned, well-thought-out, safe systems. A paradigm that gives more credence for bottom-up, operational thinking can provide better, less judgmental explanations of performance variations. It can generate interpretations sensitive to the situatedness of the performance they attempt to capture. Such a paradigm sees no errors, but rather performance variations—inherently neutral changes and adjustments in how people deal with complex, dynamic situations.

Such a paradigm also acknowledges that safety cannot just be imported, hierarchically controlled, and ordained from above with smarter planning, smarter organization, tighter monitoring and supervision, and better-protected layers of defense. In dynamic, complex systems, safety gets made and broken the whole time—actively, under changing conditions. This is what turns safety from the absence of something into the presence of something. The presence of an adaptive capacity, an ability to recognize, absorb, and adapt to changes and disruptions that fall both inside and outside what the system is designed or trained to handle. It is about identifying and then enhancing the positive capacities of people and organizations that allow them to adapt effectively and safely under varying and resource-constrained circumstances. The most important sources of knowledge and expertise for this come not from the top-down, in Taylorist fashion, but from the bottom-up. In risky judgments and managing safety-critical processes, there is no substitute for deferring to technical/operational expertise. Problem-solving processes need to engage those who are operationally practiced at recognizing and dealing with anomalies, not (just) those who understand how to bureaucratically demonstrate regulatory compliance to an outside party. This is a commitment to seeing people as a solution or a resource to harness, not as a problem to control. But how can we train for that?

TRAINING FOR RESILIENCE: FIDELITY AND VALIDITY

There are limits on our ability to prepare operators in detail for the precise operational problems they may encounter. These limits stem from the intersection between our finite knowledge and the infinite configurations of problems-to-be-solved in any complex, dynamic domain. If that is the case, then what are the implications for our training of operators in domains like that? The evolution of training simulation has often been technology-driven, helped along by ever-enhanced computing capacity, visual systems, and computer graphics. But does this increase in fidelity also improve training quality? Or does greater simulation fidelity mean greater validity of that which is trained in the simulator? There seems to be an often taken-for-granted assumption that incremental quantitative progress in simulation technology (e.g., more computing power, higher resolution, greater visual angles) adds up to a positive qualitative difference. In other words, as a simulated environment becomes ever more photorealistic, so does the yield that simulations have for crews and staff. This link between maximum fidelity to maximum training transfer is taken on faith. If it looks real, it will provide good training (Dahlström et al. 2009). However, over time, the continually increased demand for higher levels of fidelity to make simulations look "real" increases cost and lowers availability of training simulators (one has to keep in mind the huge capital investment a commitment to this style of simulation requires). This goes for industries ranging from maritime transport, to mining, nuclear power, medicine, and the military.

The focus on fidelity has muted perspectives on simulation styles and use that could allow a more subtle analysis of cognitive and group interaction aspects to form the base of training. This is particularly true for training of skills related to the management of resources, to communication, decision making, information integration, and other "soft" skills like it (Dekker et al. 2008). It is in unusual, unanticipated, and escalating situations where such skills are most needed. Highly dynamic situations involving underspecified problems, time pressure constraints, and complex group interaction are situations that cannot be resolved through procedural guidance. Training crews for them through technically anchored simulations that require close procedural compliance to even work may not be the way to create resilience.

Conventional thinking about this might suggest that people should recognize and appropriately respond to simulated challenges. For example, an emergency may call for more rapid decision making and consequently a leadership style that facilitates speed and decisiveness. Either way, an accurate account of context is a critical decision-making device. But what is missing from such theory is where that persuasive rendition of context comes from in the first place. An emergency is not just "out there"; it is recognized and described as such by the people facing it—they are the ones who "construct" the threat as an emergency. Indeed, even generically, there is disagreement on how much immediate threat there is in an emergency, as there is almost always more time than we think. But once an operator has declared an emergency, or simply told her or his fellow operators that they now face an "emergency," this legitimizes a particular mode of engagement with that situation (and perhaps a different style of team interaction and even leadership). This mode and style may not have been legitimate before the explicit construction of that situation as an "emergency." In short, operators are not just recipients of challenges that push them beyond the routine. They are constitutive of them. This contradicts the simple linear idea that there is first a challenge and then a response (appropriate or not). Rather, it is the response that helps construct the threat. This construction in turn sanctions a particular repertoire of plausible or desirable countermeasures.

Lower-fidelity simulations (ones that do not attempt to mimic directly the target technical environment) could be a cost-effective alternative and may actually improve many aspects of learning that help people deal with unanticipated situations.

This can enhance the sort of "soft" skills that are not directly anchored in the technical specifics of a simulation, but that are critical for an actionable co-construction of the situation faced by the operators. Also, there are already data to suggest that the sense of "environmental presence" experienced in simulated environments is determined more by the extent to which it acknowledges and reacts to the participant than by the simulation's physical fidelity (Dahlström 2006). In other words, high levels of technologically driven fidelity can be wasteful in terms of costs and time for whole families of tasks and competencies. The emphasis on photorealism in visual and task contexts may retard or limit the development of skill sets critical for creating safety in domains where not all combinations of technical and operational failure can be foreseen or formalized (and for which failure strategies then cannot be proceduralized and simulated).

RESILIENCE AND A NEW ERA

Competencies that are recognized as important for the creation of resilience and which are echoed in the high-performance and high-reliability organizational literatures (e.g., communication, coordination, problem solving, management of unanticipated and escalating situations) do not emerge automatically from context-fixed or photorealistic simulator training. But they matter. The research shows that high-performance teams, and resilient organizations, are those that develop and harness people's competencies and capabilities. They do not rely on control, constraint, and the eradication of deficits and negatives as much as they build on opportunity, diversity, and positive capacities to make things go right.

In a new era for safety, we should perhaps let go of the idea—as a colleague of mine, Corrie Pitzer, likes to say—that we can lead people into safety and that we, and they, should be risk averse. The Monderman experiment of shared spaces in traffic shows that this may actually get people to take more risks, because we falsely suggest to them that they are safe. If we are honest about it, we might acknowledge that in some operational worlds, we lead people into danger. And rather than making them risk *averse*, we should make them risk *competent*. This is what resilience tries to do:

- Give people the freedom and opportunity to keep a discussion about risk alive even when everything looks safe. When incident numbers are low, and everything looks in order, there still are the frustrations and workarounds that mark out real work. Regular briefings about what did not go well and what did are part and parcel of highly dynamic operational organizations and high-performing teams.
- Offer up real possibilities to say no, to trade acute production pressures for chronic safety concerns. This does not get done by putting cards around people's necks with a "stop work authority" on it, or by putting a page in the crew manual that says that copilots are entitled to take over if the captain is behaving unsafely. It is done by creating relationships of trust and confidence up and down operational hierarchies that respect operational expertise both for what it is and for what it is not.

- Create a space where honesty and learning is possible, where accountability means being given the opportunity to tell one's account to people who understand, intimately, the messy details of what it means to get work done under multiple goal conflicts and resource constraints.

These may be among the early heralds of a new era in safety, even in your organization.

STUDY QUESTIONS

1. What are some of the advantages and disadvantages of declaring a "zero vision" on safety?
2. How can you explain that organizations with a low incident rate (e.g., in construction or aviation) actually have a higher fatality risk?
3. How have people in your organization seen "safety culture"—as an emergent property that can only be influenced, or a possession that can be manipulated? Does the view of safety culture in your organization see people as a problem to control, or as a resource to harness? How?
4. Is it possible to accommodate the concerns about "prima donnas" while deferring to expertise? If so, how?
5. Consider the "shared spaces" concept developed for roads, where conflict resolutions between different kinds of traffic are encouraged to emerge from their interaction in an open environment. Is this something that can inspire new approaches to safety in your organization—a bottom-up approach, without imposing rules or protections from the top-down?
6. How is resilience a way to see people as a solution to harness, rather than a problem to control, and why would that approach be characteristic of human factors for a new era?
7. If safety is not the absence of negatives, but the presence of capacities, capabilities, and competencies, then what would those be for your organization or industry?
8. What is the difference between the validity and fidelity of a simulator or simulation? Which of these two characteristics is more important for training compliance, and which one for resilience?
9. What are some of the key obstacles to a new era in safety in your own organization or industry?

References

Adams, J. (1995). *Risk*. London: Routledge.

Aeronautica Civil. (1996). Aircraft accident report: Controlled flight into terrain, American Airlines flight 965, Boeing 757–223, N651AA near Cali, Colombia, 20 December 1995. Bogota, Colombia: Aeronautica Civil de Colombia.

Aiken, L. H., Clarke, S. P., Sloane, D. M., Sochalski, J., and Silber, J. H. (2002). Hospital nurse staffing and patient mortality, nurse burnout, and job dissatisfaction. *JAMA, 288*(16), 1987–1993.

Albright, C. A., Truitt, T. R., Barile, A. B., Vortac, O. U., and Manning, C. A. (1996). *How Controllers Compensate for the Lack of Flight Progress Strips*. Arlington, VA: National Technical Information Service.

Althusser, L. (1984). *Essays on Ideology*. London: Verso.

Amalberti, R. (2001). The paradoxes of almost totally safe transportation systems. *Safety Science, 37*(2–3), 109–126.

Amalberti, R., Auroy, Y., Berwick, D., and Barach, P. (2005). Five system barriers to achieving ultrasafe healthcare. *Annals of Internal Medicine, 142*(9), 756–764.

Amer-Wåhlin, I., Bergström, J., Wahren, E., and Dekker, S. W. A. (2010). *Escalating obstetrical situations: An organizational approach*. Paper presented at the Swedish Obstetrics & Gynaecology Week, Visby, Gotland.

Amer-Wåhlin, I., and Dekker, S. W. A. (2008). Fetal monitoring—A risky business for the unborn and for clinicians. *Journal of Obstetrics and Gynaecology, 115*(8), 935–937; discussion 1061–1062.

Angell, I. O., and Straub, B. (1999). Rain-dancing with pseudo-science. *Cognition, Technology and Work, 1*, 179–196.

Anon. (2012, 5 January). David Cameron: Businesses have a 'culture of fear' about health and safety, *The Daily Telegraph*.

Anon. (2013, 12 April). Jail for safety manager for lying about injuries, *Associated Press*.

Arbous, A. G., and Kerrich, J. E. (1951). Accident statistics and the concept of accident-proneness. *Biometrics, 7*, 340–432.

ASW. (2002, 6 May). Failure to minimize latent hazards cited in Taipei tragedy report, *Air Safety Week*.

ATSB. (1996). *Human Factors in Fatal Aircraft Accidents*. Canberra, ACT: Australian Transportation Safety Bureau (formerly BASI).

Bader, G., and Nyce, J. M. (1995). When only the self is real: Theory and practice in the development community. *Journal of Computer Documentation, 22*(1), 5–10.

Baiada, R. M. (1995). ATC biggest drag on airline productivity. *Aviation Week and Space Technology, 31*, 51–53.

Bainbridge, L. (1987). Ironies of automation. In J. Rasmussen, K. Duncan and J. Leplat (Eds.), *New Technology and Human Error* (pp. 271–283). Chichester, UK: Wiley.

Baker, G. R., Norton, P. G., Flintoft, V., Blais, R., Brown, A., Cox, J., and Tamblyn, R. (2004). The Canadian Adverse Events Study: The incidence of adverse events among hospital patients in Canada. *Canadian Medical Association Journal, 170*(11), 965–968.

Baker, J. A. (2007). *The Report of the BP U.S. Refineries Independent Safety Review Panel*. Washington, DC: Baker Panel.

Barnett, A., and Wang, A. (2000). Passenger mortality risk estimates provide perspectives about flight safety. *Flight Safety Digest, 19*(4), 1–12.

Batteau, A. W. (2001). The anthropology of aviation and flight safety. *Human Organization, 60*(3), 201–211.

Beck, U. (1992). *Risk Society: Towards a New Modernity*. London: Sage Publications.

Benner, P. E., Malloch, K., and Sheets, V. (Eds.). (2010). *Nursing Pathways for Patient Safety*. St. Louis, MO: Mosby Elsevier.

Bertin, O. (1997, 24 January). Managing dispute: Pilots cast off prima-donna image, *The Globe and Mail*, p. B.9.

BFU. (2004). Investigation Report: Accident near Ueberlingen/Lake of Constance/Germany to Boeing B757-200 and Tupolev TU154M, 1 July 2002. Braunschweig, Germany: Bundesstelle fur Flugunfalluntersuchung/German Federal Bureau of Aircraft Accidents Investigation.

Billings, C. E. (1996). Situation awareness measurement and analysis: A commentary. In D. J. Garland and M. R. Endsley (Eds.), *Experimental Analysis and Measurement of Situation Awareness* (pp. 1–5). Daytona Beach, FL: Embry-Riddle Aeronautical University Press.

Billings, C. E. (1997). *Aviation Automation: The Search for a Human-Centered Approach*. Mahwah, NJ: Lawrence Erlbaum Associates.

Björklund, C., Alfredsson, J., and Dekker, S. W. A. (2006). Shared mode awareness in air transport cockpits: An eye-point of gaze study. *International Journal of Aviation Psychology*, 16(3), 257–269.

Boeing. (1996). *Boeing Submission to the American Airlines Flight 965 Accident Investigation Board*. Seattle, WA: Boeing Commercial Airplane Group.

Bosk, C. (2003). *Forgive and Remember: Managing Medical Failure*. Chicago, IL: University of Chicago Press.

BP. (2010). Deepwater Horizon accident investigation report. London: British Petroleum.

Brown, A. D. (1997). Narcissism, identity, and legitimacy. *Academy of Management Review*, 22(3), 643–686.

Brown, A. D. (2000). Making sense of inquiry sensemaking. *Journal of Management Studies*, 37(1), 45–75.

Bruner, J. (1990). *Acts of Meaning*. Cambridge, MA: Harvard University Press.

Bruxelles, S. de (2010, 28 April). Coroner criticises US as he gives 'friendly fire' inquest verdict, *The Times*, p. 1.

Burian, B. K., and Barshi, I. (2003). *Emergency and abnormal situations: A review of ASRS reports*. Paper presented at the 12th international symposium on aviation psychology, Dayton, OH.

Burnham, J. C. (2009). *Accident Prone: A History of Technology, Psychology and Misfits of the Machine Age*. Chicago, IL: University of Chicago Press.

Byrne, G. (2002). *Flight 427: Anatomy of an Air Disaster*. New York: Springer-Verlag.

CAIB. (2003). Report Volume 1, August 2003. Washington, DC: Columbia Accident Investigation Board.

Campbell, R. D., and Bagshaw, M. (1991). *Human Performance and Limitations in Aviation*. Oxford, UK: Blackwell Science.

Campbell, W. K., Goodie, A. S., and Foster, J. D. (2004). Narcissism, confidence, and risk attitude. *Journal of Behavioral Decision Making*, 17(4), 297–311.

Campbell, W. K., Hoffman, B. J., Campbell, S. M., and Marchisio, G. (2011). Narcissism in organizational contexts. *Human Resource Management Review*, 21, 268–284.

Capra, F. (1982). *The Turning Point*. New York: Simon & Schuster.

Cilliers, P. (1998). *Complexity and Postmodernism: Understanding Complex Systems*. London: Routledge.

Cilliers, P. (2002). Why we cannot know complex things completely. *Emergence*, 4(1/2), 77–84.

Cilliers, P. (2005). Complexity, deconstruction and relativism. *Theory, Culture and Society*, 22(5), 255–267.

Cilliers, P. (2010). Difference, identity and complexity. *Philosophy Today*, 26(2), 55–65.

Clarke, L., and Perrow, C. (1996). Prosaic organizational failure. *American Behavioral Scientist*, 39(8), 1040–1057.

Connolly, J. (1981). Accident proneness. *British Journal of Hospital Medicine, 26*(5), 470–481.

Cook, R. I., and Nemeth, C. P. (2010). Those found responsible have been sacked: Some observations on the usefulness of error. *Cognition, Technology and Work, 12*, 87–93.

Cook, R. I., Potter, S. S., Woods, D. D., and McDonald, J. S. (1991). Evaluating the human engineering of microprocessor-controlled operating room devices. *Journal of Clinical Monitoring, 7*(3), 217–226.

Cordesman, A. H., and Wagner, A. R. (1996). *The Lessons of Modern War, Vol. 4: The Gulf War*. Boulder, CO: Westview Press.

Croft, J. (2001). Researchers perfect new ways to monitor pilot performance. *Aviation Week and Space Technology, 155*(3), 76–77.

Cronon, W. (1992). A place for stories: Nature, history, and narrative. *The Journal of American History, 78*(4), 1347–1376.

Crozier, M. (1964). *The Bureaucratic Phenomenon* (English Translation). Chicago, IL: University of Chicago Press.

Dahl, O., and Olsen, E. (2013). Safety compliance on offshore platforms: A multi-sample survey on the role of perceived leadership involvement and work climate. *Safety Science, 54*(1), 17–26.

Dahlström, N. (2006). *Training of collaborative skills with mid-fidelity simulation*. Paper presented at the human factors and economic aspects on safety, Linköping, Sweden.

Dahlström, N., Dekker, S. W. A., van Winsen, R., and Nyce, J. M. (2009). Fidelity and validity of simulator training. *Theoretical Issues in Ergonomics Science, 10*(4), 305–315.

Dawkins, R. (1986). *The Blind Watchmaker*. London: Penguin.

De Vries, R., and Lemmens, T. (2006). The social and cultural shaping of medical evidence: Case studies from pharmaceutical research and obstetrics. *Social Science and Medicine, 62*(11), 2694–2706.

Degani, A., Heymann, M., and Shafto, M. (1999). *Formal aspects of procedures: The problem of sequential correctness*. Paper presented at the 43rd Annual Meeting of the Human Factors and Ergonomics Society, Houston, TX.

Dekker, S. W. A. (2001). Follow the procedure or survive. *Human Factors and Aerospace Safety, 1*(4), 381–385.

Dekker, S. W. A. (2002). *The Field Guide to Human Error Investigations*. Bedford, UK: Cranfield University Press.

Dekker, S. W. A. (2003). Failure to adapt or adaptations that fail: Contrasting models on procedures and safety. *Applied Ergonomics, 34*(3), 233–238.

Dekker, S. W. A. (2009). Just culture: Who draws the line? *Cognition, Technology and Work, 11*(3), 177–185.

Dekker, S. W. A. (2011a). The criminalization of human error in aviation and healthcare: A review. *Safety Science, 49*(2), 121–127.

Dekker, S. W. A. (2011b). *Drift into Failure: From Hunting Broken Components to Understanding Complex Systems*. Farnham, UK: Ashgate Publishing.

Dekker, S. W. A. (2011c). *Patient Safety: A Human Factors Approach*. Boca Raton, FL: CRC Press.

Dekker, S. W. A. (2012). *Just Culture: Balancing Safety and Accountability* (2nd ed.). Farnham, UK: Ashgate Publishing.

Dekker, S. W. A. (2013). *Second Victim: Error, Guilt, Trauma and Resilience*. Boca Raton, FL: CRC Press/Taylor & Francis.

Dekker, S. W. A., Cilliers, P., and Hofmeyr, J. (2011). The complexity of failure: Implications of complexity theory for safety investigations. *Safety Science, 49*(6), 939–945.

Dekker, S. W. A., Dahlström, N., van Winsen, R., and Nyce, J. M. (2008). Crew resilience and simulator training in aviation. In E. Hollnagel, C. P. Nemeth and S. W. A Dekker (Eds.), *Resilience Engineering Perspectives: Remaining Sensitive to the Possibility of Failure*. Aldershot, UK: Ashgate Publishing.

Dekker, S. W. A., Nyce, J. M., van Winsen, R., and Henriqson, E. (2010). Epistemological self-confidence in human factors research. *Journal of Cognitive Engineering and Decision Making, 4*(1), 27–38.

Dekker, S. W. A., and Schaufeli, W. B. (1995). The effects of job insecurity on psychological health and withdrawal: A longitudinal study. *Australian Psychologist, 30*(1), 57–63.

Dekker, S. W. A., and Woods, D. D. (1999). To intervene or not to intervene: The dilemma of management by exception. *Cognition, Technology and Work, 1*(2), 86–96.

Dekker, S. W. A., and Woods, D. D. (2002). MABA-MABA or Abracadabra? Progress on human-automation co-ordination. *Cognition, Technology and Work, 4*(4), 240–244.

Dickerson, C. (2001). No more prima donnas. *Infoworld, 23,* 15.

Dismukes, K., Berman, B. A., and Loukopoulos, L. D. (2007). *The Limits of Expertise: Rethinking Pilot Error and the Causes of Airline Accidents.* Aldershot, UK; Burlington, VT: Ashgate.

Donaldson, C. (2013). Zero harm: Infallible or ineffectual. *OHS Professional,* March, 22–27.

Dörner, D. (1989). *The Logic of Failure: Recognizing and Avoiding Error in Complex Situations.* Cambridge, MA: Perseus Books.

Downer, J. (2013). Disowning Fukushima: Managing the credibility of nuclear reliability assessment in the wake of disaster. *Regulation and Governance, 7*(4), 1–25.

Drife, J. O., and Magowan, B. (2004). *Clinical Obstetrics and Gynaecology.* Edinburgh; New York: Saunders.

Dubois, L. (2010). Dealing with a prima donna employee. *Inc., 20,* 21–22.

Edwards, M., and Jabs, L. B. (2009). When safety culture backfires: Unintended consequences of half-shared governance in a high tech workplace. *The Social Science Journal, 46,* 707–723.

Ehrenreich, B., and English, D. (1973). *Witches, Midwives and Nurses: A History of Women Healers.* London: Publishing Cooperative.

Elkind, P., and Whitford, D. (2011, 24 January). BP: 'An accident waiting to happen', *Fortune,* pp. 1–14.

Endsley, M. R., Mogford, M., Allendoerfer, K., Snyder, M. D., and Stein, E. S. (1997). *Effect of Free Flight Conditions on Controller Performance, Workload and Situation Awareness: A Preliminary Investigation of Changes in Locus of Control Using Existing Technologies.* Lubbock, TX: Texas Technical University.

Engwicht, D. (1999). *Street Reclaiming: Creating Livable Streets and Vibrant Communities.* Philadelphia, PA: New Society Publishers.

Farmer, E. (1945). Accident-proneness on the road. *Practitioner, 154,* 221–226.

Farrington-Darby, T., and Wilson, J. R. (2006). The nature of expertise: A review. *Applied Ergonomics, 37,* 17–32.

Feyerabend, P. (1993). *Against Method* (3rd ed.). London: Verso.

Feynman, R. P. (1988). *What Do You Care What Other People Think?: Further Adventures of a Curious Character.* New York: Norton.

Fischhoff, B. (1975). Hindsight ≠ foresight: The effect of outcome knowledge on judgment under uncertainty. *Journal of Experimental Psychology: Human Perception and Performance, 1*(3), 288–299.

Fischhoff, B., and Beyth, R. (1975). I knew it would happen: Remembered probabilities of once-future things. *Organizational Behavior and Human Performance, 13*(1), 1–16.

Fitts, P. M. (1951). *Human Engineering for an Effective Air Navigation and Traffic Control System.* Washington, DC: National Research Council.

Fitts, P. M., and Jones, R. E. (1947). *Analysis of Factors Contributing to 460 "Pilot Error" Experiences in Operating Aircraft Controls.* Dayton, OH: Aero Medical Laboratory, Air Material Command, Wright-Patterson Air Force Base.

Flach, J. M. (1995). Situation awareness: Proceed with caution. *Human Factors, 37*(1), 149–157.

Flach, J. M., Dekker, S. W. A., and Stappers, P. J. (2008). Playing twenty questions with nature: Reflections on the dynamics of experience. *Theoretical Issues in Ergonomics Science, 9*(2), 125–155.

Flores, F., Graves, M., Hartfield, B., and Winograd, T. (1988). Computer systems and the design of organizational interaction. *ACM Transactions on Office Information Systems, 6*(2), 153–172.

Foucault, M. (1977). *Discipline and Punish: The Birth of the Prison* (1st American ed.). New York: Pantheon Books.

Foucault, M. (1980). Truth and power. In C. Gordon (Ed.), *Power/Knowledge* (pp. 80–105). Brighton, UK: Harvester.

Frederick, J., and Lessin, N. (2000). The rise of behavioural-based safety programmes. *Multinational Monitor, 21*, 1–7.

Freud, S. (1950). Project for a scientific psychology. In *The Standard Edition of the Complete Psychological Works of Sigmund Freud* (vol. I). London: Hogarth Press.

GAIN. (2004). *Roadmap to a just culture: Enhancing the safety environment.* Global Aviation Information Network (Group E: Flight Ops/ATC Ops Safety Information Sharing Working Group).

Galison, P. (2000). An accident of history. In P. Galison and A. Roland (Eds.), *Atmospheric Flight in the Twentieth Century* (pp. 3–44). Dordrecht, The Netherlands: Kluwer Academic.

Galster, S. M., Duley, J. A., Masolanis, A. J., and Parasuraman, R. (1999). Effects of aircraft self-separation on controller conflict detection and workload in mature free flight. In M. W. Scerbo and M. Mouloua (Eds.), *Automation Technology and Human Performance: Current Research and Trends* (pp. 96–101). Mahwah, NJ: Lawrence Erlbaum Associates.

GAO. (2012). *Workplace Safety and Health: Better OSHA Guidance Needed on Safety Incentive Programs (Report to Congressional Requesters).* Washington, DC: Government Accountability Office.

Gawande, A. (2008). *Better: A Surgeon's Notes on Performance.* New York: Picador.

Gawande, A. (2010). *The Checklist Manifesto: How to Get Things Right* (1st ed.). New York: Metropolitan Books.

Geertz, C. (1973). *The Interpretation of Cultures.* New York: Basic Books.

Geller, E. S. (2001). *Working Safe: How to Help People Actively Care for Health and Safety.* Boca Raton, FL: CRC Press.

Gergen, K. J. (1999). *An Invitation to Social Construction.* Thousand Oaks, CA: Sage.

Giddens, A. (1984). *The Constitution of Society: Outline of the Theory of Structuration.* Cambridge, UK: Polity Press.

Girard, N. J. (2005). Dealing with perioperative prima donnas in your OR. *AORN Journal, 82*(2), 187–189.

Godkin, L., and Allcorn, S. (2009). Institutional narcissism, arrogant organization disorder and interruptions in organizational learning. *The Learning Organization, 16*(1), 40–57.

Golden-Biddle, K., and Locke, K. (1993). Appealing work: An investigation of how ethno-graphic texts convince. *Organization Science, 4*(4), 595–616.

Graham, B., Reilly, W. K., Beinecke, F., Boesch, D. F., Garcia, T. D., Murray, C. A., and Ulmer, F. (2011). *Deep Water: The Gulf Oil Disaster and the Future of Offshore Drilling (Report to the President).* Washington, DC: National Commission on the BP Deepwater Horizon Oil Spill and Offshore Drilling.

Gray, G. C. (2009). The responsibilization strategy of health and safety. *British Journal of Criminology, 49*, 326–342.

Green, R. G. (1977). The psychologist and flying accidents. *Aviation, Space and Environmental Medicine, 48*(10), 922–923.

Hall, J. L. (2003). Columbia and challenger: Organizational failure at NASA. *Space Policy, 19*, 239–247.

Hallowell, M. R., and Gambatese, J. A. (2009). Construction safety risk mitigation. *Journal of Construction Engineering and Management, 135*(12), 1316–1323.

Hamilton-Baillie, B. (2008). Towards shared space. *Urban Design International, 13*(2), 130–138.

Harvey, D. (1990). *The Condition of Postmodernity: An Enquiry into the Origins of Cultural Change*. Oxford, UK: Blackwell.

Harvey, P., and Martinko, M. J. (2008). An empirical examination of the role of attributions in psychological entitlement and its outcomes. *Journal of Organizational Behavior, 30*(4), 459–476.

Hayes, J. (2012). Operator competence and capacity: Lessons from the Montara blowout. *Safety Science, 50*(3), 563–574.

Heft, H. (2001). *Ecological Psychology in Context: James Gibson, Roger Barker and the Legacy of William James's Radical Empiricism*. Mahwah, NJ, Lawrence Erlbaum Associates.

Heinrich, H. W., Petersen, D., and Roos, N. (1980). *Industrial Accident Prevention* (5th ed.). New York: McGraw-Hill.

Helmreich, R. L. (2000a). On error management: Lessons from aviation. *British Medical Journal, 320*(7237), 781–785.

Helmreich, R. L. (2000b). Culture and error in space: Implications from analog environments. *Aviation, Space, and Environment Medicine 71*(9 Suppl), A133–A139.

Henriqson, E. (2013). *Safety culture*. Paper presented at the Learning Lab on Safety Leadership, Brisbane, Australia.

Heylighen, F. (1999). Causality as distinction conservation: A theory of predictability, reversibility and time order. *Cybernetics and Systems, 20*, 361–384.

Heylighen, F., Cilliers, P., and Gershenson, C. (2006). *Complexity and Philosophy*. Brussels, Belgium: Vrije Universiteit Brussel: Evolution, Complexity and Cognition.

Hollnagel, E. (1999). From function allocation to function congruence. In S. W. A. Dekker and E. Hollnagel (Eds.), *Coping with Computers in the Cockpit* (pp. 29–53). Aldershot, UK: Ashgate.

Hollnagel, E. (2003). *Handbook of Cognitive Task Design*. Mahwah, NJ: Lawrence Erlbaum Associates.

Hollnagel, E. (2004). *Barriers and Accident Prevention*. Aldershot, UK: Ashgate.

Hollnagel, E., and Amalberti, R. (2001). *The emperor's new clothes: Or whatever happened to 'human error'?* Paper presented at the 4th international workshop on human error, safety and systems development, Linköping, Sweden.

Hollnagel, E., Nemeth, C. P., and Dekker, S. W. A. (2008). *Resilience Engineering: Remaining Sensitive to the Possibility of Failure*. Aldershot, UK: Ashgate Publishing.

Hollnagel, E., Nemeth, C. P., and Dekker, S. W. A. (2009). *Resilience Engineering: Preparation and Restoration*. Aldershot, UK: Ashgate Publishing.

Hollnagel, E., Woods, D. D., and Leveson, N. G. (2006). *Resilience Engineering: Concepts and Precepts*. Aldershot, UK: Ashgate Publishing.

Hoven, M. J. V. (2001). *Moral Responsibility and Information and Communication Technology*. Rotterdam, The Netherlands: Erasmus University Center for Philosophy of ICT.

Hugh, T. B., and Dekker, S. W. A. (2009). Hindsight bias and outcome bias in the social construction of medical negligence: A review. *Journal of Law and Medicine, 16*(5), 846–857.

Hughes, J. A., Randall, D., and Shapiro, D. (1993). From ethnographic record to system design: Some experiences from the field. *Computer Supported Cooperative Work, 1*(3), 123–141.

Hutchins, E. L. (1995). How a cockpit remembers its speeds. *Cognitive Science, 19*(3), 265–288.

Hutchins, E. L., Holder, B. E., and Pérez, R. A. (2002). *Culture and Flight Deck Operations*. San Diego, CA: University of California San Diego.

IOM. (2003). *Patient Safety: Achieving a New Standard for Care*. Washington, DC: National Academy of Sciences, Institute of Medicine.

JAA. (2001). *Human Factors in Maintenance Working Group Report*. Hoofddorp, The Netherlands: Joint Aviation Authorities.

James, W. (1890). *The Principles of Psychology* (Vol. 1). New York: Henry Holt & Co.

Janis, I. L. (1982). *Groupthink* (2nd ed.). Chicago, IL: Houghton Mifflin.

Jensen, C. (1996). *No Downlink: A Dramatic Narrative about the Challenger Accident and Our Time* (1st ed.). New York: Farrar, Straus, Giroux.

Jordan, P. W. (1998). Human factors for pleasure in product use. *Applied Ergonomics, 29*(1), 25–33.

Kern, T. (1998). *Flight Discipline*. New York: McGraw-Hill.

Kerstholt, J. H., Passenier, P. O., Houttuin, K., and Schuffel, H. (1996). The effect of a priori probability and complexity on decision making in a supervisory control task. *Human Factors, 38*(1), 65–78.

Khatwa, R., and Helmreich, R. L. (1998). Analysis of critical factors during approach and landing in accidents and normal flight. *Flight Safety Digest, 17*, 256.

Khurana, R. (2007). *From Higher Aims to Hired Hands: The Social Transformation of American Business Schools and the Unfulfilled Promise of Management as a Profession*. Princeton, NJ: Princeton University Press.

Klein, G. A. (1993). A recognition-primed decision (RPD) model of rapid decision making. In G. A. Klein, J. Orasanu, R. Calderwood and C. E. Zsambok (Eds.), *Decision Making in Action: Models and Methods* (pp. 138–147). Norwood, NJ: Ablex.

Klein, G. A. (1998). *Sources of Power: How People Make Decisions*. Cambridge, MA: MIT Press.

Kohn, L. T., Corrigan, J., and Donaldson, M. S. (2000). *To Err is Human: Building a Safer Health System*. Washington, DC: National Academy Press.

Krokos, K. J., and Baker, D. P. (2007). Preface to the special section on classifying and understanding human error. *Human Factors, 49*(2), 175–177.

Kuhn, T. S. (1962). *The Structure of Scientific Revolutions*. Chicago, IL: University of Chicago Press.

Langewiesche, W. (1998). *Inside the Sky: A Meditation on Flight* (1st ed.). New York: Pantheon Books.

Lanir, Z. (1986). *Fundamental Surprise*. Eugene, OR: Decision Research.

LaPorte, T. R., and Consolini, P. M. (1991). Working in practice but not in theory: Theoretical challenges of "High-Reliability Organizations". *Journal of Public Administration Research and Theory: J-PART, 1*(1), 19–48.

Lautman, L., and Gallimore, P. L. (1987). Control of the crew caused accident: Results of a 12-operator survey. *Boeing Airliner*, (1), 1–6.

Law, J. (2004). *After Method: Mess in Social Science Research*. London: Routledge.

Lazarus, R. S. (1966). *Psychological Stress and the Coping Process*. New York: McGraw-Hill.

Leape, L. L. (1994). Error in medicine. *JAMA, 272*(23), 1851–1857.

Lee, T., and Harrison, K. (2000). Assessing safety culture in nuclear power stations. *Safety Science, 34*, 61–97.

Leveson, N. G. (2002). *System Safety Engineering: Back to the Future*. Boston, MA: MIT Aeronautics and Astronautics.

Leveson, N. G. (2012). *Engineering a Safer World: Systems Thinking Applied to Safety*. Cambridge, MA: MIT Press.

Levin, A. (2010, 9 September). BP blames rig explosion on series of failures, *USA Today*, p. 1.

Lovell, J., and Kluger, J. (1994). *Lost Moon: The Perilous Voyage of Apollo 13*. San Francisco, CA: Houghton Mifflin.

Lund, I. O., and Rundmo, T. (2009). Cross-cultural comparisons of traffic safety, risk perception, attitudes and behavior. *Safety Science, 47*, 547–553.

Lützhoft, M. H., and Dekker, S. W. A. (2002). On your watch: Automation on the bridge. *Journal of Navigation, 55*(1), 83–96.

Lützhoft, M. H., Nyce, J. M., and Petersen, E. S. (2010). Epistemology in ethnography: Assessing the quality of knowledge in human factors research. *Theoretical Issues in Ergonomics Science, 11*(6), 532–545.

Mackay, W. E. (2000). Is paper safer? The role of paper flight strips in air traffic control. *ACM Transactions on Computer-Human Interaction, 6*, 311–340.

McCartin, J. A. (2011). *Collision Course: Ronald Reagan, the Air Traffic Controllers, and the Strike that Changed America.* Oxford, UK: Oxford University Press.

McDonald, N., Corrigan, S., and Ward, M. (2002). *Well-intentioned people in dysfunctional systems.* Paper presented at the 5th workshop on human error, safety and systems development, Newcastle, Australia.

McDonald, R., Waring, J., and Harrison, S. (2006). Rules, safety and the narrativization of identity: A hospital operating theatre case study. *Sociology of Health and Illness, 28*(2), 178–202.

Meister, D. (2003). The editor's comments. *Human Factors Ergonomics Society COTG Digest, 5*, 2–6.

Meister, D., and Farr, D. E. (1967). The utilization of human factors information by designers. *Human Factors, 9*(1), 71–87.

Miles, G. H. (1925). Economy and safety in transport. *Journal of the National Institute of Industrial Psychology, 2*, 192–193.

Mindell, D. A. (2008). *Digital Apollo: Human and Machine in Spaceflight.* Cambridge, MA: MIT Press.

Mintzberg, H. (1979). *The Structuring of Organizations: A Synthesis of the Research.* Englewood Cliffs, NJ: Prentice-Hall.

Mintzberg, H. (2004). *Managers Not MBAs: A Hard Look at the Soft Practice of Managing and Management Development.* San Francisco, CA: Berrett-Koehler.

Mokyr, J. (1992). *The Lever of Riches: Creativity and Economic Progress.* Oxford, UK: Oxford University Press.

Moray, N., and Inagaki, T. (2000). Attention and complacency. *Theoretical Issues in Ergonomics Science, 1*(4), 354–365.

Mumaw, R. J., Sarter, N. B., and Wickens, C. D. (2001). *Analysis of pilots' monitoring and performance on an automated flight deck.* Paper presented at the 11th International Symposium in Aviation Psychology, Columbus, OH.

Murray, C. A., and Cox, C. B. (1989). *Apollo, the Race to the Moon.* New York: Simon and Schuster.

Murray, S. J., Holmes, D., and Rail, G. (2008). On the constitution and status of 'evidence' in the health sciences. *Journal of Research in Nursing, 13*(4), 272–280.

Nagel, T. (1992). *The View from Nowhere.* Oxford, UK: Oxford University Press.

Naumann, S. E., Minsky, B. D., and Sturman, M. C. (2002). The use of the concept 'entitlement' in management literature: A historical review, synthesis, and discussion of compensation policy implications. *Human Resource Management Review, 12*, 145–166.

Neisser, U. (1976). *Cognition and Reality: Principles and Implications of Cognitive Psychology.* San Francisco, CA: W. H. Freeman.

Norman, D. A. (1993). *Things that Make Us Smart: Defending Human Attributes in the Age of the Machine.* Reading, MA: Addison-Wesley.

NTSB. (1990). Accident report: United Airlines Flight 232, McDonnell Douglas DC-10-10, Sioux Gateway Airport, Sioux City, Iowa, July 19, 1989. Washington, DC: National Transportation Safety Board.

NTSB. (1995). Aircraft accident report: Flight into terrain during missed approach, USAir Flight 1016, DC-9-31, N954VJ, Charlotte Douglas International Airport, Charlotte, North Carolina, July 2, 1994. Washington, DC: National Transportation Safety Board.

NTSB. (1997). Grounding of the Panamanian passenger ship Royal Majesty on Rose and Crown shoal near Nantucket, Massachusetts, June 10, 1995. Washington, DC: National Transportation Safety Board.

NTSB. (2002). Loss of control and impact with Pacific Ocean, Alaska Airlines Flight 261 McDonnell Douglas MD-83, N963AS, about 2.7 miles north of Anacapa Island, California, January 31, 2000. Washington, DC: National Transportation Safety Board.

NTSB. (2007a). Accident report: Crash of Pinnacle Airlines Flight 3701, Bombardier CL600-2-B19, N8396A, Jefferson City, Missouri, October 14, 2004. Washington, DC: National Transportation Safety Board.

NTSB. (2007b). Attempted takeoff from wrong runway Comair Flight 5191, Bombardier CL-600-2B19, N431CA, Lexington, Kentucky, August 27, 2006. Springfield, VA: National Transportation Safety Board.

O'Hare, D., and Roscoe, S. N. (1990). *Flightdeck Performance: The Human Factor* (1st ed.). Ames, IA: Iowa State University Press.

Ödegård, S. (Ed.). (2007). *I Rättvisans Namn (In the Name of Justice)*. Stockholm, Sweden: Liber.

Orasanu, J. M., and Connolly, T. (1993). The reinvention of decision making. In G. A. Klein, J. M. Orasanu, R. Calderwood and C. E. Zsambok (Eds.), *Decision Making in Action: Models and Methods* (pp. 3–20). Norwood, NJ: Ablex.

Orasanu, J. M., and Martin, L. (1998). Errors in aviation decision making: A factor in accidents and incidents. Human Error, Safety and Systems Development Workshop (HESSD), 1998. Retrieved February 2008, from http://www.dcs.gla.ac.uk/~johnson/papers/seattle_hessd/judithlynnep.

Parasuraman, R., and Manzey, D. H. (2010). Complacency and bias in human use of automation: An attentional integration. *Human Factors, 52*(3), 381–410.

Parasuraman, R., Molloy, R., and Singh, I. L. (1993). Performance consequences of automation-induced 'complacency'. *International Journal of Aviation Psychology, 3*(1), 1–24.

Parasuraman, R., Sheridan, T. B., and Wickens, C. D. (2000). A model for types and levels of human interaction with automation. *IEEE Transactions on Systems Man and Cybernetics Part A-Systems and Humans, 30*(3), 286–297.

Parasuraman, R., Sheridan, T. B., and Wickens, C. D. (2008). Situation awareness, mental workload and trust in automation: Viable, empirically supported cognitive engineering constructs. *Journal of Cognitive Engineering and Decision Making, 2*(2), 140–160.

Pellegrino, E. D. (2004). Prevention of medical error: Where professional and organizational ethics meet. In V. A. Sharpe (Ed.), *Accountability: Patient Safety and Policy Reform* (pp. 83–98). Washington, DC: Georgetown University Press.

Perrow, C. (1984). *Normal Accidents: Living with High-Risk Technologies*. New York: Basic Books.

Perry, S. J., Wears, R. L., and Cook, R. I. (2005). The role of automation in complex system failures. *Journal of Patient Safety, 1*(1), 56–61.

Pidgeon, N. F., and O'Leary, M. (2000). Man-made disasters: Why technology and organizations (sometimes) fail. *Safety Science, 34*(1–3), 15–30.

Pollock, T. (1998). Dealing with the prima donna employee. *Automotive Manufacturing and Production, 110*, 10–12.

Prigogine, I. (2003). *Is Future Given?* London: World Scientific.

Rasmussen, J. (1997). Risk management in a dynamic society: A modelling problem. *Safety Science, 27*(2–3), 183–213.

Rasmussen, J., and Svedung, I. (2000). *Proactive Risk Management in a Dynamic Society*. Karlstad, Sweden: Swedish Rescue Services Agency.

Reason, J. T. (1990). *Human Error*. New York: Cambridge University Press.

Reason, J. T. (1997). *Managing the Risks of Organizational Accidents*. Aldershot, UK: Ashgate Publishing.

Reason, J. T. (2013). *A Life in Error*. Farnham, UK: Ashgate Publishing.

Reason, J. T., and Hobbs, A. (2003). *Managing Maintenance Error: A Practical Guide*. Aldershot, UK: Ashgate.

Report on Project Management in NASA, by the Mars Climate Orbiter Mishap Investigation Board. (2000). Washington, DC: National Aeronautics and Space Administration.

Rochlin, G. I. (1999). Safe operation as a social construct. *Ergonomics, 42*(11), 1549–1560.

Rochlin, G. I., LaPorte, T. R., and Roberts, K. H. (1987). The self-designing high reliability organization: Aircraft carrier flight operations at sea. *Naval War College Review, 40*, 76–90.

Roscoe, S. N. (1997). The adolescence of engineering psychology. In S. M. Casey (Ed.), *Volume 1, Human Factors History Monograph Series* (pp. 1–9). Santa Monica, CA: Human Factors and Ergonomics Society.

Ross, G. (1995). *Flight Strip Survey Report*. Canberra, Australia: TAAATS TOI.

Sagan, S. D. (1993). *The Limits of Safety: Organizations, Accidents, and Nuclear Weapons*. Princeton, NJ: Princeton University Press.

Sagan, S. D. (1994). Toward a political theory of organizational reliability. *Journal of Contingencies and Crisis Management, 2*(4), 228–240.

Saloniemi, A., and Oksanen, H. (1998). Accidents and fatal accidents: Some paradoxes. *Safety Science, 29*, 59–66.

Sanders, M. S., and McCormick, E. J. (1993). *Human Factors in Engineering and Design* (7th ed.). New York: McGraw-Hill.

Sarter, N. B., and Woods, D. D. (1997). Teamplay with a powerful and independent agent: A corpus of operational experiences and automation surprises on the Airbus A320. *Human Factors, 39*, 553–569.

Schein, E. (1992). *Organizational Culture and Leadership* (2nd ed.). San Francisco, CA: Jossey-Bass.

Schultz, B. (1998). The prima donna syndrome. *Builder, 21*, 114.

Schulz, C. M., Endsley, M. R., Kochs, E. F., Gelb, A. W., and Wagner, K. J. (2013). Situation awareness in anesthesia: Concept and research. *Anesthesiology, 118*(3), 729–742.

Schwartz, H. S. (1989). Organizational disaster and organizational decay: The case of the National Aeronautics and Space Administration. *Industrial Crisis Quarterly, 3*, 319–334.

Schwenk, C. R., and Cosier, R. A. (1980). Effects of the expert, devil's advocate, and dialectical inquiry methods on prediction performance. *Organizational Behavior and Human Performance, 26*(3), 409–424.

Scott, J. C. (1998). *Seeing Like a State: How Certain Schemes to Improve the Human Condition Have Failed*. New Haven, CT: Yale University Press.

Shappell, S. A., and Wiegmann, D. A. (2001). Applying reason: The human factors analysis and classification system. *Human Factors and Aerospace Safety, 1*, 59–86.

Sheridan, T. B. (1987). Supervisory control. In I. Salvendy (Ed.), *Handbook of Human Factors*. New York: Wiley.

SHK. (2003). Tillbud mellan flygplanet LN-RPL och en bogsertraktor på Stockholm/Arlanda flygplats, AB län, den 27 oktober 2002 (Rapport RL2003:47) (Incident between aircraft LN-RPL and a tow truck at Stockholm/Arlanda airport, October 27, 2002). Stockholm, Statens Haverikommission (Swedish Accident Investigation Board).

Sibley, L., Sipe, T. A., and Koblinsky, M. (2004). Does traditional birth attendant training improve referral of woman with obstetric complications: A review of the evidence. *Social Science and Medicine, 59*(8), 1757–1769.

Singer, G., and Dekker, S. W. A. (2000). Pilot performance during multiple failures: An empirical study of different warning systems. *Transportation Human Factors, 2*(1), 63–77.

Singer, G., and Dekker, S. W. A. (2001). The ergonomics of flight management systems: Fixing holes in the cockpit certification net. *Applied Ergonomics, 32*(3), 247–255.

Smith, K. (2001). Incompatible goals, uncertain information and conflicting incentives: The dispatch dilemma. *Human Factors and Aerospace Safety, 1*, 361–380.

Smith, K., and Hancock, P. A. (1995). Situation awareness is adaptive, externally-directed consciousness. *Human Factors, 27*, 137–148.

Snook, S. A. (2000). *Friendly Fire: The Accidental Shootdown of US Black Hawks over Northern Iraq*. Princeton, NJ: Princeton University Press.

Snow, J. N., Kern, R. M., and Curlette, W. L. (2001). Identifying personality traits associated with attrition in systematic training for effective parenting groups. *The Family Journal: Counseling and Therapy for Couples and Families, 9*, 102–108.

Sorkin, R. D., and Woods, D. D. (1985). Systems with human monitors: A signal detection analysis. *Human-Computer Interaction, 1*(1), 49–75.

Starbuck, W. H., and Farjoun, M. (2005). *Organization at the Limit: Lessons from the Columbia Disaster*. Malden, MA: Blackwell Publishing.

Starbuck, W. H., and Milliken, F. J. (1988). Challenger: Fine-tuning the odds until something breaks. *The Journal of Management Studies, 25*(4), 319–341.

Suchman, L. A. (1987). *Plans and Situated Actions: The Problem of Human-Machine Communication*. New York: Cambridge University Press.

Thomas, G. (2007). A crime against safety. *Air Transport World, 44*, 57–59.

Tillbud vid landning med flygplanet LN-RLF den 23/6 på Växjö/Kronoberg flygplats, G län (Rapport RL 2000:38). (Incident during landing with aircraft LN-RLF on June 23 at Växjö/Kronoberg airport). (2000). Stockholm, Sweden: Statens Haverikommision (Swedish Accident Investigation Board).

Tingvall, C., and Lie, A. (2010). *The concept of responsibility in road traffic (Ansvarsbegreppet i vägtrafiken)*. Paper presented at the Transportforum, Linköping, Sweden.

Townsend, A. S. (2013). *Safety Can't be Measured*. Farnham, UK: Gower Publishing.

Transocean. (2011). *Macondo Well Incident: Transocean Investigation Report* (Vols. I and II). Houston, TX: Transocean.

TSB. (2003). Aviation investigation report: In-flight fire leading to collision with water, Swissair Transport Limited, McDonnell Douglas MD-11 HB-IWF, Peggy's Cove, Nova Scotia 5 nm SW, 2 September 1998. Gatineau, QC: Transportation Safety Board of Canada.

Tuchman, B. W. (1981). *Practicing History: Selected Essays* (1st ed.). New York: Knopf.

Turner, B. A. (1978). *Man-Made Disasters*. London: Wykeham Publications.

USAF. (1947). Psychological aspects of instrument display: Analysis of 270 'pilot error' experiences in reading and interpreting aircraft instruments. In A. M. L. E. Division (Ed.), *Memorandum Report*. Dayton, OH: US Air Force Air Material Command, Wright-Patterson Air Force Base.

USW. (2010). *Behavior-Based Safety/'Blame-the-Worker' Safety Programs: Understanding and Confronting Management's Plan for Workplace Health and Safety*. Pittsburgh, PA: United Steelworkers' Health, Safety and Environment Department.

Varela, F. J., Thompson, E., and Rosch, E. (1991). *The Embodied Mind: Cognitive Science and Human Experience*. Cambridge, MA: MIT Press.

Vaughan, D. (1996). *The Challenger Launch Decision: Risky Technology, Culture, and Deviance at NASA*. Chicago, IL: University of Chicago Press.

Vaughan, D. (1999). The dark side of organizations: Mistake, misconduct, and disaster. *Annual Review of Sociology, 25*, 271–305.

Vaughan, D. (2005). System effects: On slippery slopes, repeating negative patterns, and learning from mistake? In W. H. Starbuck and M. Farjoun (Eds.), *Organization at the Limit: Lessons from the Columbia Disaster* (pp. 41–59). Malden, MA: Blackwell Publishing.

Vicente, K. J. (1999). *Cognitive Work Analysis: Toward Safe, Productive, and Healthy Computer-Based Work*. Mahwah, NJ: Lawrence Erlbaum Associates.

Vincent, C. (2006). *Patient Safety*. London: Churchill Livingstone.

Wallerstein, I. (1996). *Open the Social Sciences: Report of the Gulbenkian Commission on the Restructuring of the Social Sciences*. Stanford, CA: Stanford University Press.

Watson, R. I. (1978). *The Great Psychologists* (4th ed.). Philadelphia, PA: Lippincott.

Webb, W. B. (1956). The prediction of aircraft accidents from pilot-centered measures. *Journal of Aviation Medicine, 27*, 141–147.

Weick, K. E. (1993). The collapse of sensemaking in organizations: The Mann Gulch disaster. *Administrative Science Quarterly, 38*(4), 628–652.

Weick, K. E. (1995). *Sensemaking in Organizations*. Thousand Oaks, CA: Sage Publications.

Weick, K. E., and Sutcliffe, K. M. (2007). *Managing the Unexpected: Resilient Performance in an Age of Uncertainty* (2nd ed.). San Francisco, CA: Jossey-Bass.

Weingart, P. (1991). Large technical systems, real life experiments, and the legitimation trap of technology assessment: The contribution of science and technology to constituting risk perception. In T. R. LaPorte (Ed.), *Social Responses to Large Technical Systems: Control or Anticipation* (pp. 8–9). Amsterdam, The Netherlands: Kluwer.

Wiener, E. L. (1988). Cockpit automation. In E. L. Wiener and D. C. Nagel (Eds.), *Human Factors in Aviation* (pp. 433–462). San Diego, CA: Academic Press.

Wiener, E. L. (1989). *Human Factors of Advanced Technology ('Glass Cockpit') Transport Aircraft*. Moffett Field, CA: NASA Ames Research Center.

Wilkin, P. (2009). The ideology of ergonomics. *Theoretical Issues in Ergonomics Science, 11*(3), 230–244.

Woods, D. D. (1993). Process-tracing methods for the study of cognition outside of the experimental laboratory. In G. A. Klein, J. M. Orasanu, R. Calderwood and C. E. Zsambok (Eds.), *Decision Making in Action: Models and Methods* (pp. 228–251). Norwood, NJ: Ablex.

Woods, D. D. (1996). Decomposing automation: Apparent simplicity, real complexity. In R. Parasuraman and M. Mouloua (Eds.), *Automation Technology and Human Performance*. Mahwah, NJ: Lawrence Erlbaum Associates.

Woods, D. D., and Dekker, S. W. A. (2001). Anticipating the effects of technological change: A new era of dynamics for human factors. *Theoretical Issues in Ergonomics Science, 1*(3), 272–282.

Woods, D. D., Dekker, S. W. A., Cook, R. I., Johannesen, L. J., and Sarter, N. B. (2010). *Behind Human Error*. Aldershot, UK: Ashgate Publishing.

Woods, D. D., and Hollnagel, E. (2006). *Joint Cognitive Systems: Patterns in Cognitive Systems Engineering*. Boca Raton, FL: CRC/Taylor & Francis.

Woods, D. D., Johannesen, L. J., Cook, R. I., and Sarter, N. B. (1994). *Behind Human Error: Cognitive Systems, Computers and Hindsight*. Dayton, OH: CSERIAC.

Woods, D. D., Patterson, E. S., and Roth, E. M. (2002). Can we ever escape from data overload? A cognitive systems diagnosis. *Cognition, Technology and Work, 4*(1), 22–36.

Woods, D. D., and Shattuck, L. G. (2000). Distant supervision-local action given the potential for surprise. *Cognition, Technology and Work, 2*(4), 242–245.

Wright, L. (2009). Prima donnas cause conflict. *Business NH Magazine, 26*, 15.

Wright, P. C., and McCarthy, J. (2003). Analysis of procedure following as concerned work. In E. Hollnagel (Ed.), *Handbook of Cognitive Task Design* (pp. 679–700). Mahwah, NJ: Lawrence Erlbaum Associates.

Wynne, B. (1988). Unruly technology: Practical rules, impractical discourses and public understanding. *Social Studies of Science, 18*(1), 147–167.

Xiao, Y., and Vicente, K. J. (2000). A framework for epistemological analysis in empirical (laboratory and field) studies. *Human Factors, 42*(1), 87–102.

Yerkes, R. M., and Dodson, J. D. (1908). The relation of strength of stimulus to rapidity of habit-formation. *Journal of Comparative Neurology and Psychology, 18*(5), 459–482.

Zaccaria, A. (2002, 7 November). Malpractice insurance crisis in New Jersey, *Atlantic Highlands Herald*.

Zwetsloot, G. I. J. M., Aaltonen, M., Wybo, J. L., Saari, J., Kines, P., and Beek, R. (2013). The case for research into the zero accident vision. *Safety Science, 58*, 41–48.

Index

Page numbers followed by f and t indicate figures and tables, respectively.

Printed in the United States
by Baker & Taylor Publisher Services